Springer Series in Materials Science

Volume 288

The Springer Series in Materials Science covers the complete spectrum of materials research and technology, including fundamental principles, physical properties, materials theory and design. Recognizing the increasing importance of materials science in future device technologies, the book titles in this series reflect the state-of-the-art in understanding and controlling the structure and properties of all important classes of materials.

More information about this series at http://www.springer.com/series/856

Lars Rebohle · Slawomir Prucnal ·
Denise Reichel

Flash Lamp Annealing

From Basics to Applications

Springer

Lars Rebohle
Institute of Ion Beam Physics
and Materials Research
Helmholtz-Zentrum Dresden Rossendorf
Dresden, Germany

Slawomir Prucnal
Institute of Ion Beam Physics
and Materials Research
Helmholtz-Zentrum Dresden Rossendorf
Dresden, Germany

Denise Reichel
AKAD University
Stuttgart, Germany

ISSN 0933-033X ISSN 2196-2812 (electronic)
Springer Series in Materials Science
ISBN 978-3-030-23298-6 ISBN 978-3-030-23299-3 (eBook)
https://doi.org/10.1007/978-3-030-23299-3

This Springer imprint is published by the registered company Springer Nature Switzerland AG
The registered company address is: Gewerbestrasse 11, 6330 Cham, Switzerland

Preface

The idea to write a book about flash lamp annealing (FLA) was born when we collected material for recent review papers [1–3]. It appeared that on the one hand sufficient information is available in different publications, but that on the other hand a comprehensive presentation of the topic is still missing. For example, details about flash lamps can be found in the literature dedicated to technical issues of laser technology or to plasma physics, but without the aspect of annealing. FLA emerged first in the field of semiconductor science and technology with the aim to reduce diffusion and segregation of dopants. The proceedings of the International Conference on Advanced Thermal Processing of Semiconductors have been a valuable source of information for a long time, but unfortunately, the conference was discontinued in 2010. A first review about the aspects of subsecond thermal processing of advanced materials in book form was published in 2014 which was basically a collection of extended papers presented at a workshop in Dresden in 2011 [4]. However, in recent times, new user communities have been developed with a rather limited exchange of ideas among each other. In the field of printed and flexible electronics, FLA is called intense pulsed light processing, flashlight sintering or photonic curing. The main motivation here is the possibility to treat temperature-sensible substrates like plastic foils and to use short process times which meet the requirements of roll-to-roll applications very well. Another community uses FLA to anneal thin films on glass, mainly for photovoltaics and display applications. Possibly, the latest development is flash-assisted deposition where FLA serves as a suitable method for in situ annealing.

Most of the users have a pragmatic approach to FLA as a valuable tool. However, FLA differs in several aspects from conventional annealing like furnace or rapid thermal annealing which sometimes gives rise to unexpected results: (1) it is not performed in thermal equilibrium, (2) there are steep temperature gradients, (3) the resulting temperature profile strongly depends on the material system, and (4) temperature measurement is challenging due to the very short processing times.

However, the non-equilibrium approach has several advantages: it allows the treatment of temperature-sensible substrates, it allows specific dynamic processes like explosive crystallization or hyperdoping, and it suppresses unwanted processes like diffusion or segregation. In addition, it saves energy, costs and process time, and it is suitable for roll-to-roll applications. The book is addressed to readers who already or potentially use FLA, to engineers and scientists who deal with materials where FLA might be an option or simply to all who are interested in the thermodynamics of materials on the subsecond timescale.

To do so, the book is divided into two parts. In the first part, Chap. 2 addresses various aspects of FLA including a brief introduction to short-time annealing, the physical background, the structure and functionality of individual components as well as the configuration of FLA tools. Chapter 3 illuminates different facets of process management such as temperature determination, homogeneity or thermal stress minimization. The second part gives an overview of the different fields of materials science in which FLA is engaged (Chaps. 4 and 5). This overview is based on a literature search which intrinsically tends to be imperfect: the result depends on topicality and completeness of literature data banks, on the keywords used to describe a specific topic and on the definition of FLA compared to other flash lamp applications. Nevertheless, we hope that this book can give a comprehensive overview, helps to get familiar with the topic and may stimulate one or the other (or better: many) novel and interesting idea(s).

Acknowledgements

The book strongly profits from decade-long experience and research with FLA which began already around 1980 at the former Zentralinstitut für Kernforschung, now Helmholtz-Zentrum Dresden-Rossendorf. Since this time, numerous colleagues have contributed to the field, and we strongly want to acknowledge all their valuable work. Thereby, special thanks go to Wolfgang Skorupa who safely navigated FLA through the tumultuous times of political changes after 1989 up to now and who made it to a unique feature of our division, the Institute of Ion Beam Physics and Material Research. We also express our thanks to Springer-Verlag who gave us the possibility to write this book and needed some patience before the book could be successfully completed.

Dresden, Germany — Lars Rebohle
Dresden, Germany — Slawomir Prucnal
Stuttgart, Germany — Denise Reichel

References

1. L. Rebohle, S. Prucnal, W. Skorupa, A review of thermal processing in the subsecond range: semiconductors and beyond. Semicond. Sci. Technol. **31**(10) (2016). https://doi.org/10.1088/0268-1242/31/10/103001
2. S. Prucnal, L. Rebohle, W. Skorupa, Doping by flash lamp annealing. Mater. Sci. Semicond. Process. **62**, 115–127 (2017). https://doi.org/10.1016/j.mssp.2016.10.040
3. W. Skorupa, T. Schumann, L. Rebohle, Millisecond thermal processing using flash lamps for the advancement of thin layers and functional coatings. Surf. Coat. Technol. **314**, 169–176 (2017). https://doi.org/10.1016/j.surfcoat.2016.08.010
4. W. Skorupa and H Schmidt (eds.), in *Subsecond Annealing of Advanced Materials*, Springer Series in Mater. Science, vol. 192 (Springer, Cham, Heidelberg, 2014)

Contents

Acronyms

ALD	Atomic layer deposition
AZO	Aluminium-doped zinc oxide
CIGS	Copper indium gallium selenide
CIGSu	Copper indium gallium sulphide
CMOS	Complementary metal–oxide–semiconductor
CVD	Chemical vapour deposition
DLTS	Deep-level transient spectroscopy
DMS	Diluted magnetic semiconductors
EB	Electron beam
EDX	Energy-dispersive X-ray spectroscopy
EELS	Electron energy loss spectroscopy
EL	Electroluminescence
ELPE	Explosive liquid-phase epitaxy
ELPN	Explosive liquid-phase nucleation
EOR	End-of-range
ERDA	Elastic recoil detection analysis
ESPE	Explosive solid-phase epitaxy
ESPN	Explosive solid-phase nucleation
FA	Furnace annealing
FET, FinFET	(Fin) field-effect transistor
FLA	Flash lamp annealing
FLASiC	Flash lamp annealing of thin 3C-SiC on silicon
FWHM	Full width at half maximum
HAADF-STEM	High-angle annular dark-field imaging STEM
IGBT	Insulated-gate bipolar transistor
IMT	Insulator–metal transition
IPL	Intense pulsed light
IR	Infrared
ITO	Indium tin oxide
LA	Laser annealing

LPC, LPE	Liquid-phase crystallization/liquid-phase epitaxy
MBE	Molecular beam epitaxy
MISFET	Metal–insulator–semiconductor field-effect transistor
MLD	Monolayer doping
MOS	Metal–oxide–semiconductor
MOSFET	Metal–oxide–semiconductor field-effect transistor
NC	Nanoclusters
NIR	Near-infrared
NP	Nanoparticles
NRA	Nuclear reaction analysis
NW	Nanowire
PECVD	Plasma-enhanced chemical vapour deposition
PET	Polyethylene terephthalate
PFN	Pulse-forming network
PIII	Plasma immersion ion implantation
PL	Photoluminescence
PLA, PLM	Pulsed laser annealing/pulsed laser melting
PV	Photovoltaics
R2R	Roll-to-roll
RBS (R/C)	Rutherford backscattering spectrometry (random/channelled)
RE	Rare earth
r-FLA	Rear-side FLA
RT	Room temperature
RTA, RTP	Rapid thermal annealing/rapid thermal processing
S2S	Sheet-to-sheet
SEM	Scanning electron microscope
SNR	Signal-to-noise ratio
SOI	Silicon on insulator
SPC, SPE	Solid-phase crystallization/solid-phase epitaxy
STEM, TEM	(Scanning) transmission electron microscopy
TCO	Transparent conducting oxide
TFT	Thin film transistor
USJ	Ultra-shallow junction
UV	Ultraviolet
VLS	Vapour–liquid–solid
XRD	X-ray diffraction

Symbols

a, $a(x, t)$	Absorptivity, absorption profile of a flash in a sample
A	Absorbed energy density
B	Bulk modulus
c	Speed of light
c_P	Specific heat capacity
C	Capacitance
C_0	Annealing point (Sect. 3.4)
C_{ijkl}	Young's modulus as a fourth-order tensor (Sect. 3.4)
d	Substrate thickness, penetration depth
d_0	Bore diameter of the lamp
d_a	Arc diameter
d_c	Critical arc diameter (maximum arc extension)
d_i	Layer thickness
D, D_0	Diffusion coefficient
D_L, D_S	Diffusion coefficient in the liquid/solid phase
D_{th}	Thermal diffusivity
e	Reduced explosion energy
E	Electric field, Young's modulus (Sect. 3.4)
E_0	Electrical input energy stored in the PFN
E_A	Activation energy
E_D	Energy density of the flash at the sample surface
E_{ex}	Explosion energy of a lamp
E_i	Ionization energy
E_{lim}	Critical input energy
E_{opt}	Optical output energy of a flash lamp
E_R, E_T	Wave amplitude of a reflected or transmitted electromagnetic wave
E_W	Light intensity due to thermal radiation of a wafer (Sect. 3.1)
f	Pulse frequency
$f(E, t)$	Flash lamp spectrum
f_S	Substitutional fraction

$g(E)$	Time-independent flash lamp spectrum
$\widehat{G}_{\mathrm{i}}$	Interface matrix (Sect. 3.2)
h	Planck's constant
$h(t)$	Pulse shape in time
h_{s}	Specific heat of evaporation
I, I_{max}	(Maximum) current, intensity
I_{L}	Light intensity signal emitted from a lamp (Sect. 3.1)
I_{SD}	Source–drain current
I_{W}	Light intensity signal from a wafer (Sect. 3.1)
j, j_{max}	(Maximum) current density
k	Extinction coefficient, segregation coefficient (Sect. 4.4), dielectric constant (Sect. 5.1.1)
k_{B}	Boltzmann constant
K_0	Lamp impedance
ℓ	Arc length
L	Inductance, line factor (Sect. 2.5)
$\widehat{L}_{\mathrm{i}}$	Layer matrix (Sect. 3.2)
N	Number of flash pulses (Sect. 2.6)
m	Lifetime coefficient
m^*	Effective mass
m_{e}	Electron mass
M	Mass
M_{bb}^{ν}, $M_{\mathrm{bb}}^{\lambda}$	Spectral radiant emittance of a black body (Sect. 3.1)
n, $\tilde{n}$	Real, complex refractive index
n	Charge carrier concentration (Chap. 4)
n_0, n_{arc}	Concentration of atoms and ions at room temperature and in the arc
n_{a}, n_{e}, n_{i}	Concentration of atoms, electrons, ions
p, p_0	Current pressure, filling pressure
p	Failure probability (Sect. 2.6.2)
p_{b}	Burst pressure
p_{R}	Reference pressure
P	Input power
P_{ave}	Average power
P_{circ}, P_{lamp}	Power dissipated in the external circuitry or the flash lamp
P_{lim}	Power limit for evaporation
$P_{\mathrm{sur}}^{\mathrm{e}}$, $P_{\mathrm{sur}}^{\mathrm{o}}$	Electrical/optical power density at the inner side of the flash lamp body
q	Elementary charge
q_{rad}	Power density of the generated radiation of a flash lamp
Q	Charge stored in a capacitor, energy input to a system by FLA
r, r_0	Radius, inner radius of the lamp
r_{D}	Ratio between dead and arc volume
r_{i}	Radius of an atom or ion (Chap. 2)
$r_{\mathrm{i,i+1}}$	Complex reflection coefficient of an interface

R	Ohmic resistance, reflectance
R_C, R_S	Reflectance of the chamber, the sample (Chap. 3)
R_L, R_U	Reflectance of the lower and upper half space of the chamber (Chap. 3)
R_P	Projected range of implantation
R_S	Sheet resistance
S_i, S_O	Inflow and outflow of energy of a system, given as volumetric power density
t_0	Pulse length
t_A	Annealing time
T	Temperature, transmittance
$T(x, t)$	Temperature profile of a flash in a sample
T_0	Starting temperature, characteristic time constant of a capacitor–coil combination (Chap. 2), softening point (Sect. 3.4)
T_A, T_{equ}	Annealing temperature, equilibrium temperature
T_C	Curie temperature
$t_{i,i+1}$	Complex transmission coefficient through an interface
T_{max}	Temperature in the centre of the arc
T_W	Temperature at the inner side of the flash lamp body
U	Elastic energy of deformation
v	Reordering velocity (Chap. 4)
v_{th}	Thermal velocity
V_0	Charging voltage
V_{arc}	Arc volume
V_{dead}	Dead volume
V_F	Voltage drop over the flash lamp
V_{min}	Minimum voltage required to fire the next flash pulse
V_{SD}	Source–drain voltage
V_t	Minimum trigger voltage
x_D	Diffusion length
x_J	Junction depth
x_{th}	(Thermal) diffusion length
Z_0	Impedance of a capacitor–coil combination
α	Absorption coefficient, thermal expansion coefficient (Sect. 3.4)
α_D	Damping constant
β	Incidence or emission angle
Γ	Correction factor between E_D and the really absorbed energy density A
$\widehat{\Gamma}$	Virtual interface matrix
δ	Volume expansion
ε	Emissivity (Chap. 3)
ε	Dielectric constant (Chap. 4)
ε_B, ε_T, ε_{ij}	Strain caused by wafer bowing, by thermal expansion, stress tensor
$\varepsilon_t(t)$	Energy released into a flash lamp up to the time t
η, η_{elec}	Conversion efficiency, electrical transfer efficiency

θ	Relative diameter (or relative radius) of the discharge column, fraction of energy consumed by evaporation (Sect. 2.6.2), propagation angle (Sect. 3.2)
κ_T	Spherical curvature
λ	Wavelength
λ_e	Mean free path of an electron
λ_T	Thermal conductivity
μ	Mobility, viscosity (Sect. 3.4)
μ_0	Strain point (Sect. 3.4)
ν	Poisson's ratio (Sect. 3.4)
ξ	Input power fraction which is used to heat up the gas
Π	Power density of an electromagnetic wave
ρ	Mass density
σ	Conductivity
σ_b	Burst stress
σ_g	Geometric cross section of an atom or ion
σ_{ij}	Stress tensor
τ^*	Critical pulse time separating the short-time from the long-time regime
τ_D	Delay between current and light maximum
τ_{th}	Thermal response time
χ, χ_R	Minimum yield in an aligned or random RBS spectrum
ω_P	Plasma frequency

Chapter 1
Introduction

Abstract This chapter gives a brief history of thermal treatments in general and flash lamp annealing in particular. It introduces the basic thermodynamic values thermal diffusivity, thermal diffusion length and thermal response time, discusses the concept of thermal budget, and provides a comparison of the three main methods of short time annealing, namely rapid thermal annealing, flash lamp annealing, and laser annealing.

What is annealing?

Annealing is a diffuse term which can have different meanings depending on the context. Originally, the verb *to anneal* came into use around the year 1000 with the simple meaning of "to set on fire" [1]. Today it is used in biology and material science, and in the latter case it is defined as

> to heat and then cool (a material, such as steel or glass) usually for softening and making less brittle. [1]

If we search for a more precise definition for material science, the European Standard EN 10052 defines annealing as:

> Heat treatment consisting of heating and soaking at a suitable temperature followed by cooling under conditions such that, after return to ambient temperature, the metal will be in a structural state closer to that of equilibrium. [2]

This definition already appears quite suitable, but it contains two limitations: the material is restricted to metals and annealing aims to bring the material more close to equilibrium. The first limitation is due to the fact that this standard is focused on the heat treatment of ferrous products, and there is no reason not to extend this definition to other material classes. However, the second limitation is more restricting. Indeed, furnace annealing (FA) and rapid thermal annealing (RTA) have annealing times long enough to fulfil this criterion, but the hallmark of ultra-short annealing techniques with annealing times in the range of ms and shorter is the possibility to anneal in non-equilibrium. It seems that *heat treatment* is a more appropriate term which is defined as:

L. Rebohle et al., *Flash Lamp Annealing*, Springer Series in Materials Science 288,
https://doi.org/10.1007/978-3-030-23299-3_1

> Series of operations in the course of which a solid ferrous product is totally or partially exposed to thermal cycles to bring about a change in its properties and/or structure. [2]

If we now replace "ferrous product" by "material", heat treatment seems to be a very suitable term which includes non-equilibrium processing steps, too. In addition, it also goes beyond the basic scheme of heating–soaking–cooling and allows more advanced thermal cycles with preheating, bimodal pulse shapes, multiple flashing etc. However, in most cases the different heat treatments are called *annealing* regardless of the restrictions mentioned above. Furthermore, the terms *thermal treatment* and *thermal processing* are also frequently used in literature. Thus, we regard these terms as synonyms for *heat treatment* and will use, depending on the context, the terms *annealing* or *thermal processing* throughout the book.

1.1 Historical Remarks

One can go as far as to say that the thermal processing of materials is a cornerstone of technological development and one of the fundamentals of our civilization. If the etymological meaning of annealing is interpreted literally, the roots of thermal processing reach back into a very distant past. Archaeological findings in Schöningen, Germany, indicate that homo heidelbergensis, a close ancestor of modern mankind, has hardened wooden spear tips by using an open fire about 300,000 BC [3]. This example already fulfils the most basic attributes of thermal processing: a material was subjected to elevated temperatures, and a specific material property, namely surface hardness, was improved. This archaic technique even exists today, but is now called flame hardening and is mostly applied to metals. Moreover, the open fire is usually replaced by a blowtorch or a similar device.

Even if this example appears controversial, the invention of advanced pottery using a kiln or furnace was undoubtedly one of the most important achievements of the early advanced civilizations. In the 4th millennium BC two-chamber kilns were used in Mesopotamia in which the kiln run (e.g. some ceramics) was separated from the fuel [4]. These furnaces already had the basic characteristics which are needed for annealing: the kiln run could be heated with a certain rate, the firing temperature could be fixed for a long time in a much more precise way than with an open fire, and the cooling rate could be adjusted within a certain range. Parallel to pottery, glass manufacture and metallurgy made important contributions. So in order to smelt copper, bronze, and finally iron; higher and higher temperatures were needed. This is even more significant for glass manufacture, although the admixture of soda can strongly decrease the melting point.

The art of annealing had been steadily improved over the centuries, but the development began to accelerate at the latest in the 18th century. At that time the first fundamental parts of thermodynamics were developed, although the onset of modern thermodynamics probably dates to the middle of the 19th century where mechanics and caloric theory was merged into thermodynamics [5]. Several temperature scales and thermometers with hitherto unequalled accuracy were invented including the well-known Fahrenheit and Celcius scale in 1724 and 1742, respectively. This development peaked out with the introduction of an absolute thermodynamic scale in 1848 by William Thomson, the later Lord Kelvin. First thermal engines were built like the Newcomen atmospheric engine in 1712 [6] and the Watt steam engine in 1776 [7]. In 1778, the potter Josiah Wedgwood invented the first pyrometer based on the calibrated shrinkage of porcelain pieces which were put together with the kiln run into the furnace [8]. In general, temperature measurements for short-time annealing are challenging, and if performed, pyrometry is the dominating technique (Sect. 3.1). In the 19th century, coal gradually became the dominant fuel, and at the end of the century electric power was another alternative source to produce heat. So the first induction furnace went into operation in 1899 in Sweden [9], and the first disappearing filament pyrometer was built by Holborn and Kurlbaum in 1901 [10]. In the last case the colour of a glowing, electrically heated filament is compared with the colour of a glowing object whose temperature has to be determined.

After this very brief journey through the early history of thermal processing we will take a closer look to flash lamp annealing (FLA). At the beginning there are two lines of development to be mentioned: namely flash annealing and that of flash lamps. The first one has nothing to do with flash lamps but was used as a synonym for fast annealing processes; in some sense it can be regarded as a precursor of RTA. This technique evolved in the 30s and 40s of the 20th century when there was a need for new solutions to produce light metal alloys with fine grains on a large industrial scale. To get fine grains, the heating rate and the annealing temperature has to be as fast as and as high as possible, respectively, with a corresponding reduction of the annealing time [11]. To meet the requirements of a large industrial scale flash annealing had to be either integrated into a conveyer belt process line or optimized for very short loading and soaking times. In any case the effective annealing time was in the range of seconds and minutes. An overview of these early solutions can be found in [11] and references therein. Until today, the term flash annealing is used from time to time to describe uncommon methods of short time annealing. One example is the use of current pulses to heat up materials on a time scale of ms [12–15].

Open-air, electrical-arc discharges were used at the beginning of the 20th century to photograph fast moving objects. For this purpose, Harold Edgerton invented the mercury-arc flash tube at the beginning of the 1930s and introduced the flash lamp as an important tool for science and industry [16]. A more detailed description of the circuitries and electrical characteristics of these early flash lamps can be found in [17, 18]. Later, the bandwidth of applications expanded and flash lamps were used to homogeneously initiate explosions, to trigger polymerization, and for short-time spectroscopy [19]. The latter point was of special scientific interest as flash lamps allowed the spectroscopic investigation of many metastable chemical compounds or fast chemical reactions. Another application is the measurement of thermodynamic properties of a material [20]. To do so, a plate with a suitable thickness is flashed from the front side whereas the temperature of the backside is recorded in a time-resolved way. The time delay between the flash and the temperature rise on the backside was used to deduce the thermal conductivity of the material.

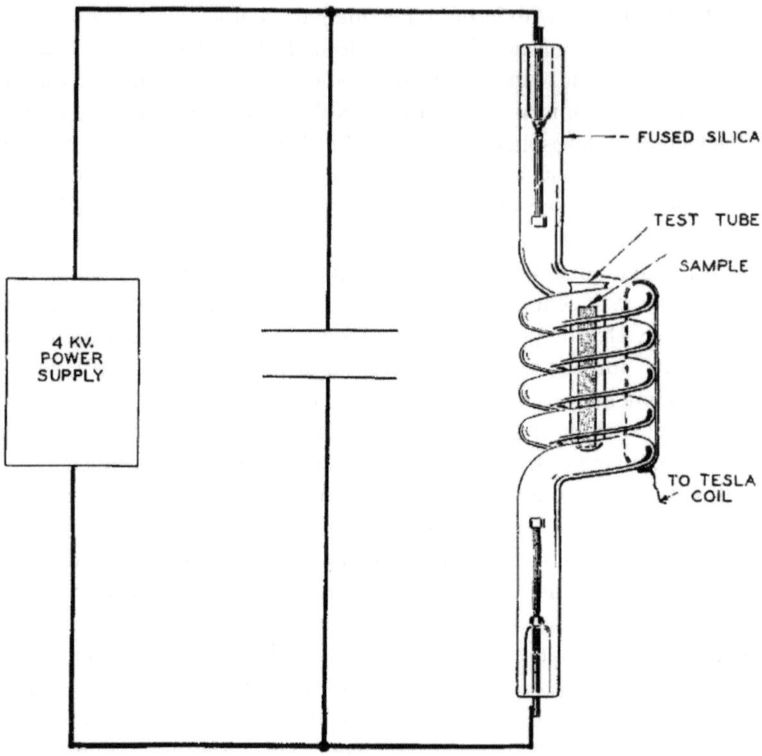

Fig. 1.1 Experimental arrangement of probably the first FLA setup in the late 50s. Reprinted by permission from Springer Nature [22]

Maybe the first FLA in a wider sense was performed by Lundberg and Nelson [21]. In this case they placed the sample to be annealed along the axis of a spiral quartz discharge lamp as shown in Fig. 1.1 [22]. Using a capacitor of 648 μF and a flash voltage of 4 kV they were able to produce an intense flash pulse with a duration of ca. 2.5 ms. The original study focused on polymers, mineral oils and the question how absorption can be maximized by using small absorbers with dimensions below 1 mm. The main advantage of FLA at that time was the possibility to heat up any material to several 1000s K within ms without the need of an explosion [23]. This means, flash light was used to modify the properties of materials, but probably not in the strict sense of defect annealing.

The invention of the laser in 1960 [24] suddenly increased the interest in powerful flash lamps which were able to pump lasers efficiently. In fact, the number of publications devoted to physics and technological aspects of flash lamps significantly increased in the following two decades. However, it was not until 1975 when researchers became aware of the high potential of short time annealing in general and laser annealing (LA) in particular. There have been a few isolated publications before which reported about the use of lasers for semiconductor applications. In 1968 phosphorous was deposited by a paint-on technique onto a p-type Si wafer followed by illumination of a ruby laser in order to form a p-n junction [25]. Shortly thereafter, a Nd:YAG laser was used to irradiate an Al layer on top of a n-type Si wafer with the result of alloying and, once again, p-n junction formation [26]. However, these techniques were not called laser annealing at this time.

Then there was a short pioneering age, approximately between 1975 and 1980, where most of the techniques of short time annealing were developed. The starting signal was given by a series of articles published by researchers from the former USSR in which they report about the recrystallization and defect annealing of Si and GaAs after ion implantation by a pulsed ruby laser [27, 28] or a continuous wave Ar laser scanned over the sample [29]. The idea became popular, and several research groups started own investigations as in Poland [30], East Germany [31], Italy [32, 33] and in the US [34–38].

Once again, the research group in Novosibirsk did pioneering work in the field of FLA [39]. The legend tells that an implanted semiconductor sample which was intended to get processed by LA had accidentally fallen into the laser mirror [40]. When the sample was found researchers were astonished that the sample was nevertheless properly annealed. The technique of FLA spread even faster than in case of LA. Thereby, the high conversion efficiency of electrical into optical energy was an especially attractive feature, even if the achievable absolute power density could not compete with that of a focused laser beam. In 1978, Cohen et al. were able to perform solid phase epitaxy on ion-implanted Si and Ge by using FLA [41]. In the same year Bomke et al. successfully annealed B-implanted Si with 15 μs short pulses using a free arc configuration as shown in Fig. 1.2 [42]. The chamber was filled with Xe or Ar, and the sample was directly exposed to the radiation of the plasma arc including the deep ultraviolet (UV) with wavelengths shorter than 200 nm. A spherical back reflector made of aluminium was used to guide more light onto the sample. A similar system was used by Lue et al. using 50 μs long pulses to anneal Si implanted with BF_2 [43]. Two years later a configuration was published in which the wafer target holder itself served as the cathode of the plasma arc [44]. However, this type of FLA was costly and somewhat impractical, which is why it has not established itself. In addition, still at the beginning it became clear that the flash needs thermal assistance to reach sufficiently high temperatures at the surface [41]. In [45] two subsequent pulses with a duration of 800 μs were applied in which the first one served as a preheating for the second pulse. There were further early reports about FLA [46, 47], and also the long-term experience in FLA of our own institute dates back to this time [48–50]. There were other methods of transient annealing which included electron beams [51, 52], the use of a graphite strip heater [53], current pulse heating [12] or pulsed microwaves [54]. However, none of these techniques has been implemented on a large scale so far.

RTA was also developed in those pioneering days when feature sizes in microelectronics came close to 1 μm. People recognized the need for shorter annealing times, but hesitated to reduce annealing time by so many orders of magnitude. A main concern about the general use of ms annealing at that time was the occurrence

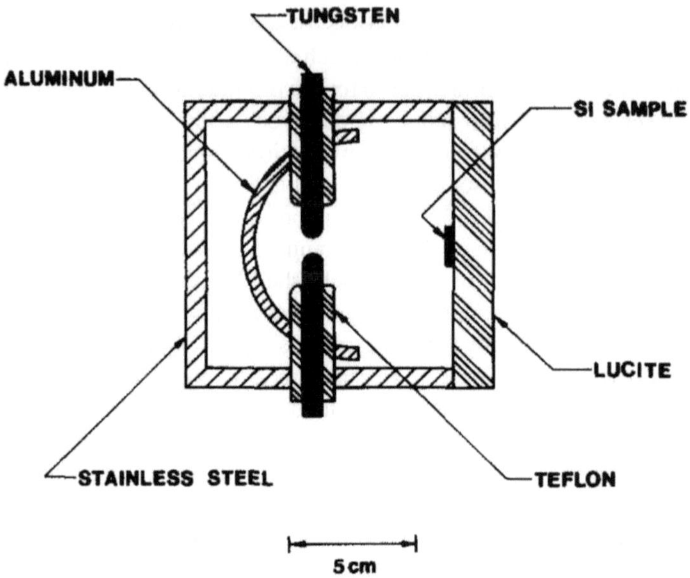

Fig. 1.2 Early FLA configuration with a free plasma arc as used in [42]. Reprinted from [42], with the permission of AIP Publishing

of thermal stress. According to [55] the first RTA using halogen lamps was performed in 1980 [56] with an annealing time of a few seconds. Alternatively, Xe flash lamps were also used working in a continuous mode [57]. From this time onwards, RTA was widely used in microelectronics, but with constantly decreasing feature size the trend was to shorten annealing times further. Even with the introduction of spike RTA, the ultimate limit was in the order of 0.5 s [58]. The turn of the millennium was a turning point in several respects. FLA was gradually introduced in microelectronics industry, and Moore's law underwent a paradigm shift. Up to the 90 nm node introduced around 2003, Moore's law was mainly driven by device shrinking with the corresponding reduction in tolerable diffusion lengths [59]. However, in order to continue to follow Moore's law, new materials like SiGe or high-k materials were introduced. These materials had to tolerate FLA, or FLA was even needed as in case of some high-k materials where interface reaction between Si and the high-k material had to be avoided.

Not even 10 years later, Moore's law required another change, namely the introduction of new architectures on the nanoscale [59]. This change is usually dated to the introduction of the tri-gate architecture at the 22 nm node around 2011. Thus, one of the current challenges of FLA is the annealing of nanostructures. However, FLA has also developed away from microelectronics and has moved into new areas of application such as photovoltaics, flexible electronics or the combination with deposition techniques.

1.2 The Thermal Cycle

Usually, an annealing process is characterized by the annealing temperature T_A and the annealing time t_A. However, in some cases the whole temperature profile over time, called the thermal cycle, is important, and the knowledge of only T_A and t_A is not sufficient. Figure 1.3a displays a typical RTA cycle consisting of heating, soaking and cooling [55]. This case is relatively simple as t_A is the soaking time and T_A is the temperature during soaking. However, if t_A becomes shorter and shorter, the impact of heating and cooling becomes more and more dominant. Heating and cooling are described by a heating and cooling rate given in K s^{-1}, respectively. In fact, the heating rate is a distinctive feature of the individual techniques of short time annealing. For annealing in the sub-second time range, the shape of the temperature profile approaches that of an asymmetric pulse where heating and cooling is described by the rising or falling edge of the pulse, respectively (Fig. 1.3b). A pulse doesn't have the equivalent of soaking, and t_A is now approximated by the width of the pulse, whereas T_A is its peak temperature. In addition, the width of an asymmetric peak is not always unambiguously defined. Typically, temperature starts from a value T_0—room temperature (RT) or a preheating temperature—increases up to the maximum T_A and decreases to an equilibrium (or quasi-equilibrium) temperature T_{equ} which is often higher than T_0. T_{equ} does not mark an equilibrium in the strict sense but indicates the region where further cooling takes place on a much longer time scale. In most cases the full width at half maximum (FWHM) is a good approximation, whereby the temperature difference $T_A - T_{equ}$ is a more appropriate measure to define the half maximum than the difference $T_A - T_0$.

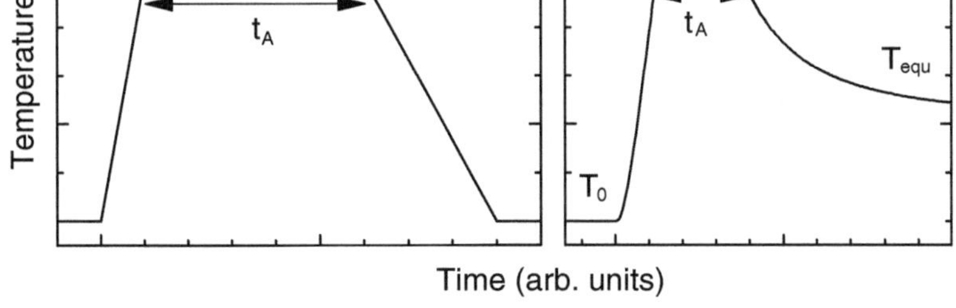

Fig. 1.3 Schematic temperature profile during a longer RTA cycle (**a**) and surface temperature profile during FLA (**b**). The latter one is the simulation for a 20 ms flash pulse applied to bulk Ge

Quantitative analysis often employs the concept of thermal budget which is defined as the area under the temperature profile [60]. This definition is, in a courteous manner, misleading. The shortcoming becomes obvious in the following thought experiment: imagine that T_A of the RTA cycle in Fig. 3.1a is doubled, whereas the soaking time has been halved. Neglecting the heating and cooling phase, both cycles have the same thermal budget wrongly implying that the physical impact would be similar. In contrast, the physical impact can be very different if e.g. the higher temperature exceeds the melting temperature whereas the lower one does not. Thermally activated processes often obey an exponential dependence on temperature, whereas diffusion processes are known for their square root-like dependence on time. If there is an interaction of different physical processes the impact of T_A and t_A becomes even more complex. Thus, any quantification strongly depends on the specific mechanisms of interest. In fact, the term "thermal budget" is almost exclusively used in a qualitative or figurative way in literature. A common subject of discussion is the reduction of the thermal budget in microelectronics. This typically means the optimization of a thermal cycle which suppresses unwanted processes like diffusion without affecting desired processes like dopant activation. This may not only include a change of T_A and t_A, but also an adjustment of heating, cooling, the introduction of preheating or the assembly of several thermal cycles. The thermal budget can also be used for qualitative comparisons if only one annealing parameter is different. Another, slightly ambiguous case is the statement that the thermal budget decreases in the order of RTA, FLA and pulsed LA. To get a similar physical effect, T_A is usually increased by 100–200 K, whereas t_A is reduced by orders of magnitude. One can argue that this annealing time reduction dominates over the temperature increase, and in many cases this is reasonable. Nevertheless, it depends on the specific conditions, and comparisons should always be considered carefully. Throughout the book we use the term thermal budget only in a general and figurative sense and refer to literature for more details [60].

There are few thermodynamic values which are helpful for imagination and discussion. The thermal diffusivity describes how fast a material can transport heat via heat conduction in relation to its ability to store heat. It is given by

$$D_{\text{th}} = \frac{\lambda_T}{\rho c_p} \tag{1.1}$$

where λ_T is the thermal conductivity, ρ is the mass density, and c_P is the specific heat capacity. Consequently, the thermal diffusion length x_{th} is defined as

$$x_{\text{th}} = \sqrt{D_{\text{th}} t} \tag{1.2}$$

x_{th} is the characteristic length scale over which heat can be transported in a given time t. Or otherwise, if the temperature at the surface of a material is suddenly increased at $t = 0$ and heat starts to propagate, x_{th} is the exponential decay coefficient with which a temperature profile slopes with increasing distance from the surface. Finally, the thermal response time τ_{th} of a physical system is obtained by:

$$\tau_{\mathrm{th}} = \frac{s^2}{D_{\mathrm{th}}} \tag{1.3}$$

where s is the relevant extension of the system. τ_{th} is the characteristic time for temperature equalization: for much shorter times heat transport is still negligible, and for much longer times the temperature has equalized. This can be illustrated by the following example. For silicon, the thermal diffusivity ranges between 0.9 cm^2 s^{-1} at RT and 0.09 cm^2 s^{-1} at 1600 K which equals a thermal response time τ_{th} of 3 and 30 ms, respectively, for a wafer thickness of 525 μm. This means, annealing times of several tens of ms are sufficient to heat up the entire wafer. For fused silica, thermal diffusivity drops down to 0.009 cm^2 s^{-1} and shows only slight changes with temperature. Thus, for a 525 μm thick fused silica wafer τ_{th} is about 300 ms, whereas for a 100 μm thin foil τ_{th} reduces to 11 ms.

1.3 Short Time Annealing

Today, the three most common methods of short time annealing are RTA, FLA and LA. They are all based on the absorption of light, and their annealing time covers more than 11 orders of magnitude: from 10^{-8} to 100 s [60] with a flexible lower limit to encompass even shorter laser pulses. The light is usually absorbed in a narrow region close to the surface, and heat conduction heats up the rest of the sample. One exception is the case of samples which are transparent or semi-transparent with respect to the spectrum of the light source. Typical examples for this are glassy substrates or SiC wafers which absorb only in the UV spectral region. Thus, the annealing time strongly determines whether only the surface, significant parts of the sample or the entire sample is heated up to high temperatures. These three processes are called adiabatic, thermal flux or isothermal [55, 61], and they are used to classify annealing methods. One exception is the CO_2 laser with a wavelength of about 10 μm which is mainly absorbed by free charge carriers. Depending on the doping level, it can therefore deeply penetrate into semiconductor materials.

The typical annealing times of RTA are in the range of 1–100 s with heating rates up to 10^3 K s^{-1}. As this is much longer than τ_{th}, RTA is an isothermal process in which the front and backside of the substrate are at the same temperature. The energy is typically supplied by halogen lamps which emit a broad spectrum in the visible and near infrared (NIR) spectral range with a maximum in the NIR. Interference and pattern effects can be neglected as the annealing time is long enough to level out temperature differences which may arise due to inhomogeneous absorption. As there are no temperature gradients, thermal stress only arises due to different thermal expansion coefficients in a heterogeneous material system. Finally, RTA is a substrate-by-substrate process.

In case of FLA annealing times mainly cover the range between 10 μs and 100 ms, although shorter times were also reported for special applications. τ_{th} is now of the same order of magnitude as the annealing time and thus in the thermal flux regime

[61]. However, the scope of potential annealing times is so large that the thermal flux regime can turn into an isothermal process like RTA for long pulse times or into an adiabatic process like LA for very short pulse times. In the thermal flux regime only a narrow region close to the surface is annealed at maximum temperature, and there is a decaying temperature profile towards the backside whose temperature is somewhere between RT and the surface temperature. Depending on whether the flash pulse is short or long the backside temperature is closer to RT or to the surface temperature, respectively. The energy is usually supplied by Xe flash lamps which also emit a broad spectrum, but now with a maximum in the blue-violet spectral region. The resulting temperature profiles depend on the thermal and optical properties of the material system, i.e. applying the same flash pulse to two different material systems may result in two different temperature profiles. Interference and pattern effects matter, but they are reduced by averaging over the broad flash lamp spectrum. Additional thermal stress is generated due to the temperature gradients, but it can be managed by choosing adequate pre-heating conditions. Similarly to RTA, FLA is a substrate-by-substrate process.

In case of LA, distinction must be made between continuous and pulsed operation. In the first case, the laser beam is moved over the sample (or the sample is passed underneath) which results in dwelling times in the ms range depending on spot size and the speed of movement. Thus, the effective annealing times are similar to those of FLA, and the process runs in the thermal flux regime. In contrast, the annealing times of pulsed LA are in the order of several tens of ns. Thus, the annealing is an adiabatic process in which only a narrow surface region with a thickness comparable to the penetration depth of the light is really heated. Heat conduction will warm up the neighbouring regions after the laser pulse, but the backside and the major part of the sample will stay close to RT. As a laser is monochromatic, interference and pattern effects are serious, and temperature depends even more on the thermal and optical properties of the material system. The problems with thermal stress can be more severe than in case of FLA, but are limited to the surface region. Thus, pulsed LA will have fewer problems with wafer bowing, but the risk of ablation or delamination is much higher. In contrast to RTA and FLA, the annealing of a whole wafer requires the scanning of the laser beam over the wafer area. The different annealing methods are compared to each other in Table 1.1.

Each of these annealing methods has its own pros and cons, but there are a couple of technical and economic effects which should also be considered. Whereas RTA and FLA are substrate-by-substrate processes, LA requires the scanning of the laser beam over the wafer area which is disadvantageous with respect to process management. Generally, all methods can be integrated in larger process lines, although RTA may delay the speed of transportation or may require long annealing zones depending on the specific annealing time. Regarding the scalability to large areas or to roll-to-roll (R2R) production lines FLA has probably the highest potential. There is no physical limit for RTA, but tools may require a very large footprint and have high energy costs. FLA is technologically more challenging, but there are flash lamps with an arc length of 3.7 m [62] which may process substrate widths up to ~3 m. LA has the problem that the two-dimensional scanning of a large area is simply inefficient.

Table 1.1 Comparison of different short time annealing methods. T_{FS} and T_{BS} denote the maximum temperature on the front and backside, respectively

	RTA	FLA	cw LA	Pulsed LA
Temperature profile	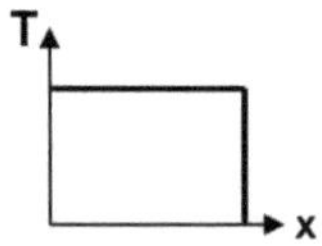	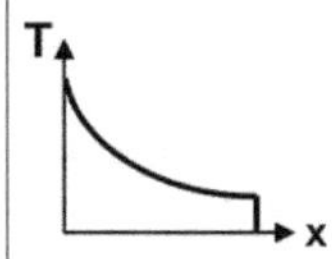		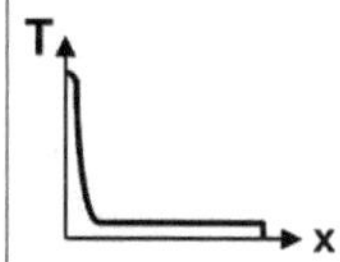
Annealing regime	Isothermal	Thermal flux		Adiabatic
Annealing times	1–100 s	10 μs–100 ms		1–1000 ns
Backside temperature	$T_{BS} \approx T_{FS}$	$T_{BS} < T_{FS}$		$T_{BS} \ll T_{FS}$
Temperature depends on material?	No	Yes		
Temperature measurement	Easy	Difficult	Medium	Very difficult
Heating rates (K s^{-1})	10–10^3	10^4–10^7		>10^8
Light source	Halogen lamps	Xe flash lamps	Laser	
Spectrum	Broad max. ~ NIR	Broad max. ~ blue	Discrete lines from UV to MIR	
Interference effects	None	Yes		
Process type	Substrate by substrate		Scanning required	
R2R compatibility	Medium	High	Medium	Low

There is the possibility to widen the laser spot to a thin line or to assembly a line scan by a large number of laser diodes, but these methods are limited in width and achievable power density. In addition, in case of indium tin oxide (ITO) annealing on large glass panels it was found that LA is much more expensive than FLA [63]. In [64], the processing cost of a specific FLA tool was projected to be roughly $0.11 US-$/m^2.

References

1. https://www.merriam-webster.com/dictionary/annealing. Accessed 19 Sept 2017
2. European Standard 10052:1993, *Vocabulary of Heat Treatment Terms of Ferrous Products*
3. D. Richter, M. Krbetschek, The age of the lower Paleolithic occupation at Schoningen. J. Hum. Evol. **89**, 46–56 (2015). https://doi.org/10.1016/j.jhevol.2015.06.003

4. G. Schneider, A technological study of North-Mesopotamian stone ware. World Archael. **21**, 30 (1989)
5. A. Bejan, Evolution in thermodynamics. Appl. Phys. Rev. **4**(1), 011305 (2017). https://doi.org/10.1063/1.4978611
6. http://www.bbc.co.uk/history/historic_figures/newcomen_thomas.shtml. Accessed 19 Sept 2017
7. http://www.bbc.co.uk/history/historic_figures/watt_james.shtml. Accessed 19 Sept 2017
8. http://www.bbc.co.uk/history/historic_figures/wedgwood_josiah.shtml. Accessed 19 Sept 2017
9. http://www.lokodex.de/IO/doku.php?id=geschichte. Accessed 19 Sept 2017
10. L. Holborn, F. Kurlbaum, Über ein optisches Pyrometer. Sitzungsberichte d. preussischen Akademie d. Wiss. Teil 1, 712 (1901) (in German)
11. R.T. Staples, Flash annealing of light alloys. J. Inst. Met. **80**(6), 323 (1951)
12. L.T. Kabacoff, M. Wun-Fogle, Magnetomechanical properties of flash annealed Metglas 2605 SC. J. Appl. Phys. **57**(8), 3499–3501 (1985). https://doi.org/10.1063/1.335040
13. A. Mitra, M. Vazquez, K. Mandal, S.K. Ghatak, Influence of stress and stress-flash annealing on the magnetic properties of amorphous $Fe_{77.5}Si_{7.5}B_{15}$ wire. J. Appl. Phys. **70**, 4455 (1991). https://doi.org/10.1063/1.349102
14. C. Morón, M.C. Sánchez, E. López, P. Sánchez, C. Aroca, Effect of high-current flash annealing on amorphous samples. J. Magn. Magn. Mater. **101**, 59–61 (1991)
15. T. Horubala, R. Masiak, H. Matyja, Microstructural changes in Fe-Si-B and Co-Si-B amorphous alloys upon flash annealing. Key Eng. Mater. **81–83**, 323–328 (1993). https://doi.org/10.4028/www.scientific.net/KEM.81-83.323
16. P.E. Gray, H.E. Edgerton, Phys. Today **44**(4), 126–128 (1991). https://doi.org/10.1063/1.2810095
17. H.E. Edgerton, K.J. Germeshausen, The mercury arc as an actinic stroboscopic light source. Rev. Sci. Instrum. **3**, 535 (1932). https://doi.org/10.1063/1.1748867
18. P.M. Murphy, H.E. Edgerton, Electrical characteristics of stroboscopic flash lamps. J. Appl. Phys. **12**(12), 848–855 (1941). https://doi.org/10.1063/1.1712876
19. M.I. Christie, G. Porter, Photochemical studies with the gaseous flash discharge. Proc. R. Soc. London A **212**(1110), 390–397 (1952)
20. W.J. Parker, R.J. Jenkins, C.P. Butler, G.L. Abbott, Flash method of determining thermal diffusivity, heat capacity, and thermal conductivity. J. Appl. Phys. **32**(9), 1679–1684 (1961). https://doi.org/10.1063/1.1728417
21. J.L. Lundberg, L.S. Nelson, High-intensity flash irradiation of polymers. Nature **179**(4555), 367 (1957)
22. J.L. Lundberg, L.S. Nelson, Initiation of thermal reactions by the flash illumination of absorbing bodies. Nature **183**(4675), 1560 (1959). https://doi.org/10.1038/1831560a0
23. L.S. Nelson, Intense rapid heating with flash discharge lamps. Science **136**, 296 (1962)
24. T.H. Maiman, Stimulated optical radiation in ruby. Nature **187**(4736), 493 (1960)
25. J.M. Fairfield, G.H. Schwuttke, Silicon diodes made by laser irradiation. Solid State Electron. **11**(12), 1175 (1968). https://doi.org/10.1016/0038-1101(68)90008-7
26. F.E. Harper, M.I. Cohen, Properties of Si diodes prepared by alloying Al into n-type Si with heat pulses from a Nd:YAG laser. Solid State Electron. **13**, 1103 (1970)
27. G.A. Kachurin, N.B. Pridachin, L.S. Smirnov, Annealing of radiation defects by laser radiation pulses. Sov. Phys. Semicond. **9**(7), 946 (1975)
28. I.B. Khaibullin, V.V. Titov, E.I. Shtyrkoy, M.M. Zaripov, V.P. Stashko, K.P. Kuzmin, Laser annealing of implanted layers, in *Proceedings of International Conference on Ion Implantation in Semiconductors*, Budapest, 1975, p. 212 (in Russian)
29. G.A. Kachurin, E.V. Nidaev, A.V. Khodyachikh, L.A. Kovaleva, Annealing of implanted layers by a scanning laser beam. Sov. Phys. Semicond. **10**(10), 1128 (1975)
30. J. Krynicki, J. Suski, S. Ugniewski, R. Grötzschel, R. Klabes, U. Kreissig, J. Rüdiger, Laser annealing of arsenic implanted silicon. Phys. Lett. **61A**(3), 181 (1977)

31. H.D. Geiler, G. Götz, K.D. Klinge, N. Triem, Investigation of laser induced diffusion and annealing processes of arsenic-implanted silicon crystals, Phys. Stat. Sol. (a) **41**, K171 (1977)
32. P. Baeri, S.U. Campisano, G. Foti, E. Rimini, Arsenic diffusion in silicon melted by high-power nanosecond laser pulsing. Appl. Phys. Lett. **33**(2), 137–140 (1978). https://doi.org/10.1063/1.90283
33. G. Battaglin, G. Della Mea, A.V. Drigo, G. Foti, G.G. Bentini, M. Servidori, Two-stage laser annealing of lattice disorder in phosphorus implanted silicon. Phys. Stat. Sol. (a) **49**, 347 (1978)
34. G.K. Celler, J.M. Poate, L.C. Kimerling, Spatially controlled crystal regrowth of ion-implanted silicon by laser irradiation. Appl. Phys. Lett. **32**(8), 464–466 (1978). https://doi.org/10.1063/1.90109
35. H.J. Leamy, G.A. Rozgonyi, T.T. Sheng, G.K. Celler, Periodic regrowth phenomena produced by laser annealing of ion-implanted silicon. Appl. Phys. Lett. **32**(9), 535–537 (1978). https://doi.org/10.1063/1.90119
36. A. Gat, J.F. Gibbons, A laser-scanning apparatus for annealing of ion-implantation damage in semiconductors. Appl. Phys. Lett. **32**(3), 142–144 (1978). https://doi.org/10.1063/1.89960
37. R.T. Young, C.W. White, G.J. Clark, J. Narayan, W.H. Christie, M. Murakami, P.W. King, S.D. Kramer, Laser annealing of boron-implanted silicon. Appl. Phys. Lett. **32**(3), 139–141 (1978). https://doi.org/10.1063/1.89959
38. J.A. Golovchenko, T.N.C. Venkatesan, Annealing of Te-implanted GaAs by ruby laser irradiation. Appl. Phys. Lett. **32**(3), 147–149 (1978). https://doi.org/10.1063/1.89962
39. G.A. Kachurin, E.V. Nidaevj, Fiz. Tekh. Poluprovod. **11**, 2012 (1977) (in Russian)
40. M. Voelskow, Historical aspects of subsecond thermal processing, in *Subsecond Annealing of Advanced Materials*. Springer Series in Material Science, vol. 192, ed. by W. Skorupa, H. Schmidt (Springer, Cham, 2014), pp. 1–13
41. R.L. Cohen, J.S. Williams, L.C. Feldman, K.W. West, Thermally assisted flash annealing of silicon and germanium. Appl. Phys. Lett. **33**(8), 751–753 (1978). https://doi.org/10.1063/1.90528
42. H.A. Bomke, H.L. Berkowitz, M. Harmatz, S. Kronenberg, R. Lux, Annealing of ion-implanted silicon by an incoherent light pulse. Appl. Phys. Lett. **33**(11), 955–957 (1978). https://doi.org/10.1063/1.90232
43. J.T. Lue, Arc annealing of BF^{+2} implanted silicon by a short pulse flash lamp. Appl. Phys. Lett. **36**(1), 73–76 (1980). https://doi.org/10.1063/1.91279
44. J.T. Lue, A multichannel flash tube implemented for large area annealing of ion implanted semiconductors. Vacuum **32**(12), 713–718 (1982)
45. L. Correra, L. Pedulli, Incoherent-light-flash annealing of phosphorus-implanted silicon. Appl. Phys. Lett. **37**(1), 55–57 (1980). https://doi.org/10.1063/1.91846
46. J.M. Poate et al., Liquid and solid phase regrowth of Si by laser irradiation and thermally assisted flash annealing. Radiat Eff. **48**(1–4), 167–174 (1980). https://doi.org/10.1080/00337578008209249
47. J.T. Lue, Y.C. Liu, W.J. Shen, Light-flash induced metallic silicides from titanium films on silicon. Appl. Phys. Lett. **38**(5), 372–374 (1981). https://doi.org/10.1063/1.92342
48. R. Klabes, J. Matthäi, M. Voelskow, G.A. Kachurin, E.V. Nidaev, H. Bartsch, Flash lamp annealing of arsenic implanted silicon. Phys. Stat. Sol. (a) **66**, 261 (1981)
49. R. Klabes, M. Voelskow, H. Woittennek, E.V. Nidaev, L.S. Smirnov, Dopant redistribution after flash lamp annealing. Phys. Stat. Sol. (a) **71**, K127 (1982)
50. K.H. Heinig, K. Hohmuth, R. Klabes, M. Voelskow, H. Woittennek, Flash lamp annealing of ion implanted silicon. Radiat. Eff. **63**(1–4), 115–123 (1982). https://doi.org/10.1080/00337578208222831
51. A.R. Kirkpatrick, J.A. Minnucci, A.C. Greenwald, Silicon-solar cells by high-speed low-temperature processing. IEEE Trans. Electron Dev. **24**(4), 429 (1977)
52. R.A. McMahon, H. Ahmed, Electron-beam annealing of ion-implanted silicon. Electron. Lett. **15**, 45 (1979). https://doi.org/10.1049/el:19790032
53. B.Y. Tsaur, J.P. Donnelly, J.C.C. Fan, M.W. Geis, Transient annealing of arsenic-implanted silicon using a graphite strip heater. Appl. Phys. Lett. **39**(1), 93–95 (1981). https://doi.org/10.1063/1.92529

54. P. Chenevier, J. Cohen, G. Kamarinos, Pulsed annealing of semiconductors by microwave energy. J. Phys. Lett. **43**(8), 291–294 (1982). https://doi.org/10.1051/jphyslet:01982004308029100
55. R. Singh, Rapid isothermal processing. J. Appl. Phys. **63**(8), R59 (1988)
56. K. Nishiyama, M. Arai, N. Watanabe, Radiation annealing of boron-implanted silicon with a halogen lamp. Jap. J. Appl. Phys. **19**(10), L563–L566 (1980)
57. R.A. Powell, T.O. Yep, R.T. Fulks, Activation of arsenic-implanted silicon using an incoherent light source. Appl. Phys. Lett. **39**(2), 150–152 (1981). https://doi.org/10.1063/1.92642
58. P. Timans, J. Gelpey, S. McCoy, W. Lerch, S. Paul, Millisecond annealing, past, present and future. Mater. Res. Soc. Symp. Proc. **912**, 3 (2006)
59. E.J.H. Collart, P.M. Kopalidis, M. Hou, S. McCoy, P.J. Timans, A. Joshi, S. Prussin, Effects of implant temperature and millisecond annealing on dopant activation and diffusion, in *Ion Implantation Conference 2012, AIP Conference Proceedings*, vol. 1496 (2012), pp. 95–98. https://doi.org/10.1063/1.4766498
60. R. Ditchfield, E.G. Seebauer, Rapid thermal processing: fixing problems with the concept of thermal budget. J. Electrochem. Soc. **144**(5), 1842 (1997)
61. T.O. Sedgwick, Short time annealing. J. Electrochem. Soc. **130**(2), 484 (1983)
62. http://www.vonardenne.biz/fileadmin/user_upload/druckschriften/Flash_Lamp_Annealing_english.pdf. Accessed 21 Mar 2016
63. Y. Kim, S. Park, B.-K. Kim, H.J. Kim, J.-H. Hwang, Xe-arc flash annealing of indium tin oxide thin-films prepared on glass backplanes. Int. J. Heat Mass Transf. **91**, 543–551 (2015). https://doi.org/10.1016/j.ijheatmasstransfer.2015.07.132
64. K.A. Schroder, S.C. McCool, W.F. Furlan, Broadcast photonic curing of metallic nanoparticle films. NSTI-Nanotech 2006 **3**, 198 (2006)

Chapter 2
The Technology of Flash Lamp Annealing

Abstract This chapter addresses various technical aspects of FLA and starts with the individual FLA tool and flash lamp components, followed by a description of the plasma properties during operation. The section about the electric operation of flash lamps deals with IV characteristics of flash lamps and the corresponding circuitries, namely the type of triggering, the use of simmer tools and the design of pulse-forming networks. In addition, the optical output is discussed in terms of spectrum, radiation pattern and pulse shape. Special attention has been paid to the question of how to shape the spectral output of a flash lamp. Finally, the subject of lamp failure and lamp degradation is shortly illuminated. The chapter closes with a short discussion about the layout of FLA tools and the possibilities of flash-assisted depositions.

2.1 Basic Structure of an FLA Tool

Figure 2.1 displays a basic scheme of a flash lamp annealing (FLA) tool as used in semiconductor processing. In the simplest case an FLA tool consists of a chamber (1) or another suitable housing which comprises the sample (2) on a substrate holder and a bank of flash lamps (3) which is powered by an external circuitry. Substrates like wafers are usually supported by a few quartz pins only in order to avoid wafer breakage if the wafer warps or bulges during the flash. If the chamber is closed, the flash can be applied under vacuum or low pressure conditions, or under a constant flow of an inert gas. However, in some special applications reactive gases like forming gas (N_2:H_2) or oxygen are also used. Nearly all FLA tools use reflectors (4) in order to guide the light towards the sample and to ensure a better homogeneity of the light on the sample surface. In addition, the efficient harvesting of the flash light allows the reduction of the electrical input power and of the thermal load of the flash lamp. Typically, the flash lamps are protected by a quartz window (5) to prevent the deposition of sample material on the quartz envelopes of the flash lamps. Inversely, the sample is protected from lamp fracture. Finally, a couple of FLA tools are equipped with a preheating system which has the task to elevate the temperature of the sample in advance of the flash. This is useful if high temperatures have to be applied or if thermal stress has to be reduced. The preheating can be a bank of halogen lamps

L. Rebohle et al., *Flash Lamp Annealing*, Springer Series in Materials Science 288,
https://doi.org/10.1007/978-3-030-23299-3_2

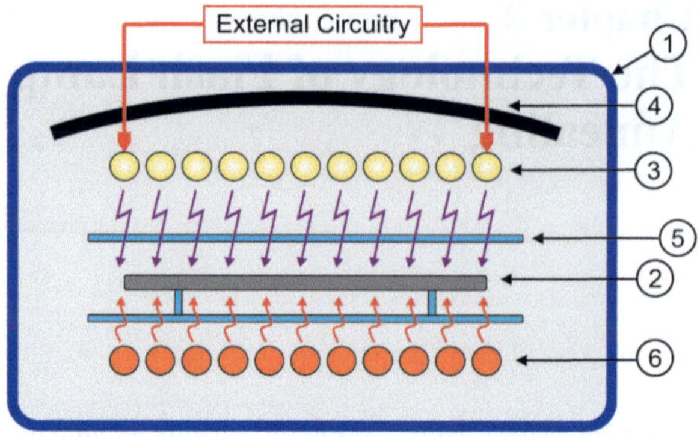

Fig. 2.1 Basic scheme of a FLA tool. Numbers are explained in the text

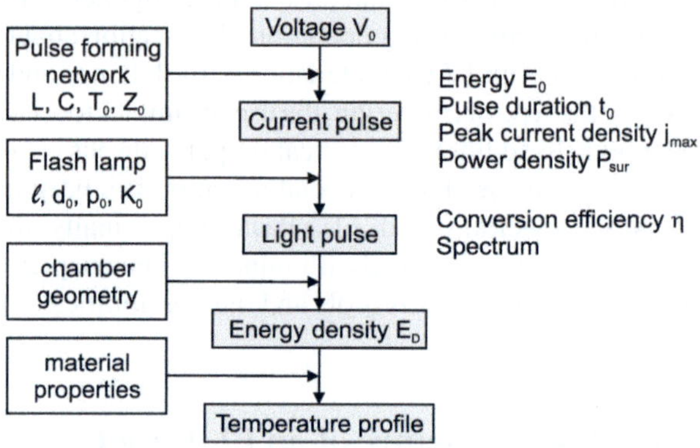

Fig. 2.2 Working scheme of a FLA tool. More details can be found in the text

(6) which heats the sample from the back side and is equivalent to a rapid thermal annealing (RTA) process. Alternatively, a hot plate can serve as well for this purpose. Finally, two synchronized flashes can be used as it was exemplarily done in [1]. In this case an 8 ms long pulse was applied to the backside to achieve a preheating temperature of 450 °C immediately followed by an 800 μs short pulse applied to the front side.

The final temperature of the sample depends in a complex manner on the FLA tool configuration, the individual process parameters, and the material properties of the sample. Figure 2.2 shows the basic working scheme from the charging voltage V_0 up to the final temperature distribution. In this picture there are four main components determining the final result of annealing: the external circuitry, the flash lamp, the chamber, and the material system. Each of these components is characterized by several parameters which are briefly introduced in the following discussion.

The first main component is the external circuitry, and for the sake of simplicity it is assumed that it consists of a single capacitor and a single coil in series which both constitute the pulse forming network (PFN). Thus, the *capacitance C*, the *inductance L* and the *charging voltage* V_0 of the capacitor are the primary parameters which define a couple of secondary parameters. Typical parameter ranges are 0.01–10 mF for C, 0.01–10 mH for L and a few kV for V_0. The usual parameter an operator is setting from the outside is V_0, and together with C it determines the *electrical input energy* E_0 which is stored in the PFN and can be released into the flash lamp:

$$E_0 = \frac{1}{2} C V_0^2 \tag{2.1}$$

In order to get high input energies for short pulses, high voltages have to be combined with sufficiently small capacitances. To give some examples, a voltage of 40 kV with $C = 0.75\ \mu F$ was used to get an energy of 600 J in a 1 μs pulse (which is enormous for that time scale) [2], and a capacitor bank with 24 kV and $C = 3$ mF was able to supply a ~360 μs pulse with an energy of 864 kJ [3]. However, it should be considered that E_0 has to be split if several flash lamps are used or if the discharge is controlled by an electronic switch. In the latter case the capacitor provides only a part of its stored energy. L and C define the *characteristic time constant* T_0, given by

$$T_0 = \sqrt{LC} \tag{2.2}$$

which is a rough measure of the time scale of the pulse. In addition, the *impedance* Z_0 of the capacitor-coil combination

$$Z_0 = \sqrt{\frac{L}{C}} \tag{2.3}$$

influences the shape of the current pulse, namely the relation between the rising and the falling edge and thus the asymmetry of the pulse. Finally, the *frequency* f of subsequent flash pulses is another parameter which is of interest for certain applications. Most of all it influences the dimensioning of the external circuitry, especially that of the charging unit, and the cooling conditions under the aspect of continuous power consumption.

The flash lamp is the second main component, and its main primary parameters are the *arc length* ℓ, the *bore diameter* d_0, the type of gas and its *filling pressure* p_0. These data are included in the calculation of the *lamp impedance* K_0 which determines the electric characteristics of the lamp. This topic is discussed in much more detail in the next section; the only thing to be said here is that both the lamp and the electric circuitry form the current pulse $I(t)$ or the current density profile $j(t)$ with

$$j(t) = \frac{4I(t)}{\pi d_a^2} \tag{2.4}$$

where $I(t)$ is the time-dependent current passing the lamp and d_a is the arc diameter. At the rising edge of the current pulse d_a is time-dependent and describes the expanding arc, whereas for later times d_a can be approximated by the bore diameter d_0. The pulse is usually characterized by the peak current I_{max} or *peak current density* j_{max}

and the *pulse length* t_0 which is pragmatically estimated as the full width at half maximum (FWHM). The pulse length is $t_0 \approx \beta T_0$ with β typically in the range of 2–3 depending on the specific conditions [4, 5]. Depending on the lamp design and operation parameters a specific part of the electric power is converted into an optical output power. A more conservative estimation of this *conversion efficiency* η, i.e. the ratio between the optical power emitted from the lamp and the electrical power pumped into the lamp, is in the order of 50%, although values above 80% have also been reported under optimized conditions [5]. Thus, the electrical power density with respect to the inner surface area of the envelope is given by

$$P_{\text{sur}}^{e} = \frac{E_0}{At_0} = \frac{E_0}{\pi d_0 \ell t_0} \tag{2.5}$$

where A is the inner lateral surface area of the cylindrical lamp around the arc. The optical counterpart is then given by

$$P_{\text{sur}}^{o} = \eta P_{\text{sur}}^{e} \tag{2.6}$$

P_{sur}^{e} is an important design parameter for dimensioning a FLA tool. Especially for short pulse times, P_{sur}^{e} can easily achieve values in the kW cm^{-2} region. However, for cooling consideration the power density averaged over several pulses, called the wall loading, is of higher interest. Thus, even for high pulse intensities cooling may not be necessary if the break between subsequent pulses is long enough. However, for very high intensities cooling can increase the lifetime of the lamp.

The optical output of the lamp is characterized by a time-dependent spectrum and a pulse shape. As shown in Fig. 2.3, the light pulse is characterized by a pulse length (FWHM) like the current pulse, but is mostly a few % shorter than the current pulse. In this example the FWHM of the current and light pulse amounts to 1.18 and 1.07 ms, respectively. However, if the current pulse is strongly asymmetric, the difference between the FWHM of the current and that of the light pulse becomes more and more significant. The optical pulse intensity is usually measured as optical energy E_{Opt} of the pulse and given by:

$$E_{\text{Opt}} = \eta E_0 \tag{2.7}$$

Although E_{Opt} is an integral measure and the spectrum changes during the flash, it is mostly sufficient to consider the time-integrated spectrum only.

The FLA chamber is the link between the light radiated from the flash lamp and the light a sample to be annealed is exposed to. Depending on the reflection properties, the chamber geometry, the lamp arrangement and the reflector design, the light is usually directed onto the sample as homogeneously as possible. The important parameter is the *energy density* E_{D} at the sample surface, given in J cm^{-2}, which is an integral measure over the spectrum and the pulse time. E_{D} can be easily estimated, and many reports about FLA and laser annealing use this value instead of a temperature which is much more difficult to estimate. However, the actually *absorbed energy density*

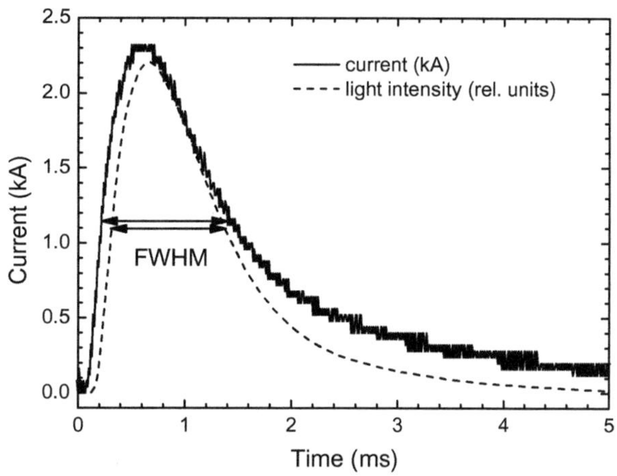

Fig. 2.3 Typical current (solid line) and light pulse (dashed line) as measured by a Rogowski coil and a photodiode, respectively. Pulses were obtained by using a 1.2 mF capacitor charged with 3.5 kV. Both signals were recorded by an oscilloscope. The light signal was digitally smoothed by a low pass filter in order to suppress high-frequency noise

A is lower due to the reflectance of the sample, although multiple reflections within the chamber may lead to additional absorption. See Sect. 3.2 for more details.

Finally, the *absorption profile* $a(x, t)$ and the *temperature distribution* $T(x, t)$ within the sample strongly depend on the geometrical, optical and thermodynamic properties of the material. While the geometrical properties mainly comprise the thicknesses within a thin film stack and the total thickness, the optical constants determine the absorption profile and which part of the incoming light is absorbed at all. The thermodynamic properties govern the heat distribution and the specific temperature increase for a given amount of heat. Further on, phase changes and other modifications which may happen during annealing will make the picture even more complex.

2.2 Components of a Flash Lamp

Flash lamps have a long history (Sect. 1.1) and their applications go well beyond the case of FLA. Thus, there is a larger body of literature about technological and physical issues which reaches back to the 1950s. The book of Marshak [6] about pulsed light sources and published in 1984 in English summarizes the knowledge of that time in a comprehensive way, of course with a focus on flash lamp activities in the former Soviet Union. Many empiric relationships about flash lamps which will be given in this chapter originate from this source. On the Western side there is a comprehensive review by ILC technology [5], a couple of scientific publications related to plasma physics inside a flash lamp [7] and several textbooks about laser physics which contain single chapters devoted to flash lamps as a pump source. More recent information can be obtained from reference manuals from industry [8, 9] and newer publications applying modern measurement techniques or simulation tools. There is also a more present study about flash lamps with a focus on electrode technology [10]. In the following we will survey these information sources and try to give a comprehensive and consistent picture about the operation and characteristics of flash lamps.

2.2.1 Dimensions and Gas Filling

Basically, flash lamps are made of two electrodes, a glass envelope and a suitable noble gas which completely fills the envelope. A basic scheme is drawn in Fig. 2.4 together with the main geometrical design parameters, namely arc length ℓ, bore diameter d_0 and wall thickness. Further design parameters include the type of electrodes, the type of sealing, the dead volume, the glass type of the envelope, the type of cooling if present, the type of gas and the gas filling pressure. These parameters will determine the electric operation conditions, especially the power range of the flash lamp. Today there is a wide variety of flash lamp applications, and consequently the design parameters vary in a large range. Most bore diameters vary between 3 and 20 mm, whereas arc lengths range from a few cm over "standard" lengths of 20–50 cm up to a few meters for large area application. One example for the latter is a flash lamp with an arc length of 3.7 m designed to treat coatings on large area glass bodies [11]. Bore diameters which are either too small or too large will cause a drop in conversion efficiency [5], which is why the specific bore diameter is often a trade-off between conversion efficiency and practical requirements. Wall thicknesses are mostly in the range of 0.5–2 mm [9] with the tendency to grow with increasing bore diameter and increasing arc length. There is a small gap between the electrodes and the envelope which allows the thermal expansion of the electrodes. This gap goes

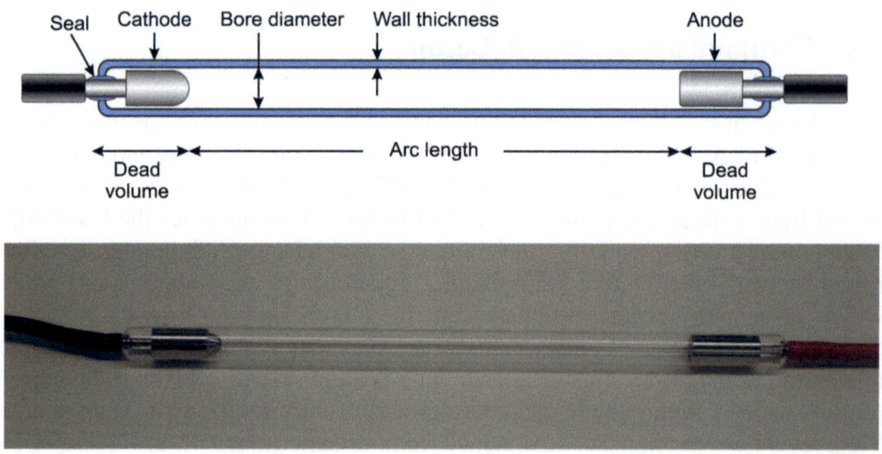

Fig. 2.4 Basic scheme of a flash lamp

down to zero if the operating temperature comes close to the shrinking temperature of the lamp during its manufacture [10].

Most flash lamps have a Xe gas filling with a filling pressure p_0 in the range of 400–600 Torr. The main reason for this is the high conversion efficiency η which increases with increasing atomic mass of the noble gas. Consequently, Xe is the most common gas filling, followed by Kr which is roughly 10% less efficient [5]. Thus, other gases than Xe are mainly used if someone wants to take advantage of special emission lines. This can be the case for certain laser applications, but it is less important for FLA. p_0 can range from values as low as 50 Torr to values exceeding the atmospheric pressure. Typically, η increases with increasing p_0, but reaches a saturation regime at a critical pressure, above which a further increase in pressure brings less benefit. For Xe lamps the critical pressure is in the range of 200–400 Torr with somewhat higher values for lower bore diameters [5]. For a comparable design Kr lamps usually have a higher p_0 than Xe lamps.

The final value of p_0 is also determined by other, electrotechnical considerations. At high pressures the lamp runs with lower currents but higher voltages. This, however, comes along with an increased risk of failure due to bulb fracture caused by enhanced stress. Low pressures foster high currents at lower voltages causing a lower conversion efficiency and the risk of enhanced electrode erosion. The dead volume V_{dead} is defined as the inner gas-filled volume between the tip of the electrode and the seal (Fig. 2.4), and it influences the gas pressure during operation which rises as the current through the lamp increases. The dead volume also lowers the working pressure by partially equalizing the pressure which finally reduces the conversion efficiency of the lamp [9]. Therefore, the dead volume is minimized in most cases.

2.2.2 Electrodes

The manufacture of electrodes requires a significant level of technical engineering as they bear the full brunt under operation. Both electrodes are often made of tungsten due to its high melting temperature, but differ in shape and the individual material modifications. The cathode should facilitate the emission of electrons for which reason it is covered with a material with a low work function. However, the negatively charged cathode attracts Xe ions of the plasma, for which reason the sputtering or evaporation of cathode material and its re-deposition on the inner wall of the envelope must be minimized. This is a challenge for engineering, and a common technique is the dispenser method in which a porous tungsten matrix is filled with materials having a low work function like barium compounds [9]. Further details about cathode materials can also be found in literature [10, 12]. The cathode is heated up by electrical resistance heating, by the absorption of light coming from the plasma and by impinging Xe ions, but is cooled by the emission of electrons. Thus, its temperature is rather high and can easily exceed a value of 2000 K [13] which further assists the emissions of electrons. Another source of material erosion is the initial streamer formation in which a hot spot at the very beginning is able to evaporate material in the order of ng to μg per pulse if typical metal evaporation rates are considered [14]. The cathode is typically formed in such a way that it guides the arc in the initial phase of lamp ignition. Depending on the specific operation conditions the shape can be more like a cone tip or a dome-like structure. The shape will determine the type of arc attachment, the plasma hot-spot temperatures, electrode fall voltages, and electrical current profiles at the electrode tips [15]. In any case, the base cross section of the cathode limits both the maximum current which can pass the lamp and the heat dissipation via thermal conduction. A more detailed discussion about the different shapes of flash lamp cathodes [8, 15] and arc attachment modes [16, 17] can be found in literature.

Both electrodes are heated up by light absorption, resistance heating and the collision of particles, but the anode gets an even higher thermal load than the cathode as the incoming electrons additionally transfer both their condensation energy (de facto the work function of the anode) and their kinetic energy to the anode. The resulting thermal problems like overheating and thermal expansion can be reduced by using a more massive anode or an anode with a cooling channel. As the thermal load scales with the current, the anode temperature is lower and higher than that of the cathode for low and high current densities, respectively. The anode is usually made of tungsten doped with rare earth oxides in order to improve its machinability [8]. The shape of the anode should keep the arc in the centre of the bore, and its surface should be large enough to evenly receive the power without the formation of hot spots. To avoid early degradation the polarity of the applied voltage should be chosen as certificated, namely the more positive potential should be applied to the anode. If the flash lamp is designed for bipolar operation, special electrodes are used which can operate both as cathode and anode.

The gas inside the glass envelope still has to be isolated from the environment, and the special challenge is to buffer the very different thermal expansion coefficients of the electrode material and the glass envelope. Expressed in figures, the thermal expansion coefficient of tungsten with 4.5×10^{-6} K^{-1} [18] at room temperature (RT) is 8 times higher than that of fused silica with 5.4×10^{-7} K^{-1} [19]. As already mentioned, high power operation of the lamp can easily heat up the electrodes to temperatures in excess of 2000 K, and at these temperatures any leakage to atmosphere will result in strongly enhanced electrode erosion, most probably by oxidation. The most common type of sealing is the tungsten-rod seal in which a highly doped borosilica glass with a thermal expansion coefficient between that of fused silica and that of tungsten serves as buffer. More details about the rod seal and other seal types can be found in [8, 9], and information about end termination and mounting is given in [5, 20].

2.2.3 The Envelope

The envelope of a flash lamp has to meet a couple of requirements which strongly reduces the number of potential candidates. The material must be highly transparent in a broad spectral region from the ultraviolet (UV) to the infrared (IR) and must be inert to the filling gas, to atmosphere and to ozone. It has to withstand high temperatures (the higher the better), high temperature gradients, high pressures and the shock wave generated during the ignition of the pulse. Finally, there are certain conditions with respect to machinability and economics.

It is therefore not surprising that in the very most cases fused silica is used as envelope material due to its excellent optical and mechanical properties. In the current language fused silica is also called quartz, although it is quite different from the mineral from which the name originally comes from. Today, there are three major groups of quartz used for flash lamps, namely clear fused quartz, doped quartz and synthetic quartz [9]. Typical production routes for clear fused quartz are electric and flame fusion [19]. In general, clear fused quartz is amorphous SiO_2 without the impurities typically known from other glass types. The trace element concentration is in the order of 1 ppm and below except that for Al, which is about 10–15 ppm [19]. Depending on the specific fabrication method, clear fused quartz may contain an OH group concentration of up to 1000 ppm which affects the transparency for certain IR wavelengths, especially for the main water absorption band at 2.73 μm. This absorption can be used to block the flash lamp light and place the diagnostic wavelength of a pyrometric temperature measurement in this band (see Sect. 3.1). Except these wavelengths, clear fused quartz is transparent between 200 nm and ca. 4 μm. It is assumed that—depending on the specific FLA parameters—a significant part of the generated light is in the deep UV below 200 nm [21] and thus absorbed by the walls. Clear fused quartz is the most commonly used and most economical form of quartz [8], but has the problem of solarisation. Doped quartz is mainly used to suppress parts of the UV emission if needed, usually by using Ti or Ce as dopants. Ce is also known to prevent solarisation which results in a very stable lamp characteristic over time [9]. Synthetic fused quartz is made from synthetically produced silicon compounds like $SiCl_4$ with an impurity concentration much lower than that of clear fused quartz. Typical trace element concentrations including that of Al range now, depending on the purity degree, between 0.001 and 0.5 ppm [19]. Synthetic fused quartz is transparent down to ca. 160 nm and much less vulnerable to solarisation, but it is also significantly more expensive than the other quartz types. Figure 2.5 compares the transmission curves of two quartz types with that of the well-known Bk7 glass for a 20 mm thick disk. This thickness was chosen for an optimum illustration of the UV absorption edge. Suprasil® 311/312 is synthetic quartz with high transparency in the UV, but with a pronounced water absorption band around 2.7 μm. In contrast, the UV absorption edge of Infrasil® 301/302 is shifted to wavelengths longer than 200 nm, but with strongly reduced water absorption. Bk7 is well-known for optical devices, but as a borosilicate glass it has a higher UV absorption and a much lower softening point which makes this material unsuitable for flash lamps.

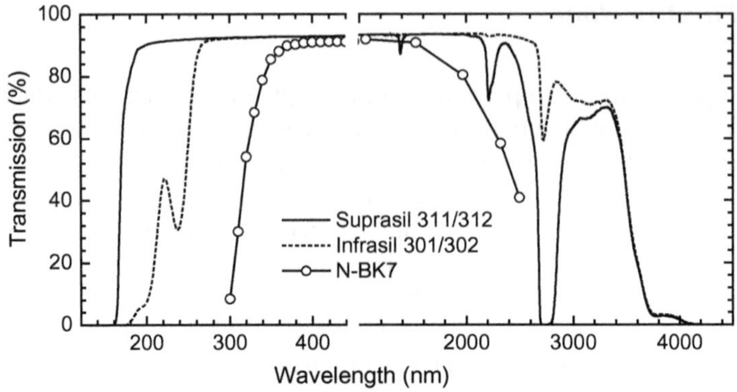

Fig. 2.5 Transmission curves of a 20 mm thick disk of Suprasil® 311/312 (solid line), Infrasil® 301/302 (dashed line) and N-Bk7 (Schott, open circles). Raw data are from [22] and [23]

As mentioned, solarisation is a degradation process which slowly but continuously lowers the transparency of the quartz glass. Solarisation comprises different processes caused by high energetic photon irradiation which induce the development of various colour centres. These colour centres can be due to impurities like Al, but also due to oxygen deficiency centres which may arise when an oxygen atom is displaced from its network site to an interstitial position [5]. The increased absorption enhances the wall loading and accelerates aging. However, high temperatures may anneal out some of these defects which eventually leads to a reduced but stable transparency after an initial drop down. Devitrification or crystallization of fused silica is another degrading mechanism which is fostered by high temperature thermal cycling and the presence of contaminants acting as catalysts on the surface, namely alkali and alkaline earth ions. The result of crystallization is cristobalite which is less transparent and has a much higher thermal expansion coefficient than fused silica, namely between ~10^{-5} K^{-1} (till 500 °C) and 1.7×10^{-6} K^{-1} (1000 °C) [24]. Opaqueness increases wall loading, whereas the higher thermal expansion coefficient fosters the development of microcracks.

The envelope thickness is a typical trade-off between high mechanical robustness and thermal durability. Thick walls are better in withstanding the shockwave of the expanding plasma, while thinner walls can better cope with thermal stress which arises from the temperature difference between the inner and outer side of the wall. Applications with a high average power but a low energy density per pulse favour thinner walls, most probably 0.5 mm. For pulses with a high energy density per pulse, thicker walls with a thickness of 1 mm and more are preferred as the risk of bulb explosion due to the shockwave is higher than the risk of failure due to thermal stress. However, wall thicknesses above 2.5–3 mm should be generally avoided as in such cases the large thermal gradients inside the wall strongly increase the risk of crack formation [9].

2.2.4 Cooling

The need for cooling mainly depends on the average wall loading at which the flash lamp is operating, but also on other lamp parameters like wall thickness, quartz type or filling gas. Although the numbers are only a rough guide, the following limits apply for Xe flash lamps [5, 8, 9]:

(i) 0–15 W cm^{-2}: Simple convection cooling by the surrounding gas is sufficient.
(ii) 15–30 W cm^{-2}: Forced gas cooling is recommended. Of course, the cooling effect depends on the flow volume, the gas temperature and the direction of flow. As the anode is expected to be hotter the gas should flow from the anode towards the cathode. In addition, air should be avoided to reduce the ozone production.
(iii) >30 W cm^{-2}: Liquid cooling. This is usually achieved by a flow tube around the flash lamp, and similar to forced gas cooling the coolant should flow from the anode towards the cathode. The upper limit strongly depends on the lamp and cooling parameters. For a wall thickness of 1 mm, upper limits of 160, 200 and 240 W cm^{-2} for doped quartz, clear fused quartz and synthetic clear quartz are given, respectively. These values must be reduced by ca. 10% if using Kr instead of Xe. For a given power density, inside wall temperatures tend to rise with increasing wall thickness (as thermal conductivity is low), decreasing filling pressure and if the plasma has direct contact with the wall (which is at least the case for "long" pulses in the ms range). With increasing inside wall temperatures the limit for wall loading decreases in an adequate way. Nevertheless, there are also reports in literature [5] that these limits can be exceeded by a larger extent for certain conditions.

The most popular coolant is deionized, purified water. Purification is required to prevent additional absorbing and etching. If the water is in contact with electrical connections deionization is needed to prevent leakage currents and short circuits. In order to keep the high quality of the coolant, only stainless steel and specific plastics compounds should be used in the water circuit. Liquid cooling may increase the problem of solarisation, as cooling also reduces the possibilities to anneal out the responsible defects. Thus, the maximum wall loading may become lower by 10–20% if solarisation is considered. Furthermore, the coolant must be in a turbulent flow and in intimate contact with the envelope as otherwise hot spots with a high risk of glass fracture may evolve. The simple standard design as shown in Fig. 2.6 is a flow tube with the flash lamp located at the central axis of the tube.

A special design of lamp cooling was developed by Mattson company [25] in which deionized water is injected at a flow rate of 60 l/min and is spiralling at the inner side of the envelope forming a water wall. Inside the water wall, Ar is additionally injected at high pressure forming a so called gas-water two-phase flow. The principle is shown in Fig. 2.7. The benefit of this lamp is that the water wall shields the quartz wall from the thermal load and the re-deposition of electrode material which increases the lifetime of the lamp and the maximum tolerable wall loading. In addition, the water absorption at 2.73 μm facilitates pyrometric temperature measurements.

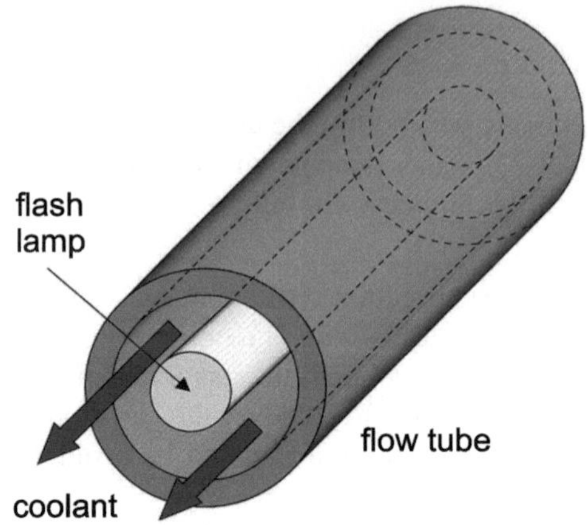

Fig. 2.6 Basic scheme of standard flow tube cooling

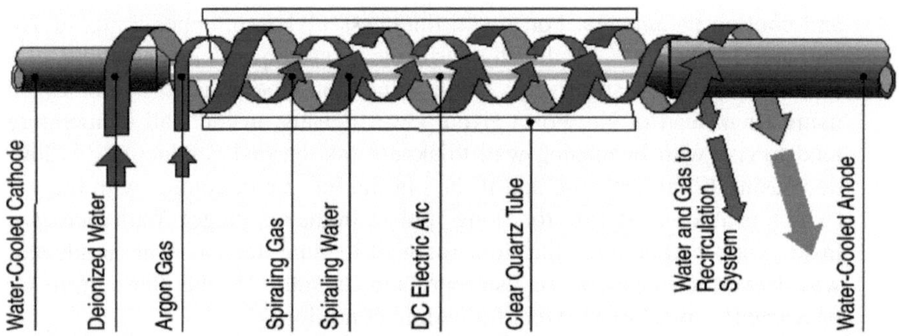

Fig. 2.7 Mattson's Vortek arc lamp—flow principle for c/w and flash arc system. Reprinted by permission from Springer Nature [25]

2.3 Plasma Discharge

2.3.1 Trigger Pulse and Initial Streamer Formation

There are two requirements in order to ignite a flash lamp: a sufficiently high number of primary electrons in the gas and a sufficiently high electric field which is able to accelerate these electrons to energies high enough to trigger impact ionization. The primary electrons can already be present as a result of previous processes like ionization by UV or particle irradiation, or they are remnants from a previous flash pulse. Otherwise, they have to be extracted from the cathode by field emission. To apply a sufficiently high electric field, a voltage higher than the self-breaking voltage of the lamp has to be applied. This can be done directly which, however, requires an electronic switch in order to supply the voltage when needed and not when the capacitor is charged. Therefore, it is more common to ignite the flash lamp by a short, high-voltage trigger pulse. There are several possibilities to feed the trigger pulse into the circuitry which will be a matter of discussion in Sect. 2.4.

It is important that the trigger pulse has to meet a couple of requirements in order to ensure a reliable operation. At first, the trigger voltage should be significantly higher than the self-breaking voltage of the lamp; a general rule states that 160% of the self-breaking voltage should be the minimum [5]. Besides the arc length, the needed voltage levels mostly depend on the filling pressure. The accelerated electrons must gain enough kinetic energy within their mean free path length which inversely scales with the atomic density and thus the gas pressure. Therefore, the minimum trigger voltage V_t increases with filling pressure and arc length. Due to the Ramsauer-Townsend-effect [26] the scattering cross section of low-energy electrons with atoms of a noble gas is quite small. This enables the efficient acceleration of electrons by electric fields without large energy losses due to scattering and leads to low V_t values. In fact, experiments performed with noble gases and an admixture of nitrogen or hydrogen showed that V_t strongly increases with the concentration of *molecular* gases [6]. More details can also be found in [10]. Secondly, the trigger pulse must be long enough to allow the initial streamer to bridge the gap between the electrodes. Recommendations in literature give values between 60 ns [5] and 100 ns per cm arc length [10]. These two requirements also imply that the current limit of the trigger source is high enough to deliver enough energy to form an initial streamer.

The build-up of an initial streamer is comprehensively discussed in [6] and can be briefly described as follows: If the applied voltage is higher than the self-breaking voltage of the lamp, the electric field is high enough to induce a couple of ionization mechanisms with a positive feedback. Probably the most important one is impact ionization by accelerated electrons which leads to an avalanche-like increase of free electrons. Nevertheless, other mechanisms like field ionization or photoionization can also give significant distributions. An initial streamer is formed in which the conductivity and the current flowing through rapidly increase by several orders of magnitude. This rapid increase is stopped by one or several of the following mechanisms:

(i) Power limitation: As the flash lamp becomes conducting and its resistance approaches that of the external circuitry, the voltage drop over the lamp and the electric field therein decrease. This limits the electric power which can be consumed in the lamp and strongly depends on the electric characteristics of both the flash lamp and the external circuitry.

(ii) Ion scattering: With increasing conductivity the fraction of ions continuously increases, but the scattering cross section of low-energy electrons with *ions* is much higher than that with *atoms*. This scattering efficiently decelerates electrons, and the concentration of electrons and ions tends to stabilize.

(iii) Thermal dissipation: With increasing conductivity the gas quickly heats up to temperatures close to 10,000 K. However, at this level the power dissipation by thermal radiation (desired) as well as by thermal conduction and convective heat transfer (usually not desired) is so strong that it compensates the input power to a large extent.

The typical timescale for the build-up of the initial streamer is a few 10 ns in correspondence with the minimum trigger pulse length mentioned above. After this time the streamer is believed to have a diameter in the order of 100 μm. Generally, there is also a time lag in the μs range between the trigger pulse and the rising edge of the real flash pulse.

The initial streamer is formed between one electrode and the inside of the lamp wall which then propagates along the wall towards the other electrode [27]. This results in a thin, conducting filament between both electrodes which is ideally oriented along the axis of the cylindrical lamp envelope [9]. However, in real cases the streamer may get caught at the wall causing a non-radial arc expansion as observed in [28, 29]. The situation becomes even more complex if the flash lamp is embedded in a lamp field. A typical arrangement as e.g. in our own FLA tools is sketched in Fig. 2.8: three flash lamps are connected in series in a meandered pattern and constitutes a lamp field (a), and four of such lamp fields are arranged in two levels and are laterally displaced relative to each other (b) to provide an homogeneous, but large area irradiation by the flash lamps. If such a lamp field is ignited, a streamer evolves in each of the lamps, but the distribution of the electric potential is inhomogeneous. Assuming that a high voltage is applied to contact 1 and 6 in Fig. 2.8a, the potential difference between contact 1 and 4 is much higher than that between contact 2 and 3 which causes a corresponding high and low electric field between the contacts, respectively. As the initial streamer formation is favoured by high electric fields, the streamers in two neighbouring flash lamps will bow to each other in such a case. Moreover, as the situation is similar for the contact pair 3–6, the streamer in the middle lamp may branch. Recently, this behaviour was experimentally verified for a lamp field containing 6 lamps with an arc length and bore diameter of 450 and

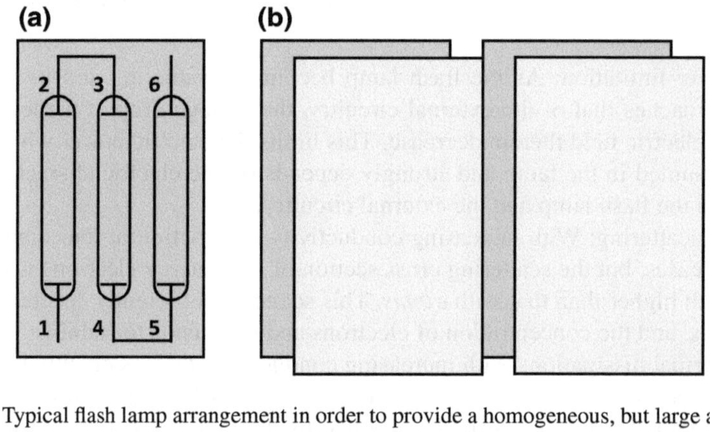

Fig. 2.8 Typical flash lamp arrangement in order to provide a homogeneous, but large area irradiation: arrangement of 3 lamps in a lamp bank (**a**) and arrangement of several banks (**b**)

48 mm, respectively [30]. It must be stated that the streamer formation introduces a significant statistical variance which affects the reproducibility of the flash process.

If a trigger pulse is applied to the lamp, both the voltage and the current peak over the lamp are delayed [10]. The voltage is at maximum when the conducting filament between both electrodes has established. Afterwards, the voltage drops due to the increasing conductivity of the lamp, whereas the current rises and has its maximum somewhere at the falling edge of the voltage pulse. The light emission follows the current pulse with a small delay caused by the time to heat up the gas to a sufficiently high temperature.

2.3.2 Arc Expansion

The very fast release of energy into the channel leads to a sharp increase of temperature and current. According to the kinetic theory of gases this causes an equivalent rise in pressure generating a shockwave which radially expands towards the inner wall of the flash envelope. The pressure compresses cold gas in front of the shockwave leading to a sharp increase of the gas density. As a result a pressure peak and a peak in gas density move outwards, but their amplitudes decrease with ongoing expansion as the energy is distributed over an increasing shell surface. Moreover, the amplitude was found to increase with decreasing filling pressure and to increase linearly with increasing input energy [31]. The shockwave and the increase of pressure push a certain amount of gas into the dead volume which lowers the overall concentration of atoms in the arc region depending on the ratio between the dead and the total volume of the lamp.

The velocity of the shockwave exhibits a similar dependency like the shockwave amplitude, namely an increase with increasing input energy and decreasing filling pressure, and it also decreases with increasing radial distance from the lamp axis. To give some figures, Lang et al. measured values up to 16 km s^{-1} [31], whereas the review in [6] mainly reports velocities between 1 and 5 km s^{-1}. Anyway, the shockwave hits the envelope wall with high momentum and can trigger, depending on the shock amplitude, the formation of circumferential and axial microcracks which widens with increasing shock values [31]. This may lead to bulb explosion and is one of the lamp degradation mechanisms.

Apart from the shockwave an ionization boundary which separates the inner, ionized region from the surrounding cold gas moves towards the walls. This gas is heated at first by thermal radiation and electronic thermal conduction which leads to a rapid increase in temperature and ionization of the surrounding gas shell. As this layer becomes conducting it is additionally heated due to the current passing through. The velocity of this boundary strongly depends on both the input energy and the time constants which are determined by the external circuitry. If the lamp is powered by a capacitance C and an inductance L in series, the typical time the plasma boundary needs to cross the lamp in radial direction can reach the order of $\sqrt{LC}$ for low input energies. With increasing input energies this time becomes shorter, but in any case the arc expansion will be slower than the shockwave. In addition, when the shockwave hits the wall the reflected wave interacts with the expanding plasma boundary and causes it to oscillate. These oscillations can be monitored as a ripple in the time profile of the voltage drop over the flash lamp [32]. Detailed investigations [29] with a high-speed camera revealed that this effect fades out for higher pulse energies, but is more present for higher filling pressures. They also found that the arc expands slower after the boundary has been hit by the reflected shockwave.

As long as the plasma boundary has not yet reached the envelope walls and is not yet influenced by its existence, the lamp is in the regime of unconfined discharge [5]. In this regime the lamp has negative impedance, i.e. the voltage drop over the lamp decreases while the current through the lamp increases. As mentioned above the time to pass this regime also depends on the input energy E_0. If this energy is sufficiently low and the power supply sufficiently limited, the arc can stabilize in the region of unconfined discharge at currents from a few tens of mA to a few A. The typical example for this behaviour is a simmer current. For a better understanding it is helpful to consider the power balance of the arc. According to this the input power P at a given time is, neglecting losses in the external circuitry and at the electrodes, balanced by the power which (1) is consumed to sustain the plasma conditions, (2) is used to heat up cold gas and (3) is dissipated through the walls. The power (1) is needed to balance the loss by radiation and rises if the diameter of the arc is increased. Thus, if P is limited the arc diameter stabilizes at a certain value which increases with increasing power.

If P is increased, the additional energy is used to heat up the cold gas in the vicinity of the arc and to partly ionize it (process 2). As a result the arc expands and consumes more and more input power which is also partly released in dynamic processes of the gas. If the arc has expanded up to the wall, P is now balanced by processes (1) and (3) only, whereas in usual cases the latter plays a minor role only. If P is now further increased, the additional energy is used to increase both the plasma temperature and the degree of ionization. Thereby, ionization usually consumes the major part of P.

The power balance also reveals that there must be a delay between the current maximum and the maximum light output as the current has first to heat up the gas before light emission can be observed. According to [10] this delay time τ_D can be roughly estimated by the energy balance

$$(p - p_0)V_{arc} = V_F I \tau_D \xi \tag{2.8}$$

where p and p_0 are the working and filling pressure, V_{arc} is the plasma volume of the arc, V_F is the voltage over the lamp, I is the current passing through, and ξ is the fraction of the input power which is used to heat up the gas. The left side of (2.8) describes the volume work to be afforded by the system, whereas the right side describes the required electrical energy under the assumption that the current suddenly rises from zero to a certain value. Equation (2.8) can be transformed to

$$\tau_D = \frac{(p - p_0)\ell}{j V_F \xi} \tag{2.9}$$

where j is the current density and ℓ is the arc length. Thus, to get a fast rise of the light pulse a high voltage has to be applied. If the current itself has a rise time in the same order, τ_D accordingly increases. Usually, this delay becomes more relevant for shorter pulses. An example is given in Fig. 2.3 where τ_D is in the order of 100 μs.

2.3.3 Plasma Properties in the Wall-Stabilized Regime

2.3.3.1 Plasma Zones

If the arc approaches the walls it enters a wall-stabilized regime in which the voltage drop over the lamp increases with increasing currents. Starting from one electrode, there is a lateral structure of the arc with different zones of different plasma conditions [10, 13, 17, 33, 34]. The zone notation is not standardized, and the schematic sketch in Fig. 2.9 follows that of [10]. In the space charge region (1), located in the close proximity of the cathode surface, the emitted electrons are repelled by the negative potential of the cathode which, on the downside, is attractive for Xe ions. As a result a space charge region with a strong electric field and an extension of a few 100's of nm evolves. Electrons are accelerated in this field, but collision events are rare, which is why the ionization level is very low and the region appears dark. Sometimes this zone is subdivided into two regions whose extensions are defined by the Debye length and the free mean path of electrons, respectively [34]. The first one is characterized by the said charge non-equilibrium, whereas in the second region ionization is not yet significant.

In the subsequent ionization zone (2), ionization and thus temperature, conductivity and emitted light power increase with distance from the cathode up to a maximum level. This hot spot is located close to the transition zone (3) and probably a few 100's of μm away from the cathode. In addition, the electron and heavy-particle temperature (Xe atoms and ions) converge to each other. The voltage drop over the space charge region and the ionization zone, called the cathode fall, is in the order of 10 V, but decreases with increasing current density [13].

In the transition zone, also called constriction zone, the arc expands up to the walls of the flash lamp, accompanied by a decrease of temperature, conductivity

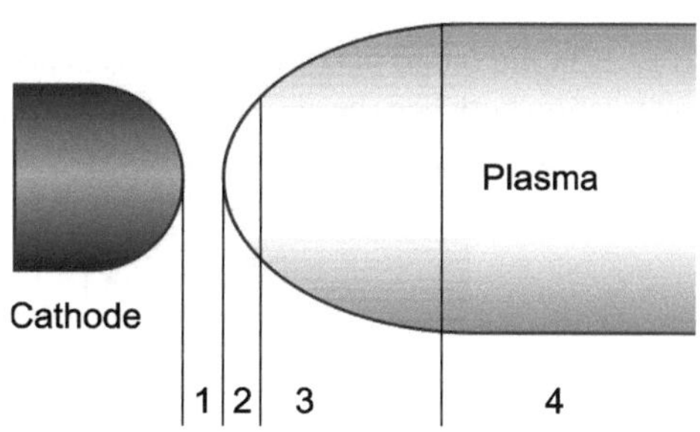

Fig. 2.9 Schematic structure of the arc in the cathode region comprising the space charge region (1), the ionization zone (2), the transition zone (3) and the fully expanded plasma column (4). Distances are not in scale

and emitted light power towards the constant values of the fully expanded plasma column (4). The reason for this is the expanding cross section at constant current which correspondingly lowers the current density. It has to be noted, that the zones (1)–(3) are not in thermal equilibrium [33, 34]. The plasma layer sequence is mirrored on the anode side, but the plasma parameters differ. Temperature and conductivity in the hot spot are higher (at least for higher currents), whereas the anode fall and the electric fields are lower than on the cathode side. Detailed plasma simulations along the lamp axis can be found in [13, 33].

Furthermore, pressure differences at the boundary regions of the arc induce plasma flows during the discharge [7, 10]. Probably the most important one is the cathode jet which arises by the action of electrodynamic forces, called the Maecker effect: the current flow induces a magnetic field which, via the Lorentz force on moved charge carriers, tends to compress the arc [35]. The effect is larger for smaller cross sections due to the higher current density which leads to a higher pressure at the endpoints of the arc. At the cathode side this leads to a plasma flow towards the anode. This in turn creates a suction which attracts cold gas from the side and finally a circulation in which gas first flows from the cathode towards the anode, drifts afterwards to the walls and from there back to the cathode. This is why evaporated material is preferably deposited in a region a few mm or cm away from the cathode. More details can be found in [7].

The fully expanded plasma (4) is in a quasi-stationary equilibrium in which most of the plasma parameters are determined by the current density j and can be derived step by step. It is assumed that most plasma parameters like ion concentration or pressure are roughly constant in radial and axial direction except for the small regions close to the wall and to the electrodes. The starting point is the question for the concentration n_{arc} of atoms and ions in the arc. n_{arc} is lower than the original atomic concentration n_0 at RT as a significant part of the gas is pressed into the dead volume V_{dead} of the lamp. If r_{D} is the ratio between the dead volume and the working volume V_{arc}:

$$r_{\text{D}} = \frac{V_{\text{dead}}}{V_{\text{arc}}}, \tag{2.10}$$

n_{arc} can be approximated by

$$n_{\text{arc}} = 0.65 \cdot n_0 \frac{1 + r_{\text{D}}}{1 + 4r_{\text{D}}} \tag{2.11}$$

if r_{D} is in the range of 0.04–1 [6]. n_{arc} is not really a constant as there is a continuous interaction between the gas in the dead volume and the plasma during the whole pulse, but changes in n_{arc} occur slowly and expression (2.11) can serve as a reasonable guess. According to kinetic theory, the mean free path of an electron λ_{e} in the plasma is

$$\lambda_{\text{e}} = \frac{1}{\sqrt{2} n_{\text{arc}} \sigma_{\text{g}}} \quad \sigma_{\text{g}} = 4\pi r_{\text{i}}^2 \tag{2.12}$$

where σ_g is the geometric cross section of an atom or ion with a radius of r_i. The electron mobility μ is given by Sze and Ng Kwok [36]:

$$\mu = \frac{q\lambda_e}{m_e v_{th}} \quad v_{th} = \sqrt{\frac{3k_B T}{m_e}} \tag{2.13}$$

where q is the elementary charge, m_e is the electron mass, and v_{th} is the thermal velocity of the electron.

For a cold Xe lamp at a pressure $p_0 = 450$ Torr the atomic concentration is $n_0 = 1.45 \times 10^{19}$ cm^{-3}, the cross section is $\sigma_g = 2.2 \times 10^{-15}$ cm^2 considering a covalent atomic Xe radius of 131 pm [37], the mean free path is $\lambda_e = 230$ nm, the thermal velocity is ca. $v_{th} = 117$ km s^{-1}, and the mobility is $\mu = 2600$ cm^2/Vs. To get the first ionization energy of 12.1 eV within the mean free path, an electric field of ~530 kV cm^{-1} has to be applied. Such high fields exist, if any, only in the ignition phase when the trigger pulse is applied. During the main discharge electric fields are much lower, which is why any drift velocities are orders of magnitude lower than the thermal velocities. Thus, atoms will be ionized by high electric fields only during ignition, whereas ionization during the main discharge is mediated by the thermal velocity of the electrons.

2.3.3.2 Ionization, Pressure and Conductivity

The electron concentration linearly depends on the current density passing the lamp with nearly no influence of the type or the pressure of the gas. Thus, the empiric formula [6]

$$n_e \approx 3.8 \times 10^{14}\ j \tag{2.14}$$

where j is the current density in A cm^{-2} and n_e is the electron concentration in cm^{-3}, is useful for a broad range of flash parameters. If the plasma is regarded as an ideal gas, the working pressure p can be derived from [38]

$$p = k_B T (n_{arc} + n_e), \tag{2.15}$$

As the temperature increases from RT to several 1000's of K, the pressure increases in the same manner and easily achieves values in the MPa range. Equation (2.15) can be also used to determine the original concentration n_0 of the gas if the filling pressure p_0 is given. Alternatively, ionization can be more correctly derived from the Saha equation [38].

$$\frac{n_e n_i}{n_a} = 2\frac{Z_i}{Z_a}\frac{(2\pi m_e k_B T)^{3/2}}{h^3}\exp\left(-\frac{E_i}{k_B T}\right) \tag{2.16}$$

where n_i/n_a is the density of ions/atoms, Z_i/Z_a is the partition function of the ion/atom, h is Planck's constant, and E_i is the ionization energy. In case that multiply charged ions can be neglected, (2.16) becomes simpler by $n_e \approx n_i$, $Z_i \approx g_i$, and $Z_a \approx g_a$ where g_i and g_a are the degeneracy of the ground states of the ion and atom, respectively.

The conductivity of the plasma is only weakly dependent on the gas and filling pressure and can be described by [5]

$$\sigma = 0.8815 \cdot \left(\frac{p_R}{p_0}\right)^{0.2} \sqrt{j} \tag{2.17}$$

where p_R is the reference pressure of the gas (see Table 2.1), σ is the plasma conductivity in $(\Omega\ \text{cm})^{-1}$ and j is given in A cm^{-2}. Neglecting the pressure term, (2.17) is in good agreement with the empiric formulas for the specific plasma resistance given in [4, 39, 40]. The dependence of σ on pressure and temperature for an Ar plasma is discussed in detail in [41]. As the electron current is much higher than the ion current, σ can also be extracted via

$$\sigma = q n_e \mu \tag{2.18}$$

in combination with (2.13) and (2.16). The current dependency of n_e and σ according to (2.14) and (2.17), respectively, is given in Fig. 2.10.

Table 2.1 Characteristic noble gas parameters relevant for flash lamp operation. Temperature was calculated using (2.25). More data can be found in [10]

Gas		He	Ne	Ar	Kr	Xe
Reference pressure (Torr)				~1100 [10]	805 [5]	450 [5]
Atomic radius (pm) [37]		46	67	96	117	131
Ionization potential (eV) [18]		24.6	21.6	15.8	14.0	12.1
Temperature (K) at 1 kA cm^{-2}		20,070	17,800	13,340	11,930	10,430
Refractive index fit constants [6]	a	1.3×10^{-24}	2.5×10^{-24}	1.0×10^{-23}	1.6×10^{-23}	2.5×10^{-23}
	b	2.3×10^{-11}	2.4×10^{-11}	5.6×10^{-11}	7.0×10^{-11}	1.0×10^{-10}

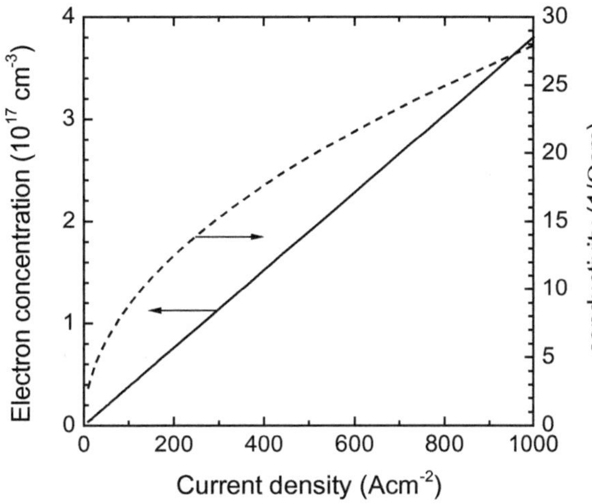

Fig. 2.10 Electron concentration (solid line, left scale) and plasma conductivity (dashed line, right scale) as a function of current density for a Xe lamp with a filling pressure of 450 Torr

2.3.3.3 Plasma Temperature

It was proposed on the basis of experimental data [6] that the power density of the generated radiation can be described by

$$q_{\mathrm{rad}} \approx 3.7 \times 10^9 p \cdot \exp\left(-\frac{E_{\mathrm{i}}}{k_{\mathrm{B}} T}\right) \tag{2.19}$$

where q_{rad} is the power density in W cm^{-3}, p is the current pressure in MPa, E_{i} is the ionization energy of the gas (see Table 2.1) in eV and T is the temperature in K. To get the temperature for a given current density, (2.19) can be combined with the energy balance of the quasi-stationary plasma of cylindrical symmetry, also known as the Elenbaas-Heller equation [42]:

$$\frac{1}{\theta}\frac{\mathrm{d}}{\mathrm{d}\theta}\left(\lambda_{\mathrm{T}}\theta\frac{\mathrm{d}T}{\mathrm{d}\theta}\right) + \sigma E^2 - q_{\mathrm{rad}} = 0 \tag{2.20}$$

where $\theta = d_{\mathrm{a}}/d_0$ is the relative diameter (or relative radius) of the discharge column, λ_{T} and σ are the thermal and electric conductivity of the plasma, respectively, and E is the electric field over the arc. The first and third term describe energy losses by thermal conduction and thermal radiation, respectively, whereas the second term describes the power released in the discharge. The radial temperature profile can be found by solving (2.20) with the start conditions [42]

$$T|_{\theta=1} = T_{\mathrm{W}} \qquad \left.\frac{\mathrm{d}T}{\mathrm{d}\theta}\right|_{\theta=0} = 0 \tag{2.21}$$

where T_W is the temperature at the wall. Usually, this will give a nearly constant temperature in the order of 10,000 K in the centre and the main part of the plasma with a sharp drop to T_W which might be in the region of 1200–1500 K [9, 43]. Alternatively, the radial temperature distribution $T(r)$ can also be approximated [10] by

$$T(r) = (T_{max} - T_W)\left(1 - \left(\frac{r}{r_0}\right)^q\right) + T_W \tag{2.22}$$

where T_{max} and T_W are the temperatures in the centre and at the wall, respectively, r_0 is the inner radius of the lamp, and q is an exponent described by

$$q \approx 1.5 + 0.063 \cdot j \tag{2.23}$$

with j given in A cm^{-2}. T_{max} is the temperature as shown in Fig. 2.11, and $T(r)$ becomes more boxlike with increasing q and thus increasing current density.

To get T_{max}, (2.20) can be further simplified by assuming that the input energy is dominantly used to generate radiation, that losses at the electrodes are negligible, and that losses in the external circuitry are out of the viewing frame:

$$\sigma E^2 \approx q_{rad} \tag{2.24}$$

Inserting (2.17) and (2.19) in (2.24) will give the approximation (please note the cube of j)

$$j^3 = 1.064 \times 10^{19}\left(\frac{p_R}{p_0}\right)^{0.4} p^2 \exp\left(-2 \times 10^4 \frac{E_i}{T}\right) \tag{2.25}$$

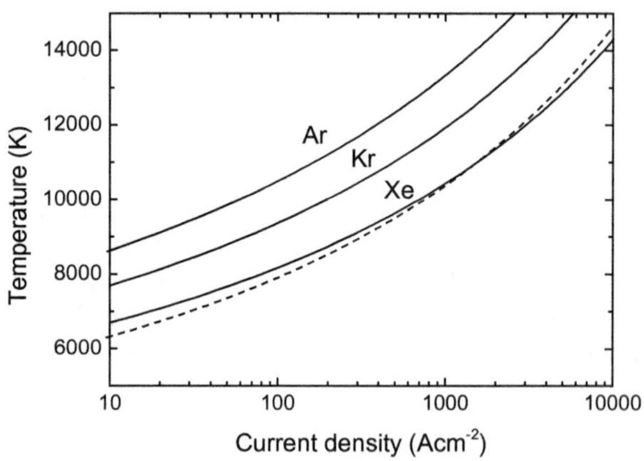

Fig. 2.11 Plasma temperature as a function of current density for different gases according to (2.25) (solid lines) and (2.26) (dashed line)

Marshak used an effective charge model to consider deviations from the ideal gas and proposed an approximation as following [6]:

$$j = 11p^{0.5}T^{1.25}E_i^{-0.5}\exp\left(-5000\frac{E_i}{T}\right) \tag{2.26}$$

Both approximations can be used to numerically derive the $T(j)$ dependence, although they are not equivalent in a strict mathematical sense. However, in most cases their results differ by a few % only as shown in Fig. 2.11. The displayed $T(j)$ dependencies for a Xe lamp with 450 Torr were obtained assuming $p \approx k_B T n_{arc}$. In addition, for a given current density the plasma temperature rises in the order Xe-Kr-Ar due to the increasing ionization energy. Further approximations of the $T(j)$ dependence [17, 44] as well as more detailed simulations of the plasma properties [13, 33, 45, 46] can be found in literature.

2.3.4 The Post-discharge Period

There are a couple of physical processes after the flash pulse which restore the original condition of the flash lamp. These processes are important for multiple flash operations as their time scale determine the maximum possible frequency of the flash pulses, provided that the charging unit can supply enough power and the flash lamp can survive this. The post-discharge period is usually described by the breakdown strength recovery characteristic which describes the development of the minimum voltage V_{min} necessary to fire the next flash pulse. According to [6] the post-discharge period can be divided into three stages.

In the first stage, the plasma cools down to a temperature were deionization becomes significant. During this time the plasma channel still exists, and the conditions in the plasma channel are similar to those during the flash, but with continuously decreasing temperature. V_{min} is low which means that it is easy to continue to release energy into the channel. The duration of this stage depends on how fast the current pulse fades out. A non-zero current density will still heat up the plasma and delays the cooling of the plasma. In order to shorten this time an electronic switch can be used which sets the current density to zero. In this case the plasma cooling time is in the order of 10–100 μs. The emitted light intensity of the lamp was observed to decline exponentially with a decay time in the same order [10]. In case of underdamped flash pulses with a duration of 10 μs it was found that the decaying light intensity is superimposed with oscillations, even if the current pulse has already expired [47]. In the second stage deionization is dominant which destroys the plasma channel. During time V_{min} strongly rises to a value which is already close to the original value. Finally, in the third stage V_{min} slowly approaches its original value. Physical processes in this stage include the pressure equalization in the ms-range and the cooling down of the electrodes and the glass envelope which may last up to a few seconds depending on geometry and cooling conditions. The complete cool down of the lamp can be regarded as the transition to single pulse operation. Finally it should be noted that the recovery characteristic can significantly vary between individual lamps.

Under multiple flash conditions subsequent flashes are fired in the last stage of the post-discharge period which means that the lamp cools down only to an elevated equilibrium temperature. The upper frequency limit, typically in the kHz range, is given by the time when the second stage has finished. If the gas is not sufficiently deionized the lamp can run into a continuous mode characterized by a permanent discharge with a slight intensity modulation instead of individual flashes. The recovery characteristic is faster for smaller input energies, shorter pulse durations, longer arc length, smaller bore diameters and smaller filling pressures [6]. If optimized for ultra-short pulses, flash pulse decay times in the order of 3 μs have been experimentally achieved [2]. A further possibility to increase the flash frequency is the use of simmer currents as the plasma does not need to be fully deionized after the flash. Furthermore, the use of lighter gases allows even higher frequencies at the expense of conversion efficiency. However, if one considers that the conversion efficiency is already quite low for pulse durations of only a few μs, the use of lighter gases is probably not a favourable solution.

There are already a couple of parameter shifts for frequencies far below the frequency limit. For 300 μs pulses it was found that the lamp impedance K_0 and thus the plasma resistivity decreases with increasing pulse frequency above 100 Hz, most probably due to residual ions [48]. In turn, a reduced lamp impedance may shift the pulse length depending on the circuit design.

2.4 Electric Operation

2.4.1 *Electric Characteristics of a Flash Lamp*

According to [5] and (2.17) the voltage drop V_F over the flash lamp during the discharge can be calculated by

$$V_\mathrm{F}(t) = \mathrm{sgn}(I(t)) \cdot K_0(t) \cdot \sqrt{|I(t)|} \tag{2.27}$$

where I is the time-dependent current through the lamp and K_0 is the arc or lamp impedance parameter given in $\Omega\sqrt{A}$. K_0 mainly depends on lamp parameters and is described by

$$K_0(t) = 1.28\left(\frac{p_0}{p_\mathrm{R}}\right)^{0.2} \frac{\ell}{d_\mathrm{a}(t)} \tag{2.28}$$

where p_0 is the filling pressure, p_R is the reference pressure of the gas (see Table 2.1), ℓ is the arc length and d_a is the current diameter of the arc. As p_R decreases in the order Ar-Kr-Xe, K_0 slightly increases and thus the electric power released in the lamp. For a given filling pressure, a given current $I(t)$ and a given ohmic load R of the external circuitry, the electrical transfer efficiency is defined by

$$\eta_{\text{elec}} = \frac{P_{\text{lamp}}}{P_{\text{lamp}} + P_{\text{circ}}} = \frac{K_0}{K_0 + R\sqrt{|I(t)|}} \tag{2.29}$$

where P_{lamp} and P_{circ} are the electric power dissipated in the lamp and the external circuitry, respectively [49]. According to (2.29), Xe lamps cause lower external losses than lamps filled with other noble gases due to the lower reference pressure and thus higher K_0 values.

For the regime of unconfined discharge, where the arc is still expanding, there is a complicated, but empirical time dependence of the arc diameter [5]:

$$d_{\text{a}}(t) = 1.5 \cdot \left(\frac{\varepsilon_{\text{t}}(t)}{\ell}\right)^{0.6} \quad \varepsilon_{\text{t}}(t) = \int_0^t I(t')V_F(t')\mathrm{d}t' \tag{2.30}$$

In this formula $\varepsilon_{\text{t}}(t)$ is the energy in J which has been released into the flash lamp up to the time t, and all lengths are given in cm. Clearly, d_{a} must not exceed a critical diameter d_{c} which is smaller but close the bore diameter. However, simple insertion of typical lamp parameters in (2.30) shows that $\varepsilon_{\text{t}}(t)$ is in the order of 10–100 J, whereas typical values for E_0 are well in the kJ range. Under these conditions the arc already expanses up to the wall during the rising edge of the flash pulse, typically after a few % of the pulse time or even less. Thus, for a basic modelling of the current pulse it is sufficient to use the critical, time-independent diameter d_{c} which is empirically described by [5]

$$d_{\text{c}} = 0.7 d_{\text{a}}^{1.15} \tag{2.31}$$

where all lengths are given in cm. However, d_{c} was also found to converge to d_{a} with increasing lamp currents [10]. Thus, in many cases it is sufficient to use the bore diameter d_0 instead of d_{c} [8].

2.4.2 Circuitries

The most basic circuitry as shown in Fig. 2.12a comprises a charging unit, a PFN composed of at least one capacitor C and one inductance L, the flash lamp F and a switch S which either connects the charging unit or the flash lamp to the capacitor. The switch is not needed but recommended in order to protect the charging unit from backstroke effects during the discharge. The dimensioning of the charging unit depends on the energy E_0 to be stored in the capacitor and the repetition rate. More details about the charging unit and related circuitries can be found in [20, 50]. The circuitries in Fig. 2.12 are sketched in such a way that the cathode is grounded, but

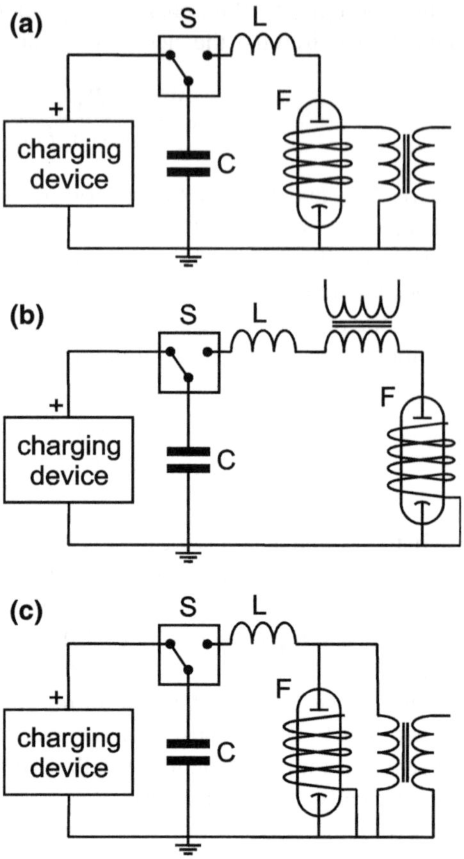

Fig. 2.12 Schematic flash lamp circuitries with different trigger modes: external triggering (**a**), series triggering (**b**), and parallel triggering (**c**)

designs with a grounded anode and a cathode being on high negative potential are also possible.

In the usual case where the charging voltage is below the self-breaking voltage the flash lamp must be triggered. Figure 2.12a displays external triggering in which the trigger pulse is supplied by an external circuitry. External triggering has the advantage to be flexible in design and to need only a relatively small transformer [5]. In order to ensure a reliable triggering, a reference plane in the form of a wire is wrapped around the flash lamp [9]. Triggering can also be designed as series triggering (Fig. 2.12b) and parallel triggering (Fig. 2.12c) in which the secondary coil of the transformer is in series or parallel to the flash lamp, respectively. In both cases the reliability is larger than for external triggering, but now the secondary coil of the transformer is part of the PFN and must be considered. Parallel triggering usually needs a smaller transformer than series triggering, but probably needs additional circuitry elements like diodes [5]. In both cases the reference plane is usually on ground. In general, nearby reflectors can also be used as a reference plane. More details about trigger circuitries are given in [5, 8, 9, 51].

However, the pulse shapes which can be achieved by circuitries as shown in Fig. 2.12 or by more complex capacitor-coil configurations are limited. For certain applications special shapes like rectangular pulses are needed or the pulse shape must be controlled in real-time. In such cases high-power, controllable switches such as gate turn-off thyristors, metal-oxide-semiconductor field-effect transistors (MOSFETs) and insulated-gate bipolar transistors (IGBTs) can be used [20]. Thus, an IGBT was used to create flash pulses with a constant light output and variable durations between 1.4 and 40 ms [52, 53]. In this case the falling edge of the pulse was determined by a trigger signal closing the IGBT. In another case the lamp current was measured in situ by a current sensor and processed in a feedback circuit which controls the base current of a MOSFET connecting the energy storage with the flash lamp [48]. Finally, this method was fine-tuned by a technique called Flat-Top Flash Annealing™ [54] in which the feedback loop between temperature measurement and lamp current control is so close that the temperature profile at the sample surface shows a broad, constant plateau instead of a peak. The critical feedback loop for such a control is described in more detail in [55].

In case of higher repetition rates or if reliability is a major concern, the circuitry can be complemented by a simmer supply. The trigger pulse is applied once at the very beginning leading to an initial arc formation, followed by a limited power input from a simmer power supply (Fig. 2.13). The typical simmer current is in the range of a few 100 mA to a few A and has to keep the arc in the regime of unconfined discharge and to prevent the self-extinction of the arc. Thus, the time profile of the lamp current is a constant current plus the high current peaks of the individual flash pulses. However, if the repetition rate is too low the power consumption of the simmer can be irresponsibly high. In this case, called the pseudo-simmer mode, the trigger signal is applied, subsequently followed by a simmer period of a few 100 ms and the real flash.

Simmer operation offers several advantages. Most importantly, the impact of the shockwave in the early arc formation is diminished. In addition, the simmer current keeps the plasma and the lamp envelope at an elevated temperature which reduces the temperature gradients during the real flash and thus the corresponding degradation effects. Finally, simmer operation reduces the fluctuations between individual flashes and ensures a higher reproducibility [56, 57]. The reason behind this is that the arc has time to stabilize avoiding perturbations in the initial phase of arc formation

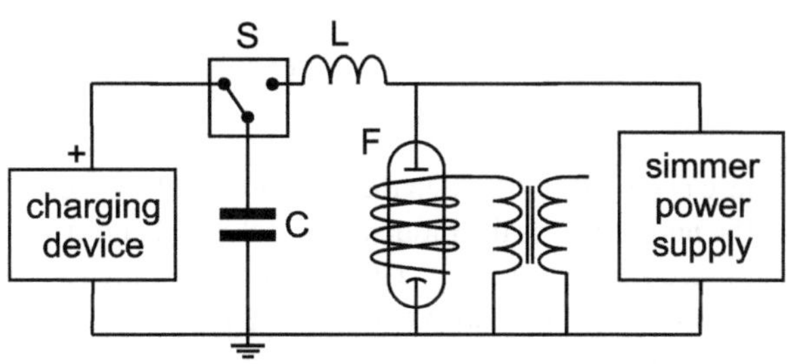

Fig. 2.13 Schematic flash lamp circuitry with simmer operation

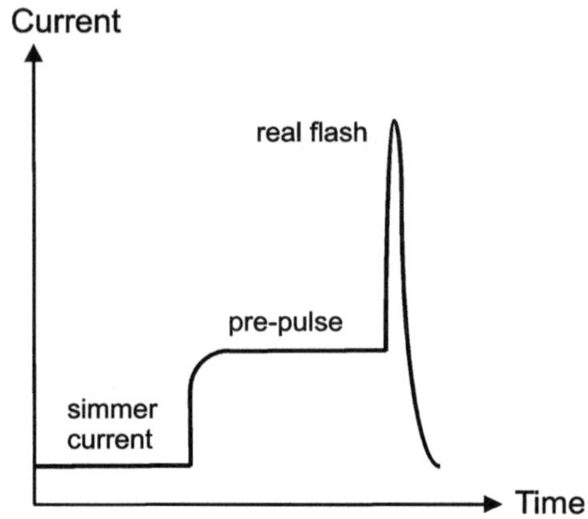

Fig. 2.14 Schematic current profile with simmer operation and the application of a pre-pulse. After [58]

like random contact points on the electrodes. However, there is also a weak but significant interaction between the magnitude of the simmer current and the flash lamp behaviour during the flash. It has been found that the lamp impedance K_0 slightly decreases with the simmer current, leading to a slight increase of the peak current, the plasma temperature and the pulse duration [32]. Furthermore, higher simmer currents suppress the oscillations of the luminous plasma boundary.

In [58] the simmer operation is performed in an even more sophisticated way as shown in Fig. 2.14: A few 100 ms before the real flash the simmer current is ramped up to a determined current value which causes the arc to expand up to the inner wall of the lamp. This pre-pulse reduces the deleterious accompanying effects of the flash even further, namely those caused by the shockwave and thermal stress. In some cases the pre-pulse can even replace the preheating of the sample. However, the pre-pulse current level should be in the range of several 10 or 100 A in order to have a real impact which requires a corresponding high-performance power supply.

2.4.3 *Modelling of a Single RLC Section Circuit*

To model the time dependence of the current pulse the simple partial differential equation for an oscillating circuit plus a lamp term can be used:

$$\frac{Q}{C} + R\frac{\mathrm{d}Q}{\mathrm{d}t} + L\frac{\mathrm{d}^2Q}{\mathrm{d}t^2} + V_\mathrm{F}(t) = 0 \tag{2.32}$$

where Q is charge stored in the capacitor with capacitance C, R is the ohmic resistance of the network, L is the inductance of the coil, and V_F is the voltage drop over the flash lamp according to (2.27). This equation assumes a series connection of all elements and that the voltage drop over the capacitor is divided among the ohmic resistance, the coil and the flash lamp. The standard oscillating circuit introduces a damping coefficient α_D which, with the introduction of a flash lamp, is dominated by K_0 and is approximately given by:

$$\alpha_D = K_0 \sqrt[4]{\frac{C}{LV_0^2}} \tag{2.33}$$

The value of critical damping is about $\alpha_D \approx 0.8$ [5] which separates weak ($\alpha_D < 0.8$) from strong damping ($\alpha_D > 0.8$). Strong damping avoids a negative undershoot of the current or multiple oscillations. Considering that a flash lamp should only work in one direction (positive polarity for the anode) strong damping is usually the preferred case. A more detailed investigation devoted to α_D for short and long pulses can be found in [47].

Figure 2.15 displays the time profiles of the voltage drop over the lamp (solid line) and the lamp current density (dashed line) obtained by simulations. As shown, the curves can be roughly divided into three different regimes. For times below 1 ms the lamp is in the regime of unconfined discharge (Sect. 2.3.2) where the arc diameter increases nearly linearly with time and an average speed of 6.6 ms^{-1}. This part of the profiles was simulated using a partial differential equation system which includes (2.27–2.32). The voltage drop over the lamp (Fig. 2.15) decreases in the same manner as the voltage drop over the coil increases (not shown). The current increases too, but as the cross section of the arc increases faster the current density decreases. During that time the charge stored in the capacitor is only insignificantly lowered. After ca. 1 ms the arc diameter and thus K_0 are assumed to be constant, and it is sufficient to use only (2.32) with the proper initial conditions. The lamp is in the wall-stabilized regime (Sect. 2.3.3), and between 1 and 39 ms the main discharge occurs with a pronounced current and voltage pulse with a pulse length of ca. 20 ms. The stored energy is mainly released in this phase, and the capacitor is nearly completely discharged. Under these operational conditions (see figure caption 2.15) the damping constant α is 0.76 and thus still in the weak damping region. Consequently, an undershoot is observed after 39 ms. For pulse lengths in the ms-range the first simulation can be skipped if the initial arc formation is not of interest. The difference to a full detailed simulation is a small time shift for the rising edge of the pulse. To give some numbers, in the maximum of the main discharge the current density reaches a value of 633 A cm^{-2} for $V_0 = 4$ kV which, according to (2.14), (2.15), (2.17), and (2.25) gives an ionization degree of ca. 1.7%, a peak pressure of 2 MPa, a plasma conductivity of 22 $(\Omega\,\mathrm{cm})^{-1}$ and a plasma temperature of ca. 9900 K. Similar current and voltage profiles for different operational conditions can be found in [40].

The shape of the current pulse strongly depends on the design of the PFN, the lamp impedance K_0 and the applied voltage V_0. Figure 2.16a displays the current pulse for different voltages and for $L = 19$ mH, $C = 6$ mF, $R = 0.1\ \Omega$ and $K_0 = 64\ \Omega\sqrt{A}$. At a voltage of 2 kV the oscillation is strongly damped with $\alpha_D = 1$. With increasing V_0 the damping becomes weaker, and at 4 kV ($\alpha_D = 0.76$) the oscillation has changed to the weakly damped regime showing a slight undershoot after the first

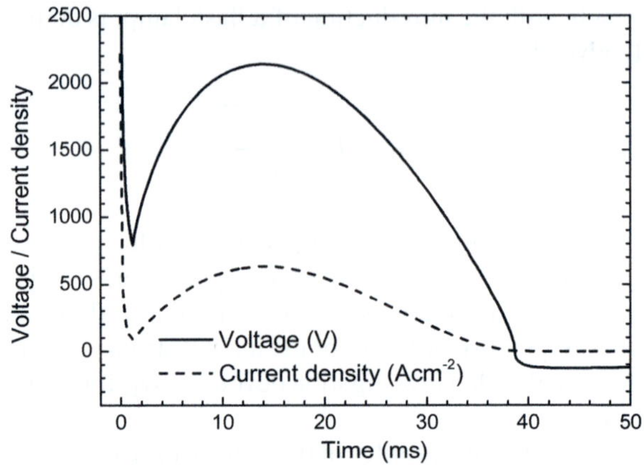

Fig. 2.15 Simulated time profiles of the absolute values of the voltage drop over the lamp (solid line) and the lamp current density (dashed line) for the following conditions: $C = 6$ mF, $H = 19$ mH, $V_0 = 4$ kV, $R = 0.1\ \Omega$, $\ell = 75$ cm, $d_0 = 1.5$ cm

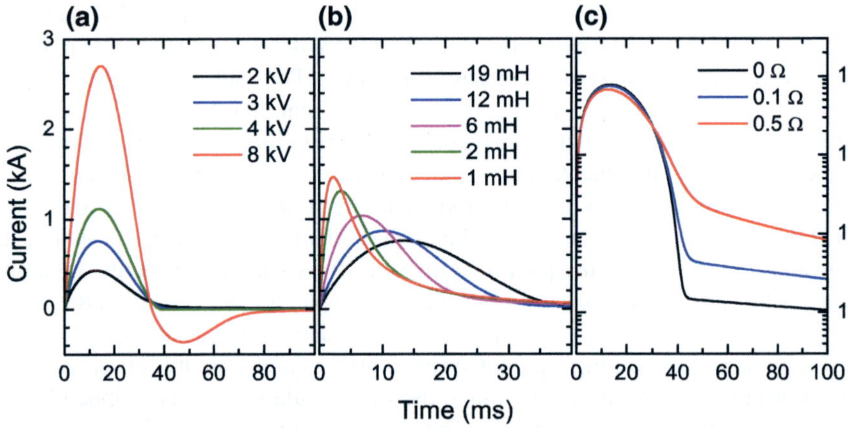

Fig. 2.16 Simulated current pulse shapes as a function of the applied voltage (**a**), the inductance (**b**) and the ohmic resistance (**c**) of the PFN. Please note the logarithmic scale in (**c**)

peak. At 8 kV ($\alpha_D = 0.54$) the oscillation is clearly weakly damped. In a similar way, the decrease of K_0 also leads to a decrease of α_D (not shown).

Figure 2.16b demonstrates that an inductance, which is too low in comparison to the capacitance, leads to a strongly asymmetric current pulse. The current pulses of this graph were obtained for $V_0 = 3$ kV and the same parameters as in Fig. 2.16a, except that the inductance is varied from 1 to 19 mH. According to (2.3) this is equivalent to an impedance running from 0.41 to 1.78 Ω. The rising edge of the pulse is determined by the inductance: the higher it is the more energy is buffered in the coil and released with delay. Thus, the inductance must be high enough to get a more symmetric time profile. This problem is most relevant in situations where the secondary coil of the trigger transformer provides the only relevant inductance of the PFN. Finally, the ohmic resistance of the PFN should be as low as possible to minimize ohmic energy losses within the circuitry. In addition, a larger resistance leads to a larger long-running tail of the current pulse (Fig. 2.16c).

2.4.4 Active Pulse Shaping

There are a couple of reasons to shape the current or light pulse of the flash lamp to a greater extend as it is possible with a series connection of a single capacitor and inductor. The most common reason is the minimization of thermal stress despite the short pulse time. This can be done by preheating the wafer to reduce the total temperature difference between front and backside or by smoothing the rising edge of the current pulse to conserve the lamp [59] and to avoid large temperature gradients within the surface region of the sample. Another reason could be to address two different physical processes requiring different annealing parameters. One example is the use of a short, intense pulse to activate dopants after ion implantation, followed by a longer pulse of lower intensity for the annealing of implantation-induced defects [60].

There are several technical ways to implement active pulse shaping which is also referred to by different names in literature like flexibly-shaped pulse FLA [60] or Flat-Top Flash Annealing™ [54]. One possibility is to use two separate lamp circuitries, probably one for preheating and one for the actual flash [1]. If there is only one lamp circuitry as in most cases, two or three pulses can be applied one after the other within a very short period of time [61] which, however, requires a sufficiently powerful charging unit. If FLA is performed in the simmer mode and the power supply is sufficient, a high simmer current can be used for preheating [58]. Another option is the use of an active switch as discussed in the previous section. The advantage of this method is the high flexibility and strict control over the pulse shape, but the approach is costly and might be limited by the power class of available switches. Finally, more complicated networks of capacitors and inductors can be used.

One type of an advanced PFN consists of a capacitor-inductor combination which is repeated several times. Figure 2.17 displays a circuit diagram of such a network with three parallel LC sections with an inductance of $L_i = 5$ mH and a resistance of $R_i = 0.1\ \Omega$ in each section. The capacitance decreases from 5 to 3 and 2 mF in the order C_1, C_2 and C_3, but all are charged with the same voltage. These values were chosen in such a way that the lamp current can be roughly approximated by a box-like profile. For a charging voltage of 2 kV the time profile of the voltage over the capacitances C_i and the current through R_i is given in Fig. 2.18a and b, respectively. The voltage curves immediately reveal that the capacitor C_3 is discharged first, subsequently followed by C_2 and C_1. However, the capacitances will not be discharged continuously, but

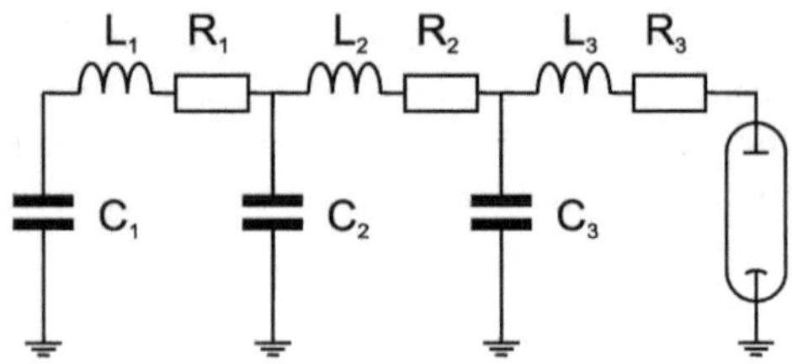

Fig. 2.17 Circuit diagram of a triple LC network with a flash lamp

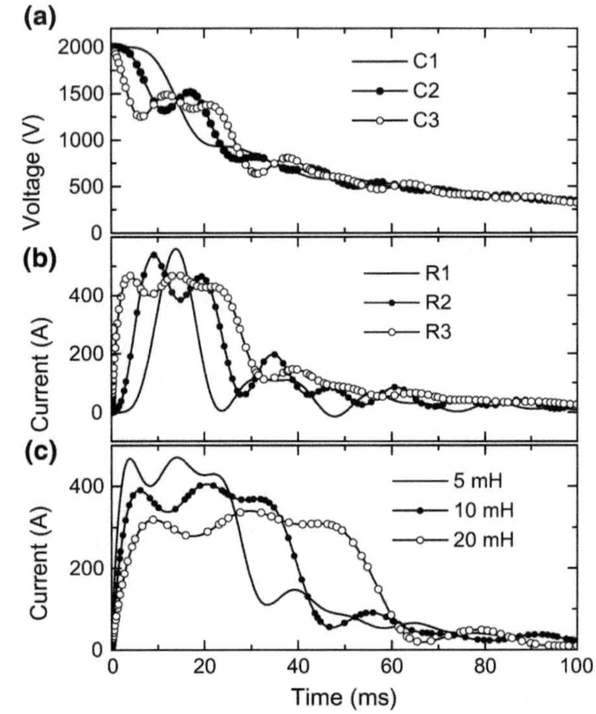

Fig. 2.18 Simulated voltage drop over the capacitors C_i (**a**) and current through the resistors R_i (**b**) of a triple LC network. Change of the flash lamp current as a function of the inductances L_i (**c**). More details are given in the text

will be partially recharged by the other capacitances. This causes oscillations of the discharge current with potential undershoots, even if they are small compared to the peak currents (Fig. 2.18b). The current flowing through R_3 and thus through the flash lamp is less effected, but there is now a long tail after the main pulse has decayed. More drastically, the capacitors will not be completely discharged, and they still have a significant residual voltage.

The individual profiles of the capacitor voltages and the lamp currents strongly depend on the specific parameters of the network components, but some general trends become apparent. In addition, the decrease of the inductance will only lead to a convergence towards a simple series connection with a very asymmetric shape (Fig. 2.16b with only 1 mH or even less). Vice versa, an increase of the inductance as shown in Fig. 2.18c will both prolong the pulse and decrease the current tail. As a consequence, if long current tails should be avoided large impedance values according to (2.3) have to be combined with high voltages.

2.5 Optical Output

2.5.1 The Plasma Spectrum

Basically, the output spectrum of a flash lamp consists of a broad continuum from the UV to the near IR and a couple of discrete spectral lines. The latter ones are so-called bound-bound transitions which are element specific and are due to discrete energy transitions within the atoms and ions. The broad spectrum is the result of free-bound transitions (recombination luminescence if an ion captures an electron) and free-free transitions [5]. Whereas the discrete lines are well visible at low flash energies, the continuum dominates the spectrum with increasing flash energy. To quantify the contribution of discrete lines in the spectrum the line factor L is defined by

$$L = \frac{P_{\text{tot}}}{P_{\text{cont}}} = \frac{P_{\text{tot}}}{P_{\text{tot}} - P_{\text{lines}}} \tag{2.34}$$

where P_{tot}, P_{lines} and P_{cont} are the total optical power, the optical power emitted by bound-bound transitions and the optical power emitted by the continuum, respectively [38]. L starts at high values for low input energies, decreases monotonically with increasing input energy and finally converges towards 1. P_{tot} and P_{lines} can be easily determined from a measured flash lamp spectrum on the energy scale by integration and by adding all peak areas of discrete emission lines. It should be noted that some of the emission lines, e.g. in the IR, are probably out of the detection range of the used spectrometer.

In a first approximation, the continuum can be modelled by blackbody radiation with a wavelength-dependent emissivity ε which, however, sensitively depends on plasma parameters. It was found that ε is a monotonically increasing function of the wavelength meaning that less light is emitted in the UV compared to the visible and near infrared (NIR) spectral range and less than expected by Planck's radiation law [5]. If the input energy is increased, the plasma temperature increases only weakly as the additional input energy is mainly compensated by additional plasma losses. However, ε increases and converges to 1, and this increase in the UV region is stronger than in the visible and the NIR range. As a consequence, the spectrum shifts toward shorter wavelengths with increasing input energy. As ε cannot exceed 1 blackbody radiation for a given plasma temperature is the intrinsic limitation for the maximum light output of the flash lamp which, however, is only reached for very high input energies.

The conversion efficiency η as defined in Sect. 2.1 decreases with decreasing atomic mass, whereas the specific amount of reduction depends on the operation conditions. η is practically independent on medium and larger arc lengths but decreases for smaller ones [6]. There is a similar dependence on filling pressure: η is constant for higher pressures and decreases only if the pressure is significantly lower than the reference pressure [62]. Thereby, this behaviour is more pronounced for higher input energies. Regarding the bore diameter, there is an optimum range for η which increases with input energy. Whereas at low bore diameters losses are attributed to losses at the wall, at large diameters the arc expansion period with its additional need of energy to heat up the cold gas may cover a significant part of the pulse time. Depending on input energies and pulse times, optimum ranges for the bore diameters were found between 3 and 6 mm [62] as well as between 10 and 20 mm [6]. Furthermore, η is seriously affected by larger dead volumes.

The arc expansion time τ_{exp} also influences the dependence of η on the pulse length: as long as τ_{exp} is small compared to the pulse length, η is high and independent of pulse length, but decreases if the pulse length becomes too short. For a bore diameter of 6 mm and an input energy of 25 J, the critical pulse length was found to be about 150 μs [62]. With increasing current density η increases at first, reaches a broad maximum and decreases for a further increase. For the specific case of a Xe lamp with $d_0 = 6.5$ mm and a 900 μs pulse this maximum was around 5–7 kA cm^{-2} [6]. For shorter pulse lengths the maximum η decreases and shifts to higher current densities.

In order to obtain the most authentic spectrum the light of a flash lamp was directly collected by a glass fibre cable and analysed by a mobile spectrometer as shown in Fig. 2.19 [63]. The glass fibre is in the direct line of sight to the flash lamp through a small opening in the reflector in order to avoid any distortion by reflection. A stack of neutral density filters with known transmission characteristics has been used to attenuate the flash lamp light to levels which are harmless to the spectrometer. Light pulse shapes can be measured in the same arrangement if the glass fibre is replaced by a photodiode being part of a measurement circuitry.

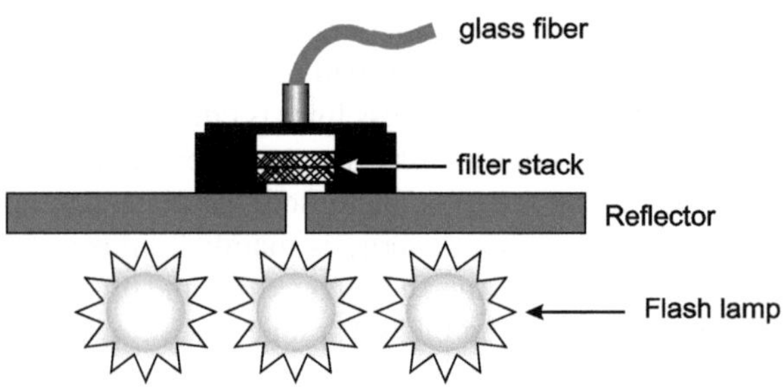

Fig. 2.19 Basic setup for the measurement of a flash lamp spectrum. After [63]

2.5.2 Transparency and Radiation Pattern

The plasma itself is only semi-transparent with absorption coefficients in the order of 1 cm^{-1} [47]. It was found that the absorption increases roughly to the square of the current density passing the lamp and with a power of 1–1.5 of the wavelength [64]. In case of uniform plasma in thermal equilibrium the emissivity can be given by [65]:

$$\varepsilon(\lambda) = 1 - \exp(-\alpha(\lambda)d) \tag{2.35}$$

where α is the wavelength-dependent absorption coefficient and d is the thickness of the plasma. Depending if $\alpha d \ll 1$ or $\alpha d \gg 1$ the plasma can be regarded to be transparent or opaque, respectively. According to (2.35) the emissivity starts at low values for low current densities and thus low absorption values, and converges towards 1 for high current densities. For medium input energies it can be all the same: transparent in the UV, semi-transparent in the visible and opaque in the NIR. This situation can cause deviations from the monotonically decreasing function of the line factor with input energy [66]. If the majority of the discrete emission lines is in the red or NIR spectral region, and if the continuum part of the flash lamp spectrum shifts more to the blue and UV, the light of the discrete emission lines originates from a larger plasma volume than the light from the continuum due to the different absorption coefficients. This in turn results in an overestimation of the line factor. However, the plasma of a flash lamp is only uniform in a quasi-stationary equilibrium in the fully expanded arc (zone 4 in Fig. 2.9), which is why (2.35) is only a rough guess. Furthermore, the time profile of the absorption coefficient of the lamp plasma roughly follows that of the current pulse in case of a damped PFN as it was verified in laser transmission measurements [47]. The increasing opacity with increasing current density has another consequence: an increasing amount of light is reabsorbed by the plasma and the absolute light intensity increases much less than expected [66]. Fortunately, this will not lead to an adequate decrease of η as reabsorption is not a dissipation of energy. Reabsorption helps to achieve a certain plasma temperature with less electrical input power. Finally, the refractive index of the plasma is approximated by the following formula [6]

$$n = 1 - 4.94 \times 10^{-14}\lambda^2 n_e + a\left(1 + \frac{b}{\lambda^2}\right)n_0 \tag{2.36}$$

where λ is the wavelength in cm, n_e is the electron concentration in cm^{-3}, n_0 is the concentration of atoms and ions in cm^{-3}, and a as well as b are coefficients which depend on the type of gas and are given in Table 2.1 (Sect. 2.3.3). The first and second term in (2.36) describes the electronic and atomic contribution to the refractive index, respectively.

Thus, the flash lamp spectrum is often composed of contributions which come from different depths of the plasma. As a consequence, the angular distribution of intensity is different from that of a Lambertian emitter. However, the data in literature give not always a consistent picture (Fig. 2.20). Due to the semi-transparent nature of the plasma it can be assumed that the real distribution is between a $\cos(\beta)$ (Lambertian emitter) and a $\cos^2(\beta)$ distribution (volume emitter). In fact, a linear combination of both was assumed in [67, 68], but experimental measurements of these authors revealed that the real distribution was even *narrower* than $\cos^2(\beta)$. This may be explained by the truncation of the distribution at larger angles caused by the passage of light rays through a thick-walled glass envelope and different refractive indices of air and plasma [69]. In contrast to this, recent measurements were in good agreement with a Lambertian emitter in [44], and in [5] an even broader far-field angular distribution was assumed and described by

$$I(\beta) = \frac{2\cos(\beta)}{1+\cos^2(\beta)} \tag{2.37}$$

In [6] it was argued, that the narrowing is mainly due to a more and more increasing opacity of the plasma leading to a distribution whose width depends on the product αd_0 of the absorption coefficient α and the arc diameter d_0. For a transparent plasma with $d_0 \to 0$, the distribution is similar to that of (2.37), whereas for an opaque plasma with $\alpha d_0 \to \infty$ the distribution is slightly narrower than a Lambertian distribution. Finally, a $\cos^q(\beta)$ distribution was found to be a good approximation for a flash lamp with a Kr:Xe ratio of 90:10 and a fill pressure of 500 Torr [70]. The coefficient q was fitted according to

$$q = 17.7 \cdot (I t_0)^{-0.34} \tag{2.38}$$

where I is the current in A in the range between 56 and 410 A and t_0 is the pulse duration in ms between 0.36 and 4.45 ms

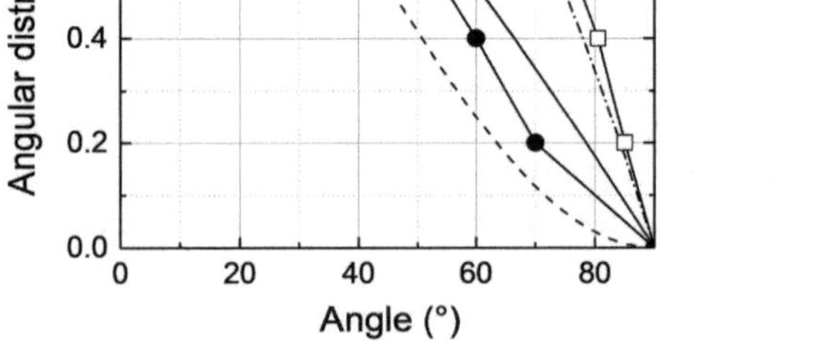

Fig. 2.20 Different models for the angular distribution of a flash lamp

The radiation originally generated by the plasma is assumed to be unpolarised. However, as it irradiates the inner wall of the quartz envelope with a certain angle distribution, reflections at the wall leads to a certain polarisation of the light emitted from the outer surface of the lamp. As the spectrum changes with time and the reflection depends on the wavelength, the polarisation of the emitted radiation is, strictly speaking, time- and wavelength-dependent. However, the polarization changes only slightly with the operation conditions, and according to [6] the general pattern is the following: For detection angles <30° with respect to the surface normal the degree of polarization is only in the order of 1–2%, but increases to a level of c. 10% if the angles increases to 60°. The degree of polarization continues to increase for even higher angles, but as the intensity strongly decreases for such high angles the impact on the overall balance is limited. Finally, it can be stated that flash lamp radiation is slightly polarised which, however, has no practical consequences in the very most cases.

2.5.3 *Time-Resolved Optical Properties*

As the flash lamp spectrum depends on the current density it should also change during the pulse. To investigate this, time-resolved spectra of a 20 ms pulse were recorded with a single spectrum integration time of 2 ms (Fig. 2.21a). The light intensity obtained by spectrum integration follows the current signal measured with a Rogowski coil and an oscilloscope (Fig. 2.21b). However, the maximum of the relative UV output is significantly delayed.

A more detailed investigation reveals that the light pulse is somewhat shorter than the current pulse. Figure 2.22 shows the voltage (blue), the current (black), the light intensity (red) and the electrical power (olivine) of a flash lamp with an initial capacitor voltage of 3.2 kV and for a 20 ms (a) and 3 ms pulse (b). In both cases the light intensity follows the profile of the electrical power which leads to significant differences at the falling edge of the pulse, i.e. the light fades out faster than the current. This finding also reveals that the conversion efficiency η is nearly constant over the pulse. In case of 20 ms the damping is close to the critical value of $\alpha_D \sim 0.8$, and there is a slight undershoot of the current. However, the converted electrical power during undershoot is negligible, and the lamp extinguishes when the current changes its polarity.

2.5.4 *Shaping the Spectral Output*

The purpose of spectral shaping is the maximization (or minimization) of the optical output at specific wavelengths or in individual optical bands. A prominent example is the laser pumped by flash lamps where the output of the flash lamp has to overlap with certain laser absorption lines. However, for FLA the output in certain

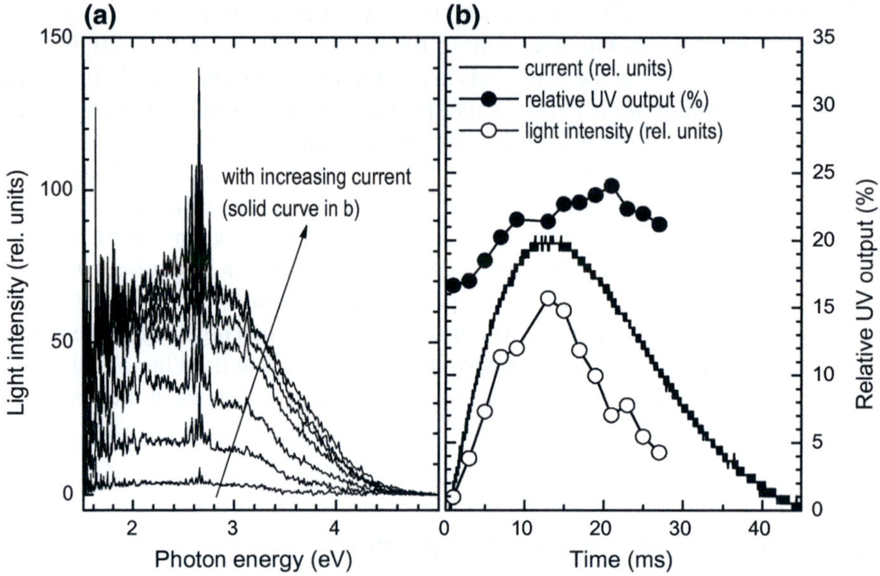

Fig. 2.21 Time-resolved spectra after 1, 3, 5, 7, 9, 11 and 13 ms (**a**) and time-resolved FLA parameters (**b**) of a 20 ms pulse. Please note that the right scale in (**b**) is different for the different curves

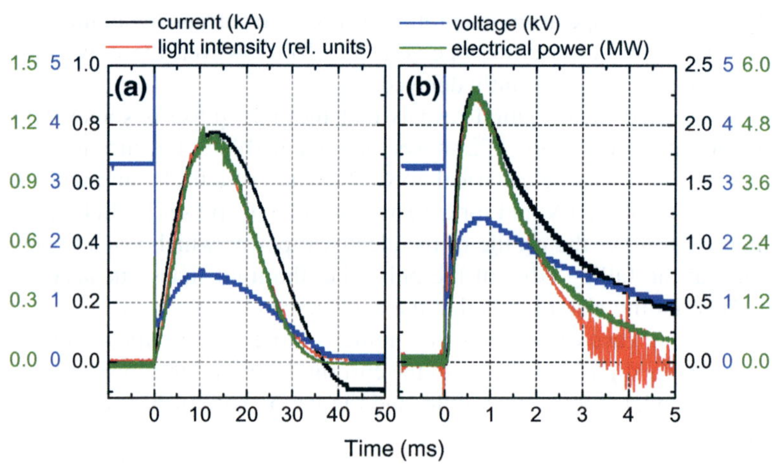

Fig. 2.22 Voltage (blue), current (black), light intensity (red) and electrical power (olivine) of a flash lamp with an initial capacitor voltage of 3.2 kV for a 20 ms (**a**) and 3 ms pulse (**b**). The current and the light intensity were measured via a Rogowski coil and a photodiode, respectively, whereas the voltage was measured with a 1:1000 oscilloscope probe

optical bands is more important. To simplify discussion the interest in a specific optical band is reduced to the question if the UV output of a flash lamp should be maximized or minimized compared to the visible and NIR spectral region. In the very most cases the absorption depth at UV wavelengths is smaller than those in the visible-NIR region which results in an accordingly shallower absorption profile. Thus, maximum UV output is of interest if shallow temperature profiles are needed as in case of temperature-sensible substrates or very thin layers. Another application is FLA of substrates which are transparent in the visible and where the major part of absorption has to occur in the UV. A typical example of this is the class of transparent conduction oxides (TCOs) which may even been deposited on a transparent substrate. However, UV irradiation can trigger photochemical reactions like polymerization or bond breaking which possibly has to be avoided in some applications. Basically, there are three possibilities to shape the spectrum of a flash lamp: (i) to choose optimum lamp parameters, (ii) to choose optimum operation conditions, and (iii) to cut parts of the spectrum by external filters or reflectors.

(i) The first option to maximize UV output is to choose synthetic fused quartz which is transparent down to 160 nm (see Sect. 2.2.3), but is also the most expensive type of quartz glass. For opposite requirements Ce-doped quartz can be used which absorbs radiation for wavelengths below 370–380 nm. Trivalent Ce, especially in an oxide environment, can be efficiently pumped by UV irradiation and emits around 460 nm [71], which is why Ce-doped quartz can transform a part of the UV lamp output into the blue spectral region. However, in this case the maximum input energy is more limited due to the additional thermal load of the quartz glass caused by absorption.
As discussed in Sect. 2.3.3 a low atomic number of the gas atoms leads to higher plasma temperatures at the expense of power conversion. Figure 2.23a compares the time-integrated flash lamp spectrum of a Xe and a Kr lamp with a filling pressure of 300 Torr during a 3 ms flash performed with a capacitance voltage of 2 kV. Due to the relative high voltage and filling pressure the spectra are dominated by the continuum, and they differ only slightly by the element-specific lines and by a higher output of Xe in the visible range. If the UV output of the spectrum should be optimized, a Kr lamp is slightly more efficient, although the overall efficiency is somewhat smaller. A similar but more unambiguous result was obtained in a study which compared different flash lamps filled with Ar, Kr or Xe at a filling pressure of 100 Torr [72]. The authors found that in the UV range of 175–250 nm, important for the photodissociation of various precursors in chemical vapour deposition, Kr lamps are superior compared to Xe and Ar lamps. A more recent publication investigated the spectra of Xe and Kr flash lamps and came to the conclusion that the maximum output in the wavelength region between 300 and 500 nm is achieved for a gas mixing ratio of 90% Kr and 10% Xe [73]. Finally, the UV output slightly increases if the filling pressure decreases.
(ii) As already discussed an increase in the maximum current density will increase the plasma temperature and the emissivity for shorter wavelengths, and both

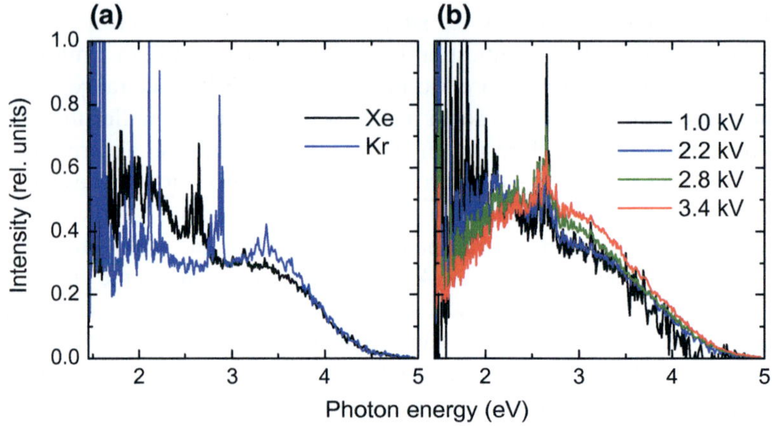

Fig. 2.23 Time-integrated spectrum of a 3 ms flash from lamp (**a**) filled by Xe or Kr at a filling pressure of 300 Torr and (**b**) filled by Xe at different capacitance voltages. All spectra are normalized to an integral of 1

effects lead to an increase of the UV part of the spectrum. This is shown in Fig. 2.23b where the spectrum of a Xe flash lamp shifts to the UV with increasing capacitance voltage. Furthermore, the decrease of the pulse time will have a similar effect if the pulse energy remains unchanged. As the same energy is now applied in shorter times, the peak current density is correspondingly higher. To quantify this, the relative UV output, defined as the integral between 3.1 and 4.96 eV (250–400 nm) on the energy scale compared to the integral over the whole spectral range (250–1100 nm), was determined for several FLA conditions. As shown in Fig. 2.24, the UV output continuously increases with increasing flash voltage V_0 and with shortening of the pulse duration from 20 to 6 ms. In case of 3 ms the capacitance was lower which results in a lower flash energy at a given V_0. As a consequence, the curves for 3 and 6 ms are close to each other as the increase in UV output caused by the shorter pulse time is balanced by the smaller flash energy.

(iii) Finally, external filters or reflectors can nearly eliminate the UV output by absorption or, the other way around, can maximize the UV output by suppressing wavelengths in the visible-NIR. Thus, an edge filter suppressing wavelengths shorter than 540 nm were used to avoid UV degradation of a polymer substrate during FLA [74]. Another example are dichroic reflectors which reflect in the UV and partly in the short-wavelength, visible region and let longer wavelengths through. However, the thermal load of these filters or reflectors has to be considered which probably includes additional cooling measures. Furthermore, cladding layers absorbing in certain wavelength regions can be used, although they require additional costs in processing and have to be considered with respect to heat distribution within a sample. Some of the possibilities to maximize the UV output are addressed in [75].

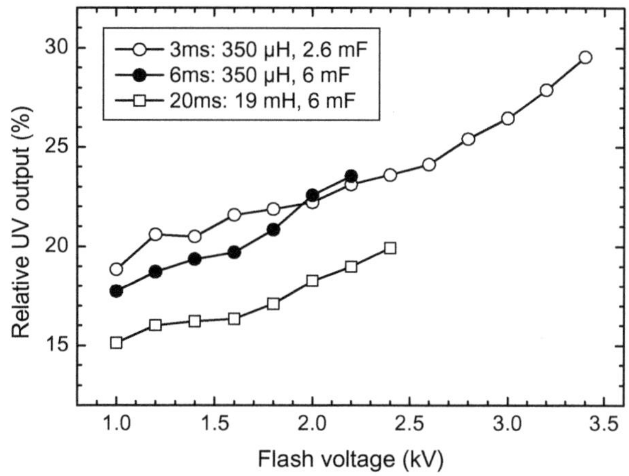

Fig. 2.24 Relative UV output of a Xe flash lamp as a function of the flash voltage and the pulse duration. Parameters of the PFN and the definition of the relative UV output are given in the legend and the text, respectively. Data extracted from [63]

2.6 Flash Lamp Lifetime

2.6.1 Single and Multiple Flash Operation

From the viewpoint of application, flash lamps should have an "excellent lifetime", and the costs for replacement should be minimal. According to [5] flash lamps which deserve this title have to fulfil the following three criteria: (i) the lifetime should be so long that all expenses which result from lamp replacement are only a small part of the overall operation costs, (ii) the end of a lamp's lifetime is not characterized by catastrophic events like lamp explosion, and (iii) the lifetime of a lamp is predictable. As some failure mechanisms include catastrophic events, one strategy to fulfil the criteria (ii) and (iii) is the recognition of early signs of an impending failure in the aging characteristics. When considering the maximal load for a flash lamp one has to distinguish between single and multiple flash operation. In case of single flash operation the break between different flashes is so long that individual flash pulses are uninfluenced by each other (see Sect. 2.3.4). In this case the load is limited by the explosion energy E_{ex} which is defined as the electrical input energy at which the lamp explodes with the first flash. Although E_{ex} is mostly given in Joule, sometimes the unit Joule per arc length is used in literature due to the linear dependence of E_{ex} on the arc length. The lifetime is usually measured by the number of flashes the lamp can complete without failure. In case of multiple flash operation flash pulses with an input energy E_0 are repeated with a frequency f, and the limit is the average power P_{ave} given by

$$P_{ave} = E_0 \cdot f \tag{2.39}$$

In contrast to single flash operation the lifetime is now measured in hours of operation. The explosion energy is a statistical value which may differ from lamp to lamp with a variation in the order of 10% [6]. The same applies for the average power limit, although the variations are assumed to be smaller.

2.6.2 *Lamp Failure*

Generally, lamp failure is an event where the lamp immediately stops working, mostly accompanied by a burst or an explosion (Fig. 2.25) with the corresponding inconveniences. Thus, users are strongly interested in the prediction of such failures. The physical processes leading to lamp failure can be assigned to one of the following categories [6]:

(i) Lamp explosion or breakage due to the shockwave when the lamp is fired.
(ii) Lamp explosion or breakage if the pressure exceeds the burst pressure of the lamp.
(iii) Thermal failure caused by thermal stress which evolves from the difference of a hot inside and a cold outside of the envelope.
(iv) Failures which originate from the different kinds of electrode erosion. This is typically the final point of a longer aging history.

The main parameter to predict failures is the explosion energy E_{ex} which primarily depends on the geometrical dimensions of the lamp. It is clear that the released energy is distributed over the inner surface area leading to a linear relationship between E_{ex} and the surface area [6]:

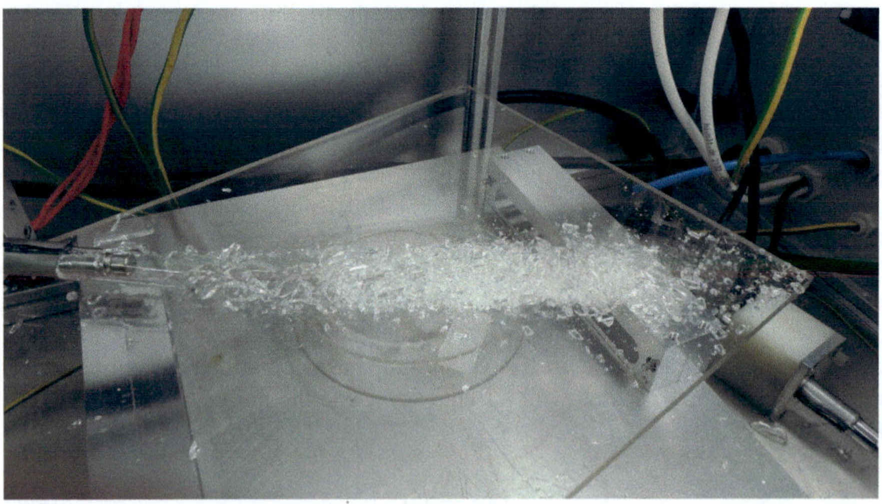

Fig. 2.25 Pulverized flash lamp in a lamp test stand after applying an energy close to the explosion energy E_{ex}. A U-shaped girder made of Al was used as lamp reflector and splinter protection. Courtesy of M. Neubert

$$E_{\mathrm{ex}} \approx 10{,}000 \cdot d_0 \ell \sqrt{t_0} \tag{2.40}$$

where d_0 is the bore diameter in cm, ℓ is the arc length in cm, t_0 is the pulse length in s, and E_{ex} is the explosion energy in J. Equation (2.40) is used for quartz glass; for glass the values are lower by a factor of 4–5 [6]. In addition, E_{ex} tends to increase with increasing wall thickness and decreasing filling pressure: (2.40) fits best to "standard" parameters, namely wall thicknesses in the range of 1–2 mm and filling pressures in the order of 400–500 Torr [6]. For a given explosion energy E_{ex} and input energy E_0, the predicted number of flash pulses N which can be applied within the lifetime of a lamp is given by

$$N = \left(\frac{E_0}{E_{\mathrm{ex}}}\right)^{-m} \tag{2.41}$$

This empiric formula is well known, but several references partly differ in the specific value of the lifetime coefficient m. In [6] a range between 8.5 and 12.5 is given, whereas more recent references give $m = 8.5$ [8, 9]. In [5] an value of $m = 8.5$ and $m = 14$ is given for bore diameters smaller than 6 mm and greater than 15 mm, respectively, with a transition region between. The two cases of $m = 8.5$ and $m = 14$ are illustrated in Fig. 2.26a.

It was also found that (2.40) fits well for pulse lengths in the ms range but deviates more and more for shorter pulse lengths. In fact, there is a critical pulse time τ^* which separates a short-time ($t_0 < \tau^*$) from a long-time ($t_0 > \tau^*$) regime [6]. τ^* depends on

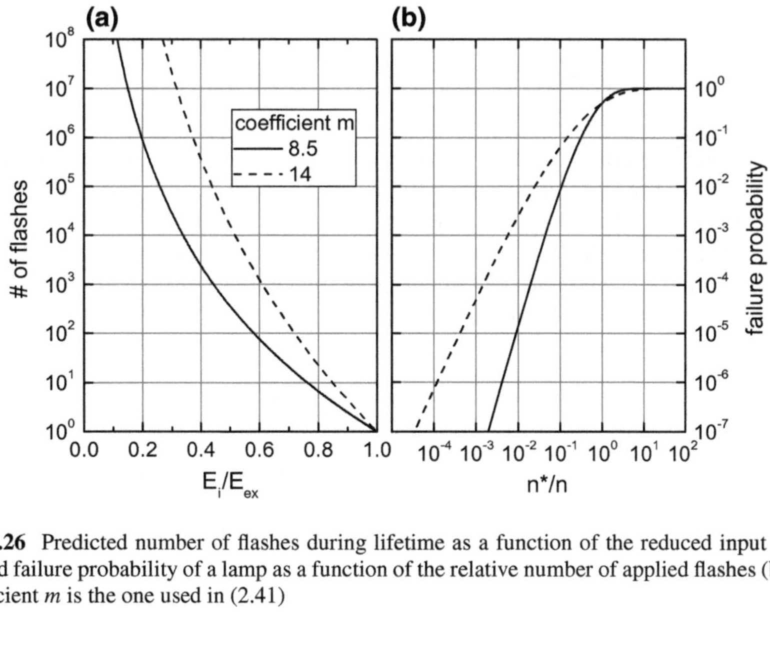

Fig. 2.26 Predicted number of flashes during lifetime as a function of the reduced input energy (**a**) and failure probability of a lamp as a function of the relative number of applied flashes (**b**). The coefficient m is the one used in (2.41)

many design parameters of the lamp and often ranges between 100 μs and 1 ms. One distinct dependence is that on pressure: τ^* becomes shorter with decreasing filling pressure. In the short-time regime, E_{ex} scales with t_0 instead with $\sqrt{t_0}$ which leads to significantly lower explosion energies than predicted by (2.40). In addition, the variation of experimental values is large and may reach values of 50% which makes the prediction of a failure more difficult. It is believed that failures in the short-time regime are due to the shockwave when the lamp is fired. As a consequence, the short time region should be avoided if possible. Even if the pulse length is long, pulse shapes with a very sharp current rise are critically. In this case the current rise time is more important than the pulse length. One recommendation in literature is that the current rise $\mathrm{d}I/\mathrm{d}t$ should not exceed a value of 10^7 A s^{-1} [10].

In contrast to this, failures in the long-time regime are more predictable with a variation of the explosion energy in the order of 10% [6]. Moreover, if the applied energy is well below E_{ex}, a possible sign of an impending failure is the increase of the required trigger voltage. According to (2.15) the internal pressure strongly rises during the flash pulse if the temperature ramps up to values of 10,000 K and more. However, a lamp is designed to withstand this pressure under normal operation conditions. For very high input energies the temperature increases only slowly, and the contribution of an increasing rate of ionization is also rather small. Thus, the burst of the flash lamp is probably not caused by the pressure increase during the discharge. Alternative models propose a significant increase in pressure due to the vaporization of material from the inner side of the envelope [6] or the development of sufficiently high thermal stresses in the quartz walls [10].

In the first scenario evaporation starts if a certain power limit P_{lim} is exceeded. P_{lim} is the power per cm^2 absorbed in the envelope wall and is approximated by [6]:

$$P_{lim} = \rho h_s \sqrt{\frac{D_{th}}{t_0}} \tag{2.42}$$

where t_0 is the pulse length and ρ, h_s and D_{th} are the mass density, the specific heat of evaporation and the thermal diffusivity of quartz glass, respectively. Equation (2.42) can be also read in the sense that evaporation starts for a given P_{lim} after a time t_0. If it is now assumed that the part $(1 - \eta)$ of the electrical energy, which does not contribute to the measureable optical output of the lamp, is absorbed in the envelope wall, (2.42) and (2.5) can be combined to

$$E_{lim} = \pi d_0 \ell \frac{\rho h_s \sqrt{D_{th} t_0}}{1 - \eta} \tag{2.43}$$

E_{lim} is now the critical input energy at which significant evaporation of quartz starts. It has to be noted that this is an underestimation as some of the losses are not due to absorption. Finally, if E_0 exceeds E_{lim} the evaporated mass M can be approximated by [6]

$$M = \frac{\theta E_0}{h_s} \tag{2.44}$$

where θ is the fraction of energy which is really consumed by evaporation. It depends on the excess energy E_0–E_{lim} and may consider the underestimation mentioned above. Based on this, (2.44) gives the number of additional O and Si ions in the plasma which originate from the quartz envelope, and the use of (2.15) can give an estimate of the final pressure. However, even if the burst pressure is not reached, the evaporation of SiO_2 material can accelerate degradation (next section). Interestingly, both the model of SiO_2 vaporization and that of thermal stress generation lead to similar empirical results, especially with respect to (2.40).

The easiest way to get a robust lamp for single flash operation is to increase the wall thickness. If σ_b is the burst stress of the material (ca. 60 MPa for quartz glass), the burst pressure is given by [6]

$$p_b = \frac{R^2 - r^2}{R^2 + r^2}\sigma_b \tag{2.45}$$

where r and R are the bore and outside diameters of the lamp, respectively. However, experimentally measured burst pressures are mostly lower than predicted by (2.45), for which reason this equation should only be regarded as a rough guide for comparing different lamps.

Equation (2.41) gives the number of flashes which can be applied in the lifetime of a flash lamp for a given E_0 and E_{ex}. However, if we assume that there is a distribution of explosion energies and E_{ex} is the median of this distribution, the failure probability of the lamp after N flashes is 0.5. Without doubt, this is unacceptable for most applications. If $f(e)$ is the distribution of the reduced explosion energy $e = E^*_{ex}/E_{ex}$, the failure probability after N^* flashes is given by

$$p = \int_0^{e^*} f(e)\mathrm{d}e \quad \text{with} \quad e^* = \left(\frac{N^*}{N}\right)^{1/m} \tag{2.46}$$

where N is the number of flashes obtained by (2.41). If $f(e)$ is a Gaussian distribution with the width σ, (2.46) can be written as

$$p = \frac{1}{2}\left(1 + erf\left(\frac{e^* - 1}{\sqrt{2}\sigma}\right)\right) \tag{2.47}$$

Figure 2.26b displays the failure probability as a function of the relative number of applied flashes. For a better illustration let us assume that someone decides to apply pulses with 20% of the explosion energy. According to (2.41) with $m = 8.5$ the predicted number of flashes till breakdown is ~9×10^5 (Fig. 2.26a). Figure 2.26b now describes the probability that the lamp fails earlier: for 9×10^5 pulses ($N^*/N = 1$) the probability is 0.5, for 9×10^4 pulses ($N^*/N = 0.1$) the probability drops down to 0.009 etc. A higher m value leads to a steeper increase of the predicted number of flashes with decreasing input energy, but also to a broader distribution of the failure probability. To continue with this example, 20% of the explosion energy would allow 6×10^9 pulses for $m = 14$, but a reduction by factor 10 ($N^*/N = 0.1$) lowers the failure probability to ca. 0.06 only.

It should be considered that the risk of failure approximately increases by N_L times if a lamp is working in a bank of N_L lamps. A failure in form of an explosion usually destroys all lamps of a bank which is why the weakest lamp determines the lifetime of all. If $1 - p$ is the survival probability of one lamp, the failure probability of N_L lamps is

$$p(N_\mathrm{L}) = 1 - (1 - p)^{N_\mathrm{L}} \approx N_\mathrm{L} \cdot p \tag{2.48}$$

The linear approximation was obtained by Tailor expansion for $p \ll 1$.

The heating of the inner side of the envelope leads to material expansion and compressive stress there. Theoretically, if this stress exceeds the burst stress σ_b, the lamp should burst. However, experimental investigations suggest that thermal stress is usually not the limiting factor [6]. However, it can be the limit in case of multiple flash operation and might be generally responsible for the formation of hairline cracks in advance of a lamp failure. Another potential failure mechanism is the confinement of the arc in a tiny spot of the electrode during arc formation. In this scenario the corresponding current density is high enough to evaporate significant amounts of material which also leads to lamp explosion. However, it was found that this will only happen under really harsh conditions like very short times combined with high voltages. Other electrode failure mechanisms may lead to a failure without lamp explosion or breakage, and the explosion energy is probably not an adequate description for these cases.

2.6.3 Lamp Degradation

In the ideal case a flash lamp will not operate up to failure but up to a time where lamp parameters leave a defined range of values. In most cases the lifetime is defined as the time where the optical output power of the lamp falls below a certain critical value like 70% or 80% of its original value. Degradation is mainly due to electrode erosion, degradation of the lamp envelope or a combination of both.

The mechanisms of electrode erosion were already mentioned in Sect. 2.2.2 and mainly include sputtering and material evaporation from the cathode and the anode. Sputtering is especially high if the lamp is switched on with a cold cathode. The reason behind this is that electron emission only occurs by field emission which requires higher electric fields than in cases where field emission is assisted by a high cathode temperature. In turn, higher electric fields increase the number and energy of Xe ions hitting the cathode surface. If sputtering is avoided, cathode material can also be evaporated depending on the cathode temperature [33]. This temperature can be controlled and limited, inter alia, by lowering the work function [13] and choosing a suitable arc attachment mode [16]. The re-deposition of cathode and anode material at the wall decreases its transparency and increases the wall loading. An additional degradation mechanism is the depletion of Ba (or another low work function material) at the cathode which increases their operation temperature and starts a phase of accelerated aging. Depletion in turn increases with temperature, and a linear relationship between the logarithm of the lifetime and the reciprocal of the cathode temperature was found in [12]. Furthermore, a leaky seal or SiO_2 evaporation may oxidize the electrodes leading to the formation of tungsten oxide and tungsten disilicide which may lead to additional deposits on the walls [10]. It is said that oxygen levels as low as 10^{-9} Torr are sufficient to "poison" a cathode and to efficiently shorten the lamp lifetime [76]. One remedy is the impregnation with osmium- or iridium-based protective coatings [76]. The argumentation for the anode is similar, but here the critical process is thermal evaporation instead of oxidation. Thus, traces of molten tungsten on the anode surface are a first sign of degradation. As thermal conductivity in noble gases decreases with atomic mass, electrode lifetimes are higher for Xe than for Kr due to the better heat isolation [10]. Electrode erosion may become the dominant lifetime limit in the long term if other failure or degradation mechanisms like lamp explosion are eliminated, e.g. by using input energies being much lower than the explosion energy.

At temperatures above 1300 °C SiO_2 starts to decompose to SiO and O_2, but in the presence of hydrogen decomposition already starts at 1200 °C [10]. SiO itself is not stable and continues to decompose, this time to Si and SiO_2. The deposition of Si or SiO leads to blackening and partly conductive paths. The first one increases the thermal load of the walls, whereas the second one leads to an increase of the required trigger voltage. The processes of evaporation and re-deposition can cause a local, but significant thinning of the envelope walls over the lifetime of the lamp. The weight loss of flash lamp envelopes made of different materials by these processes was already investigated in [77]. Furthermore, quartz becomes more plastic with increasing temperature, which can lead to lamp bending and the formation of quartz bubbles in the long term. Such problems, however, can be avoided by appropriate means of cooling. Finally, microcracks can evolve during the cooling phase and are a precursor of a lamp burst.

It was found that lifetime increases at first with increasing size of the cathode but becomes constant above a certain threshold [6]. Furthermore, lifetime increases with increasing filling pressure. In case of multiple flash operation, lifetime increases with the flash frequency as long as the power is small compared to the average power limit [6].

2.7 Layout of FLA Tools

2.7.1 From Sheet-to-Sheet to Roll-to-Roll Processes

The typical FLA tool for single substrate handling has already been sketched in Fig. 2.1 and is commonly used in semiconductor industry and research. Depending on the complexity of the machine and the throughput requirements handling can be performed by hand or by automatic handling systems. In any case such a handling, called a sheet-to-sheet (S2S) process, requires subsequently the feed-in of the substrate or wafer into the chamber, the evacuating or purging of the chamber, the FLA process itself, eventually the venting of the chamber, and the feed-out of the wafer. The required process time compared to the pulse length is long and allows large charging times. In most cases the required annealing temperatures are high and sometimes even close to the melting temperature of the semiconductor. Thus, the electrical performance is often designed for single flash operation with flash pulses with a high energy density and comparatively long pulse times of a few ms or tens of ms. In addition, relatively large lamp fields are used in order to cover the whole area of the substrate.

However, if substrates become large they cannot be flashed in one piece any longer. In this case the substrate is moved under the flash lamp with constant speed, and during the flash only a stripe-type area of the substrate is illuminated. Mostly, only a small number of flash lamps (or even only one) is used, but they can be quite long depending on the width of the substrate. The FLA tool is now rather a module of a larger system than a stand-alone machine. Basic sketches of such a module for one [74, 79–82] and more lamps [83, 84] can be found in literature. Depending on whether the substrate is rigid or flexible the transport system is like a conveyer belt or a roll-to-roll (R2R) system, respectively. The step-by-step annealing also changes the electrical requirements of the FLA tool: it uses now multiple flash operation, and the flash frequency strongly increases which is accompanied by a decrease of pulse length and energy density. Figure 2.27 gives a basic overview of the electrical performance for different kinds of application. This sketch is by far not comprehensive, but it illustrates the general trend quite well. The area of high energy density and long-pulses is the domain of semiconductors and is typical for S2S systems. The region of conveyer belt or R2R systems may start in the lower left edge of the Si and Ge domain with the crystallization of amorphous Si on large glass carriers as an exemplary application, but extends over transparent conducting oxides (TCO) to printed electronics and flexible substrates. In this order the required energy density per flash decreases, the pulse length becomes shorter, and the flash frequency increases to several Hz and beyond. The flash frequency is often a complex balance between the advance rate of the substrate, the illuminated area, the annealing requirements and the power design of the charging unit.

In conveyer belt or R2R systems the annealed area is composed piece by piece, and it is nearly impossible to have a gapless but non-overlapping coverage of the area. The first possibility to use FLA is the camera mode in which different illuminated areas are separated by a small, but untreated stripe (Fig. 2.28a). This mode is preferred if

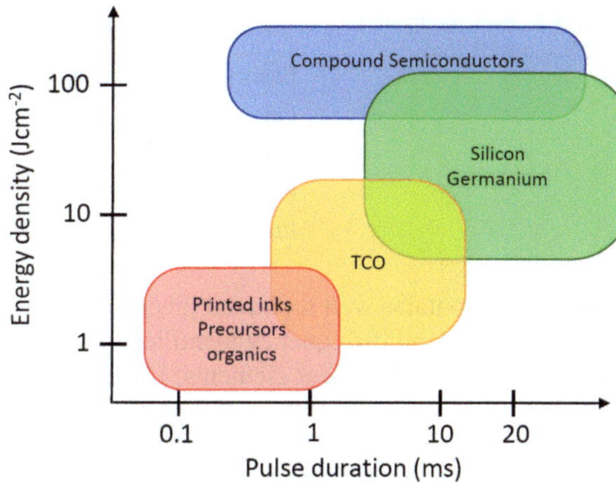

Fig. 2.27 Overview of the electrical performance of FLA tools for different kinds of application. After [78]

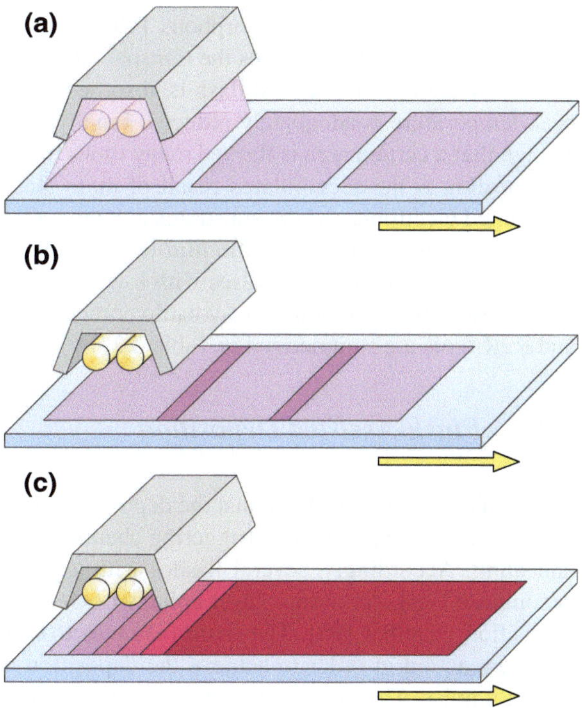

Fig. 2.28 FLA operation in the camera mode (**a**), the low frequency (**b**) and the high frequency continuous mode (**c**). The arrow indicates the movement of the substrate

Table 2.2 Basic comparison of S2S and R2R tools

	S2S	R2R
Energy density per pulse	Medium/large	Low
Pulse length	ms range	<1 ms
Flash frequency	Low	Medium/high
Number of lamps	Many	One or few
Illuminated area	Whole area	Stripe-type part

the substrate contains well separated functional areas to be flashed, such as a narrow roll with printed radio-frequency identification (RFID) tags or circuitries with large spacing between subsequent elements. If the whole area of the substrate should be annealed, FLA can work in the continuous mode in which subsequent areas overlap. In the low frequency case (Fig. 2.28b) the overlap region is exposed twice to the flash but is rather narrow. This mode can be used if the temperature has to exceed only a certain threshold and if the second flash is without impact. A typical example for this is the crystallization of amorphous TCO layers where the first flash crystallizes the layer and strongly increases the transparency in the visible and near UV region. As a consequence, the second flash is absorbed to a lesser extent and the resulting peak temperature is adequately reduced. Finally, the frequency of the flash can be so high that a certain area is flashed many times (Fig. 2.28c). In this mode the effect of annealing is the accumulated result of many flash pulses. This mode is typical for printed electronics where substrates are temperature-sensitive and the annealing also comprises the drying of the film or the sublimation of solvents. In this case the fast succession of short pulses with low energy density should also prevent the rapid, destructive evaporation of volatile compounds. The differences between S2S and R2R tools are summarized in Table 2.2.

2.7.2 Flash-Assisted Deposition

FLA is increasingly used to assist the deposition or modification of thin films, especially by cracking precursors, triggering chemical reactions or performing an in situ annealing. Accordingly, several flashes with a pulse length of 1 ms were used to decompose GeH_4 for atomic layer epitaxy of Ge in a low pressure chemical vapour deposition chamber [85]. The main idea was to separate surface adsorption from surface reaction in order to achieve the deposition of a single atomic layer of Ge per flash pulse. The method was later successful extended to SiH_4 for Si deposition [86] and to NH_3 for the nitriding of a Si(100) surface [87]. In the latter case it was found that FLA is able to enhance the nitriding rate by up to 5 orders of magnitude and to achieve higher N concentrations at the surface. In all these cases the UV part of the flash lamp spectrum was essential. According to [72] wavelengths must be shorter than 160, 185 and 206 nm in order to efficiently dissociate SiH_4, N_2O and NH_3 molecules, respectively.

A more recent development is the use of FLA for atomic layer deposition (ALD). The idea to assist ALD by light illumination already appeared in the patent of Ishizaka

[88], but according to the power range given in the design example the use of flash lamps was probably not yet intended. ALD is a low temperature process in order to limit the growth rate and to avoid the decomposition of precursors which, on the other side, often requires a post annealing step to improve the film quality [89]. This improvement may include the crystallization of amorphous layers, the growth of larger grains, defect annealing, or even the growth of quantum structures [90]. ALD becomes especially difficult if the individual steps of the ALD cycle have to run at different temperatures. An FLA step can be now introduced to decompose adsorbed precursor molecules, activate chemical reactions, or perform in situ annealing. Figure 2.29 displays a typical ALD cycle with an FLA step after the second purging (step 4) in order to perform an in situ annealing step. Due to the short time scale the chamber walls are not significantly affected by the flash, and the next ALD step can start without any delay caused by cooling. If the FLA step is intended to decompose precursor molecules or activate the oxidation reaction, it can be incorporated after or during the precursor pulse (step 1) or the oxidant pulse (step 3), respectively. More details about the process control, namely the dependence of ALD parameters like growth rate on FLA parameters, can be found in literature [89, 90].

The typical tool design of a flash-assisted ALD as shown in Fig. 2.30 closely resembles the design of S2S machines. However, special attention has to be paid to the fact that ALD is depositing thin films everywhere in the process chamber including windows in the line of sight of the flash lamps. As in most FLA tools, the flash lamps are protected by a quartz glass window. In addition, a sacrificial, simple to replace glass window can be inserted between the substrate and the quartz glass window. In order to enhance the lifetime of this second window, an inert gas purging unit is integrated just below with the task to minimize the adsorption of precursors

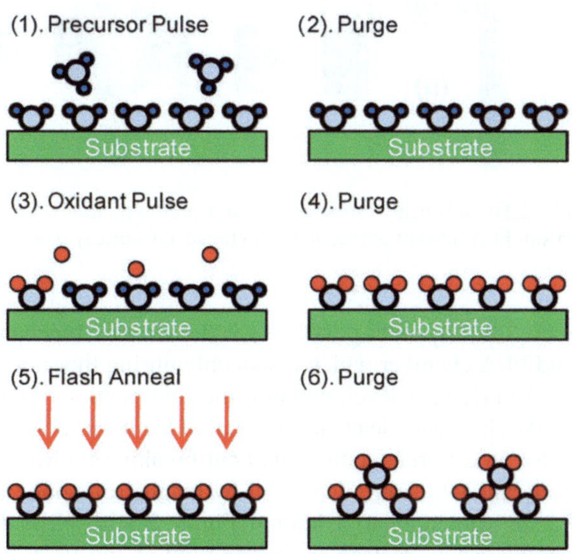

Fig. 2.29 Schematics of a modified ALD cycle with an FLA step during the oxidant purging step. Reprinted with permission from [90], Copyright (2012) American Chemical Society

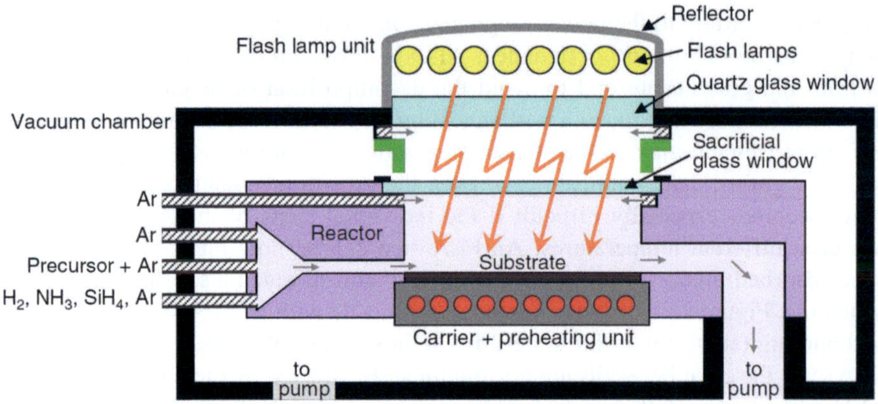

Fig. 2.30 Basic scheme of a flash-assisted ALD chamber (by Henke et al. [89])

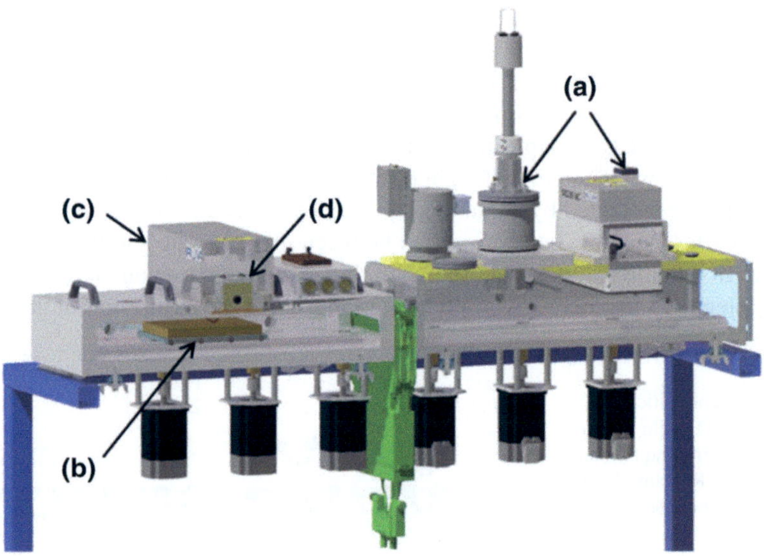

Fig. 2.31 Schematic cross section of a prototype tool for magnetron-sputtering combined with in situ FLA. Details are explained in the text. Courtesy of Rovak GmbH

there [89]. Another possibility is the use of a mechanical shutter between the ALD and FLA chamber which opens only during the short time of the flash [90].

A fairly new development is the combination of magnetron-sputtering and in situ FLA. The basic idea behind is that of a cluster tool in which a sample may commute between different stations without breaking the vacuum. This especially applies to materials which are deposited by magnetron-sputtering and needs to be annealed, but are vulnerable to oxidation. Figure 2.31 displays the basic scheme of a prototype combining magnetron-sputtering with in situ FLA. The tool is made of a chamber for FLA (left) and a chamber for magnetron-sputtering (right) equipped with two sputtering sources (a). Both chambers are separated by a valve (green) which enables, if needed, the execution of FLA and magnetron-sputtering at different vacuum or inert gas conditions. A sample carrier (b) can commute between both chambers to realize alternating cycles of deposition and annealing. The FLA unit (c) is equipped with a water-cooled flash lamp (d) and an Al reflector which results in an homogeneously irradiated, stripe-shaped area of about $30 \times 200\ \text{mm}^2$.

References

1. G.G. Bentini, L. Correra, Analysis of thermal stresses induced during xenon arc lamp flash annealing. J. Appl. Phys. **54**, 2057–2062 (1983)
2. J.C. LeBlanc, M.A. Herbert, D.W. Whillans, H.E. Johns, An intense microsecond flash lamp. Rev. Sci. Instr. **43**, 1814 (1972)
3. B.M. Kovalchuk et al., Capacitor bank module for multi megajoule energy storage, in *Pulsed Power Conference* (2007), p. 1479
4. R.H. Dishington, W.R. Hook, R.P. Hilberg, Flashlamp discharge and laser efficiency. Appl. Opt. **13**, 2300 (1974)
5. ILC Technology, An overview of flashlamps and CW arc lamps. Technical Bulletin 3 (1986)
6. I.S. Marshak, *Pulsed Light Sources* (Consultants Bureau, New York, 1984)
7. G.R. Jones, M.T.C. Fang, The physics of high-power arcs. Rep. Prog. Phys. **43**, 1415 (1980)
8. The Heraeus Noblelight Technical Reference Book for Arc and Flash Lamps. http://www.dmlittle.eclipse.co.uk/lasers/documentation/Heraeus%20Noblelight%20Flashlamp%20Catalog.pdf. Accessed 12 Nov 2018
9. High Performance Flash and Arc Lamps, Perkin Elmer. https://laser-caltech.web.cern.ch/laser-caltech/report/Flash%20lamp%20Eg&G.pdf. Accessed 12 Nov 2018
10. R. Iffländer, *Gas Discharge Lamps for Laser Excitation*, 2nd edn. (Schramberg, Iffländer, 2008)
11. http://www.vonardenne.biz/fileadmin/user_upload/druckschriften/Flash_Lamp_Annealing_english.pdf. Accessed 21 Mar 2016
12. T. Aida, H. Tanuma, S. Sasaki, T. Yaguchi, S. Taguchi, N. Koganezawa, Y. Nonaka, Emission life and surface analysis of barium-impregnated thermionic cathodes. J. Appl. Phys. **74**, 6482 (1993)
13. P. Flesch, M. Neiger, Numerical simulation of dc high-pressure discharge lamps including electrodes. J. Phys. D **35**, 1681 (2002)
14. J.D. Cobine, E.E. Burger, Analysis of electrode phenomena in the high-current Arc. J. Appl. Phys. **26**, 895 (1955). https://doi.org/10.1063/1.1722115
15. P. Flesch, M. Neiger, Time-dependent simulation of plasma and electrodes in high-intensity discharge lamps with different electrode shapes. J. Phys. D **36**, 849 (2003)
16. S. Lichtenberg, D. Nandelstädt, L. Dabringhausen, M. Redwitz, J. Luhmann, J. Mentel, Observation of different modes of cathodic arc attachment to HID electrodes in a model lamp. J. Phys. D Appl. Phys. **35**, 1648 (2002)
17. M. Redwitz, L. Dabringhausen, S. Lichtenberg, O. Langenscheidt, J. Heberlein, J. Mentel, Arc attachment at HID anodes: measurements and interpretation. J. Phys. D Appl. Phys. **39**, 2160 (2006)
18. D.R. Lide (ed.), *CRC Handbook of Chemistry and Physics*, 87th edn. (CRC Press, Boca Raton, Fl, 2006)
19. Lamp Materials, Heraeus Quarzglas. https://www.heraeus.com/media/media/hqs/doc_hqs/products_and_solutions_8/tubes/Quartz_lamp_materials_EN.pdf. Accessed 19 Sept 2017
20. W. Koechner, *Solid-State Laser Engineering* (Springer Science+Business Media, New York, 2006)

21. S. Beleznai, G. Mihajlik, A. Agod, I. Maros, R. Juhasz, Z. Németh, L. Jakab, P. Richter, High-efficiency dielectric barrier Xe discharge lamp: theoretical and experimental investigations. J. Phys. D **39.** 3777–3787 (2006)
22. https://www.heraeus.com/en/hqs/fused_silica_quartz_knowledge_base/t_calc/transmission_calculator.aspx. Accessed 19 Sept 2017
23. https://refractiveindex.info/. Accessed 5 Oct 2017
24. F. Aumento, Stability, lattice parameter, and thermal expansion of β-cristobalite. Am. Miner. **51**, 1167 (1966)
25. D.M. Camm, J.C. Gelpey, T. Thrum, G.C. Stuart, S. McCoy, Flash-assist RTP for ultra-shallow junctions. J. Miner., Metal. Mater. Soc. (TMS) **58**, 32–4
26. C. Ramsauer, Über den Wirkungsquerschnitt der Gasmoleküle gegenüber langsamen Elektronen, Annalen der Physik. **6**(369), 513–540 (1921) (in German)
27. M.J. Kushner, Arc expansion in xenon flashlamps. J. Appl. Phys. **57**, 2486 (1985). https://doi.org/10.1063/1.335434
28. G.W. LeCompte, H.E. Edgerton, Xenon arc transients, electrical and optical. J. Appl. Phys. **27**, 1427 (1956)
29. Y. Vitel, M. Skowronek, K. Benisty, M.M. Popovic, Plasma-wall interaction in flashtubes. J. Phys. D **12**, 1125 (1979)
30. J. Shenli, L. Rui, L. Jianjun, L. Xingwen, S. Xiachuan, L. Haibing, The plasma channel evolution characteristics of pulsed flashlamps working in an array. Plasma Sci. Techn. **15**, 640 (2013)
31. K.R. Lang, F.S. Barnes, Shock waves in xenon flashtubes and tube deterioration. J. Appl. Phys. **35**, 107 (1964)
32. T.G. Barton, R. Guttenberger, H.J. Foth, Effects of simmer current on flash-lamp impedance and their combined influence on the output of the Ho, Cr, Tm:YAG laser. Appl. Opt. **34**, 2004 (1995)
33. M.S. Benilov, Understanding and modeling plasma–electrode interaction in high-pressure arc discharges: a review. J. Phys. D Appl. Phys. **41**, 144001 (2008)
34. H.A. Dinulescu, E. Pfender, Analysis of the anode boundary layer of high intensity arcs. J. Appl. Phys. **51**, 3149 (1980)
35. H. Maecker, Plasmaströmungen in Lichtbögen infolge eigenmagnetischer Kompression. Zeitschrift für Physik **141**, 198 (1955) (in German)
36. S.M. Sze, K. Ng Kwok, *Physics of Semiconductor Devices*, 3rd edn. (Wiley-Interscience, Hoboken, N.J., 2007)
37. P. Pyykkö, M. Atsumi, Molecular single-bond covalent radii for elements 1–118. Chem. Eur. J. **15**, 186 (2009)
38. U. Bauder, Radiation from high-pressure plasmas. J. Appl. Phys. **39**, 148 (1968)
39. J.H. Goncz, Resistivity of xenon plasma. J. Appl. Phys. **36**, 742 (1965)
40. P.J. Walsh, W. Lama, T.J. Hammond, Voltage-current relationship for pulsed arc discharges. J. Appl. Phys. **52**, 5476 (1981)
41. R.S. Devoto, Transport coefficients of ionized argon. Phys. Fluids **16**, 616 (1973)
42. M.-R. Liao, H. Li, W.-D. Xia, Approximate explicit analytic solution of the Elenbaas-Heller equation. J. Appl. Phys. **120**, 063304 (2016). https://doi.org/10.1063/1.4960777
43. K. Günther und R. Radtke, Elektrische Leitfähigkeit von Xenon-Impulsplasmen, Beiträge aus der Plasmaphysik **12**(1), 63 (1972) (in German). https://doi.org/10.1002/ctpp.19720120109
44. Y. Wu, J. Zhu, Z. Zhang, Y. Li, Radiation model of a xenon flash lamp in a laser amplifier pump cavity. High Power Laser Sci. Eng. **3**, e31 (2015)
45. A. El-Deib, F. Dawson, D. Camm, Physics-based circuit model for a water-wall high temperature arc lamp using finite volume method. IEEE Trans. Ind. Appl. **50**, 4289 (2014)
46. H. Elloumia, G. Zissis, T. Berthier, J.J. Damelincourt, Time-dependent behavior of a pulsed high power xenon flashlamp. Application to the Mega-Joule laser pumping. J. Quant. Spectrosc. Rad. Transfer **86,** 361–369 (2004). https://doi.org/10.1016/j.jqsrt.2003.12.003
47. G.J. Linford, Time-resolved xenon flash-lamp opacity measurements. Appl. Opt. **33**, 8333 (1994)

48. S. Amano, T. Mochizuki, Dynamic characteristics of high-repetition-rate flash lamps in high-power slab lasers. Opt. Comm. **201**, 173 (2002)
49. J.R. Oliver, F.S. Barnes, A comparison of rare-gas flashlamps. IEEE J. Quantum Electron. **5**(5), 232 (1969)
50. J. Wegrzyn, G. Patonay, I. Warner, M. Ford, Versatile power supply circuit for xenon flashlamps. Rev. Sci. Instr. **60**, 90 (1989)
51. M.S. Ansari, B. Singh, S.V.G. Ravindranath, M.S. Bhatia, Multi-resolution analysis of partial discharge current in low pressure flash lamps, in *XXVI International Symposium on Discharges and Electrical Insulation in Vacuum* (2015), p. 81
52. H. Kiyama, T. Kusuda, S. Kato, T. Aoyama, Flexibly controllable sub-second flash lamp annealing, in *8th International Workshop on Junction Technology* 206 (2008)
53. H. Kiyama, S. Kato, T. Aoyama, T. Onizawa, K. Ikeda, H. Kondo, K. Hashimoto, H. Murakawa, T. Kuroiwa, Advanced flash lamp annealing technology for 22 nm and further device, in *International Workshop on Junction Technology* (2010)
54. P. Timans, G. Xing, S. Hamm, S. McCoy, J. Cibere, G. Stuart, D. Camm, Flat-top flash annealing^{TM} for advanced CMOS processing, in *12th International Workshop on Junction Technology* (2012), pp. 69–72
55. H.S. Grover, F.P. Dawson, D.M. Camm, Y. Cressault, M. Lieberer, Control of wafer temperature using a vortex water-wall argon arc. IEEE Trans. Ind. Appl. **51**, 4802 (2015)
56. R. Arya, J.M. Thomas, A.G. Bhujle, D.D. Bhawalkar, Effects of simmer current on flashlamp impedance and their combined influence on the output of a quasi-CW Nd:YAG laser. IEEE J. Quantum Electr. **36**, 872 (2000)
57. T. Gebel, L. Rebohle, R. Fendler, W. Hentsch, W. Skorupa, M. Voelskow, W. Anwand, R.A. Yankov, Millisecond annealing with flash lamps: tool and process challenges, in *14th International Conference on Advanced Thermal Processing of Semiconductors* (2006), pp. 47–55
58. T. Thrum, D. Camm, S. Dets, A. Hewett, I. Rudic, G.C. Stuart, A. Viel, Development of a powerful vortex stabilized water-wall flash lamp for RTP applications, in *IEEE Industry Application Society Annual Meeting* (2004), p. 1019
59. J.L. Emmett, A.L. Schawlow, Enhanced ultraviolet output from double-pulsed flash lamps. Appl. Phys. Lett. **2**, 204 (1963)
60. T. Onizawa, S. Kato, T. Aoyama, Y. Nara, Y. Ohji, A proposal of new concept milli-second annealing: flexibly-shaped-pulse flash lamp annealing (FSP-FLA) for fabrication of ultra-shallow junction with improvement of metal gate high-k CMOS performance, in *Symposium on VLSI Technology Digest of Technical Papers IEEE-2008* (2008), pp. 110–111
61. W. Lerch et al., Advanced activation trends for boron and arsenic by combinations of single, multiple flash anneals and spike rapid thermal annealing. Mater. Sci. Eng. B **154–155**, 3–13 (2008). https://doi.org/10.1016/j.mseb.2008.08.017
62. A. Buck, R. Erickson, F. Barnes, Design and operation of xenon flashtubes. J. Appl. Phys. **34**, 2115 (1963)
63. M. Braun, Elektrische und optische Eigenschaften von dotierten ZnO Dünnfilmschichten nach Temperung mit spektral modifizierten Blitzlampen, Master Thesis, Westsächsische Hochschule Zwickau (2016) (in German)
64. J.L. Emmett, A.L. Schawlow, E.H. Weinberg, Direct measurement of xenon flashtube opacity. J. Appl. Phys. **35**, 2601 (1964)
65. J.F. Holzrichter, J.L. Emmett, Design and analysis of a high brightness axial flash lamp. Appl. Opt. **8**, 1459 (1969)
66. R. Radtke, M. Kettlitz, On the energy balance of isothermal wall-stabilized xenon arcs. Plasma Sour Sci. Technol. **1**, 274 (1992)
67. D. Bigazzi, P. Mazzinghi, L. Ulivi, A simple model of the wavelength dependent angular radiation characteristics of flash-lamps. Opt. Comm. **110**, 321 (1994)
68. P. Mazzinghi, D. Bigazzi, Wavelength-dependent model of Kr flash lamp emission and absorption. Appl. Opt. **36**, 2473 (1997)
69. L. Ulivi, P. Mazzinghi, Single shot measurements of angular radiation characteristics of flash-lamps. Opt. Comm. **81**, 157 (1991)

70. P. Greve, H. Gross, H. Sakowski, R. Iffländer, R.A. Huber, Emission characteristic of high power Kr/Xe-flashlamps. Proc. SPIE **1021**, 84 (1988). https://doi.org/10.1117/12.950077
71. J.M. Sun, S. Prucnal, W. Skorupa, M. Helm, L. Rebohle, T. Gebel, Increase of blue electroluminescence from Ce-doped SiO_2 layers through sensitization by Gd^{3+} ions. Appl. Phys. Lett. **89**, 091908 (2006)
72. J. Flicstein, Y. Vitel, O. Dulac, C. Debauche, Y.I. Nissim, C. Licoppe, Tunable UV-flash krypton lamp array useful for large area deposition and in situ UV annealing of Si-based dielectrics. Appl. Surf. Sci. **86**, 286 (1995)
73. A. Tauchi, K. Yoshikawa, N. Takahashi, Y. Kato, M. Ohkita, T. Sakurai, Development of flash lamp having high radiant efficiency in short-wavelength range, in *IEEE International Conference on Power Electronics Systems* (2004), p. 204
74. H.S. Kim, S.R. Dhage, D.E. Shim, H.T. Hahn, Intense pulsed light sintering of copper nanoink for printed electronics. Appl. Phys. A **97**, 791–798 (2009)
75. H. Gross, M. Berendt, German Patent DE10 2012 107 816 A1
76. C.R.K. Marrian, A. Shih, The operation of coated tungsten-based dispenser cathodes in nonideal vacuum. IEEE Trans. Electr. Dev. **36**, 173 (1989)
77. R.A. Dugdale, R.C. McVickers, S.D. Ford, The bursting of tubes containing gas discharges. Brit. J. Appl. Phys. **13**, 508 (1962)
78. T. Gebel, M. Neubert, J. Weber, R. Endler, W. Skorupa, Flash lamp annealing: (sub)millisecond thermal treatment of layers on heat sensitive substrates, in *10th International Conference on Coatings on Glass and Plastics* (2014), pp. S4–05
79. W.-S. Han, J.-M. Hong, H.-S. Kim, Y.-W. Song, Multi-pulsed white light sintering of printed Cu nanoinks. Nanotechnology **22**, 395705 (2011)
80. H.-J. Hwang, W.-H. Chung, H.-S. Kim, In situ monitoring of flash-light sintering of copper nanoparticle ink for printed electronics. Nanotechnology **23**, 485205 (2012)
81. W.B. Jin, Y. Park, B.K. Kim, H.J. Kim, J.H. Hwang, H. Chung, J.H. Park, D.H. Kim, S. Park, Thermal warpage of a glass substrate during Xe-arc flash lamp crystallization of amorphous silicon thin film. Int. J. Thermal Sci. **83**, 25–32 (2014)
82. Y. Kim, S. Park, B.-K. Kim, H.J. Kim, J.-H. Hwang, Xe-arc flash annealing of indium tin oxide thin-films prepared on glass backplanes. Int. J. Heat Mass Transfer **91**, 543 (2015)
83. W.-H. Chung, H.-J. Hwang, S.-H. Lee, H.-S. Kim, In situ monitoring of a flash light sintering process using silver nano-ink for producing flexible electronics. Nanotechnology **24**, 035202 (2013)
84. S. Weller, M. Junghähnel, Flash lamp annealing of ITO thin films on ultra-thin glass. Vak. Forsch. Prax. **27**, 29–33 (2015)
85. M. Sakuraba, J. Murota, N. Mikoshiba, S. Ono, Atomic layer epitaxy of germanium on silicon using flash heating chemical vapour deposition. J. Crystal Growth **115**, 79–82 (1991)
86. J. Murota, M. Sakuraba, S. Ono, Silicon atomic layer growth controlled by flash heating in chemical vapour deposition using SiH_4 gas. Appl. Phys. Lett. **62**, 2353–2355 (1993)
87. T. Watanabe, M. Sakuraba, T. Matsuura, J. Murota, Separation between surface adsorption and reaction of NH_3 on Si(100) by flash heating. Japan. J. Appl. Phys. **38**, 515–517 (1999)
88. T. Ishizaka, J. Faguet, F.M. Cerio, Method of light enhanced atomic layer deposition, World Patent WO 2007/111781 A2
89. T. Henke, M. Knaut, C. Hossbach, M. Geidel, L. Rebohle, M. Albert, W. Skorupa, J.W. Bartha, Flash-enhanced atomic layer deposition: basics, opportunities, review, and principal studies on the flash-enhanced growth of thin films. ECS J. Solid State Sci. Technol. **4**, P277–P287 (2015)
90. M.C. Langston, N.P. Dasgupta, H.J. Jung, M. Logar, Y. Huang, R. Sinclair, F.B. Prinz, In Situ cycle-by-cycle flash annealing of atomic layer deposited materials. J. Phys. Chem. C **116**, 24177–24183 (2012)

Chapter 3
Process Management

Abstract This chapter discusses a couple of issues which are relevant for process management. Thereby, the main focus is on temperature measurement and temperature simulation. The first section deals with pyrometry as the dominating temperature measurement technique in short time annealing, and it addresses the various challenges that come with optical temperature measurements. This especially includes the problems with partly unknown emissivities and the outshining of thermal radiation by the flash itself. The section about temperature simulations gives a brief introduction into the mathematical background, discusses the influence of chamber wall reflections, and explores the question how temperature profiles change as a function of the annealing parameters, the optical characteristics of the sample and the thermodynamic material properties. Further sections cover the issues of an homogeneous irradiation, pattern effects, thermal stress, reproducibility and outgassing.

3.1 Temperature Measurement

Contactless surface temperature measurement by optical techniques, most of all pyrometry, during flash lamp annealing (FLA) has been sparely investigated. The objective of the following chapter is to study the reasons to this finding. The challenges and prerequisites for different materials used during FLA will be investigated after a short revision of the fundamentals of pyrometry will have been given.

For the sake of completeness it is important to say that contact temperature measurement is not an issue for the contamination-sensitive microelectronics and photovoltaics (PV) industry. Moreover, for FLA being a surface treatment the thermal mass of a thermocouple, for example, may significantly influence the homogeneity and the thermal budget of the thin processed layer. Nevertheless, contact temperature measurements can be used to measure or calibrate the preheating system, especially if preheating is realized by a hot plate.

L. Rebohle et al., *Flash Lamp Annealing*, Springer Series in Materials Science 288,
https://doi.org/10.1007/978-3-030-23299-3_3

3.1.1 Pyrometry

Pyrometers are radiation detectors. A sketch of their principle set-up is given in Fig. 3.1. Radiation from some source, e.g. a hot object, is collected through a set of lenses to be focused through an aperture onto a detector. There may be an additional eyepiece, a video module or a target laser to adjust the line of sight. The detector may be a semiconductor detector like Si, InGaAs or lead salts, which operates according to the principle of the inner photoelectric effect. Another important detector group are pyroelectric detectors, in which the polarization of the pyroelectric film responds to mechanical deformation brought about by temperature raise. Finally, thermopiles and bolometers are mainly used to measure the mid-infrared regime. The first are based on the Seebeck effect which creates a thermoelectric voltage between the joint of two distinct metals and a reference joint due to a thermal difference. The latter makes use of the temperature dependence of its resistivity.

The pyrometer measures the thermal radiation any object emits according to its temperature. The calibration of this radiation power density with respect to temperature is carried out using a black body source. For the measurement of non-black bodies an emissivity can be set to correct the measured radiation power density to give a true temperature reading. Emissivity is the ratio of radiation power density emitted by real bodies with respect to that emitted by a black body. This correction may be done digitally or analogously. The value of emissivity is generally a function of wavelength and temperature. For a high signal-to-noise-ratio (SNR) a detector should be chosen whose diagnostic wavelength regime covers that particular part of the material's absorption spectrum which shows high emissivity. However, the thermal radiation that can be detected must exceed the noise equivalent power by orders of magnitude to ensure a high SNR likewise and thus reliable temperature reading.

Given the latter is true, let us redirect our attention to the emissivity. A changing (and maybe unknown) emissivity may lead to large temperature errors dependent on the wavelength regime of the detector. For an emissivity change of 10% the resulting

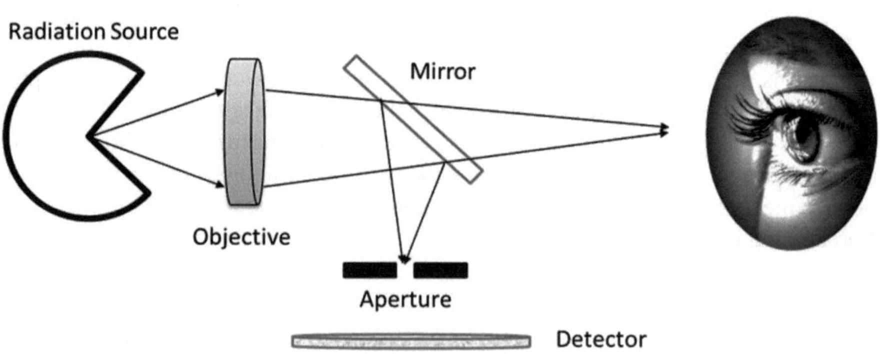

Fig. 3.1 Block diagram of a radiation pyrometer

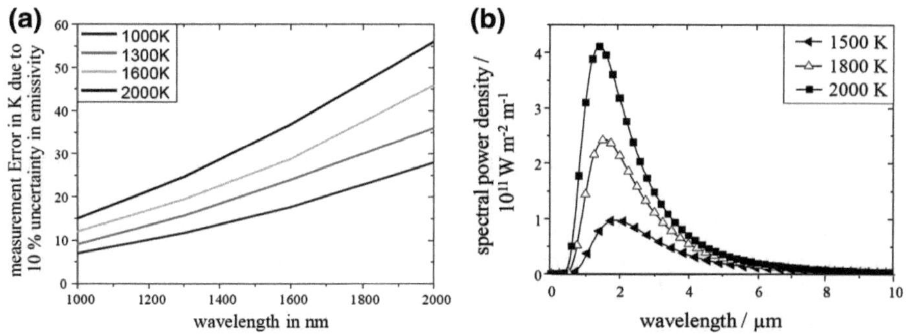

Fig. 3.2 Deviation in temperature measurement due to 10% uncertainty in emissivity (**a**) [5] and black body spectrum according to Planck's radiation law (**b**)

measurement error can be found in Fig. 3.2a. It is shown that the measurement error following a change in emissivity (either due to a lack of knowledge or due to changing properties of the material which temperature needs to be measured) increases with wavelength as well as with temperature. This arises from Planck's Law of thermal radiation in which the spectral radiant emittance of a black body M_{bb} rises steeply from its short wavelength end due to the Law's exponential component and falls smoothly at its long wavelength tail due to its power function (Fig. 3.2b):

$$M_{\mathrm{bb}}^{\nu} = \frac{8\pi h\nu^3}{c^2}\frac{1}{\exp(h\nu/k_{\mathrm{B}}T) - 1} \tag{3.1}$$

$$M_{\mathrm{bb}}^{\lambda} = \frac{8\pi c^2}{\lambda^5}\frac{1}{\exp(hc/\lambda k_{\mathrm{B}}T) - 1} \tag{3.2}$$

where c and k_{B} refer to the speed of light in vacuum and to the Boltzmann constant, respectively, and λ and T represent the diagnostic wavelength and the material temperature, respectively. For a given percentual deviation from the true material emissivity this results in a larger temperature uncertainty for longer wavelengths and higher temperatures. Further—as the measurement error is solely a consequence of the relative (!) change in emissivity—larger emissivity values should be preferred. Consequently, the best way to measure temperature is to find the balance between a detection regime at a wavelength as short as possible, where the material emissivity is preferably large, and a regime as furthest away from the flash lamp maximum at around 500 nm as possible (Fig. 3.5). Clearly, emissivity depends on the material, which is why target materials of FLA will be looked at more closely in the following.

3.1.2 *Temperature Measurement of Semiconductors*

3.1.2.1 Spectral Emissivity

Semiconductors have a distinct bandgap which decreases with temperature and limits the range of photon energies to be absorbed. Thus—according to *Kirchhoff's Law* who found in 1860 that at thermal equilibrium the wavelength-specific absorptivity equals the emissivity of the same material [1]—it also limits the spectral regime of emitted radiation. However, as to be seen, this does not mean that no radiation with a wavelength larger than the bandgap is emitted, but the probability of which is smaller and thus diminishing the emissivity. Semiconductors, therefore, show a larger emissivity for wavelengths smaller than the bandgap. Figure 3.3, for example, shows the emissivity of Si as a function of temperature and wavelength. Indeed, larger nominal values for the spectral emissivity can be found for smaller wavelengths. Above the band gap around 1100 nm the emissivity drops to smaller values before it recovers for elevated temperatures above 600 °C due to free carrier emission. The rise for larger wavelengths occurring for temperatures below 600 °C is due to enhanced lattice vibration [2–4].

Very little literature is, however, available for the emissivity of other common semiconductors regardless of the possibility to extract this information from other optical properties. Yet their spectral absorption coefficient at room temperature (RT) is often known [6]. Following *Gustav Kirchhoff* the knowledge of the first is sufficient to gain an understanding of the behaviour of the material emissivity keeping in mind

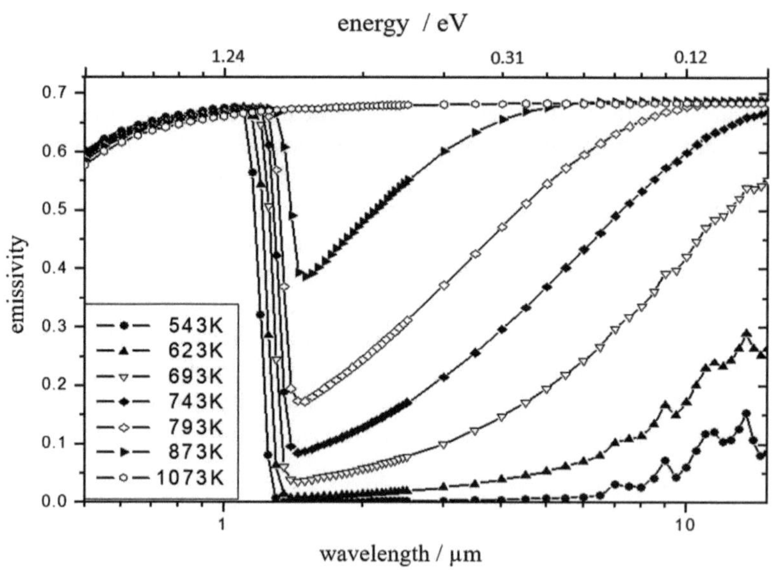

Fig. 3.3 Emissivity spectrum of Si according to [5]

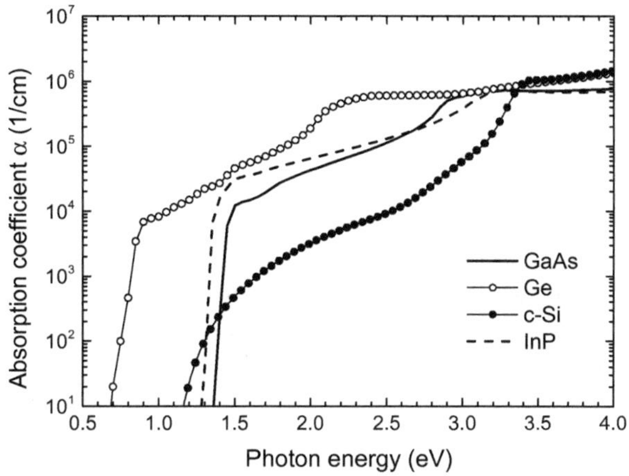

Fig. 3.4 Absorption coefficient of common semiconductors as a function of photon energy. Optical constants were taken from [8]

that—strictly speaking—thermal equilibrium is not present during FLA. Ignoring interference effects, the absorption coefficient α and the unitless absorptivity a are defined by $a = (1 - e^{-\alpha d} - R)$ where d is the material thickness and R equals its reflectivity [7].

Figure 3.4 shows a significant drop in the absorption coefficient for various common semiconductors below the band gap as expected and similar to what has been found before for Si. Moreover, a temperature dependence of the emissivity similar to Fig. 3.3 may be presumed, yet little data exists. Now following these findings the temperature measurement should take place close to the infrared (IR) edge of the flash lamp spectrum, yet still below the spectral bandgap (given in units of wavelength) of the semiconductor. In fact, many semiconductors possess a bandgap that is found even closer to the flash lamp output than Si (Fig. 3.4) shifting the immediate decrease in absorptivity towards increased background radiation of the flash lamp. Figure 3.5 shows the afore described mutual position of the Si bandgap and the flash lamp in the spectrum [9, 60]. Keeping in mind that contactless temperature measurement is radiometry, pyrometers, namely radiation detectors, are not able to distinguish between radiation from non-thermal and those from thermal sources as long as the wavelength regime of the detector is met. The demand for solutions to distinguish thermal radiation from its background is still an issue. Despite the obstacles brought about by the spectral neighbourhood of the pyrometer's diagnostic wavelength and the flash lamp maximum temperature measurement of semiconductors below the energy bandgap at larger wavelengths is an undesired alternative due the low emissivity there. However, emissivity is not the only reason to avoid pyrometry in the near- and mid-infrared regime. Semiconductors are not opaque below their bandgap (Fig. 3.4). This gives rise to many problems like temperature measurement of the bulk instead of the surface and simultaneous measurement of the preheating source.

Rapid thermal annealing or processing (RTA or RTP respectively) is faced with similar obstacles. RTA uses halogen lamps for thermal annealing in a time regime of several seconds. Approaches to solve this issue have mostly made use of local infrastructure, e.g. X-ray or mass spectroscopy. More generally applicable methods evolve around filtering the undesired background or imposing a modulation on it [10]. However, in contrast to flash lamps the power density of halogen lamp radiation that is needed in order to achieve comparable energy densities is a lot smaller due to the longer time scale of annealing.

Now let us take a closer look at the state of the art for temperature measurement during FLA. This challenge has been approached from two sides. One side monitors the thermal radiation of one source through a (*water*) *filter* which is opaque to the other source [11]. The other approach puts a modulation signal (*ripple*) on the background radiation source to distinguish it from the other [5].

3.1.2.2 Water Filter

This approach uses the first harmonic mode of the OH stretch mode in water around 1450 nm [11–13]. The use of the ground mode which is located between 2700 and 2800 nm in quartz windows is also known [14]. In either case lamps and the wafer are separated by such a window as shown in Fig. 3.6. Thus lamp radiation at this specific wavelength—represented by the red arrow—is absorbed before it hits the wafer. The remaining radiation (orange arrow) is sufficient to heat up the wafer to the desired temperature without significant impact on the process parameters. The flash lamp spectrum is very broad having its maximum in the visible region around 500 nm (Fig. 3.5) such that annealing is not significantly affected.

Now, if small band pyrometry between the water window and the wafer is used as depicted in Fig. 3.6 and if the diagnostic wavelength is (in this example) 1450 nm, it is certain that the measured radiation stems from the wafer. Further, if IR halogen lamps

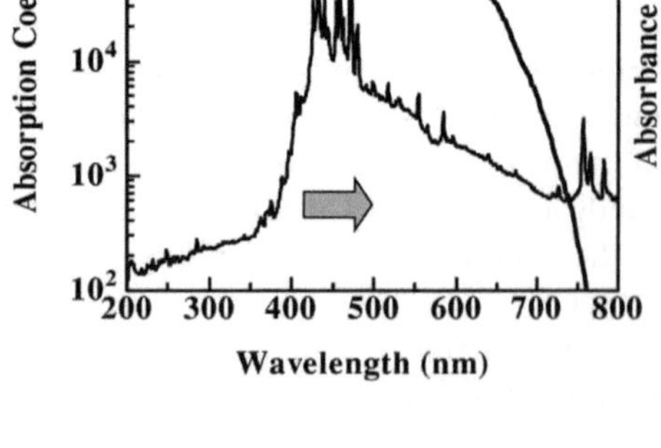

Fig. 3.5 Emission spectrum of a typical Xenon flash lamp in comparison to the Si absorption coefficient. Reproduced from [60] with permission of Electrochemical Society

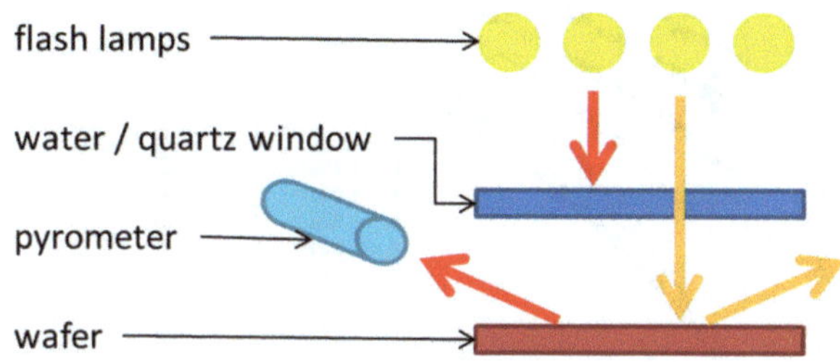

Fig. 3.6 Illustration of a set-up for temperature measurement using OH-mode filtering

Table 3.1 SNR for an opaque hypothetical material with an emissivity of 0.7 and a melting point above 2000 K during FLA

Temperature	1300 K	1500 K	1800 K	2000 K
Relative contribution in % at a central wavelength of 1 μm ± 0.1 μm	0.6	1.4	3.8	6.6

The values were obtained by simulating the energy density necessary to gain the given temperature for a 20 ms flash pulse [5]

are employed for preheating, the same precautions must be met to hinder the lamp radiation to reach the pyrometer if the wafer temperature still allows transmission for radiation at the diagnostic wavelength. The pyrometer and the wafer, however, may be only separated, if necessary, by a minimum OH-content as it is present in infrared grade quartz, which has a concentration of OH of less than 10 ppm [15]. Temperature measurement carried out by such means thus avoids the severe background of the lamp radiation. Figure 3.5 has already shown the low-energy emission spectrum for Xenon. However, emission lines in flash lamps have also been found in the far infrared regime [16] and similar results may be assumed for other noble gases used such as Argon [17]. Although on a global scale the intensity of the flash lamp spectrum in the IR regime is small with respect to the intensity at the maximum of the spectrum and further decreases with increasing current density [18] it may still be enough to compete with the small intensity of thermal radiation. An impression of what this means can be found in Table 3.1 for a diagnostic wavelength around 1000 nm. The values in Table 3.1 have been created by taking a model flash lamp spectrum between a wavelength of 200 and 1100 nm. The actual shape of the spectrum shall not be important at this moment as it is irrelevant for the point that needs to be made here: to gauge the consequences of the IR portion of the flash lamp spectrum on contactless temperature measurement. By the help of a simulation program created by [19] the incident flash energy density was found to achieve the indicated maximum temperature T if a hypothetical material with sample emissivity of 0.7 and a melting point above 2000 K was assumed. Then the thermal radiation at T was compared to the reflected lamp radiation with a supposed sample reflectivity of 0.3. Table 3.1 indicates that the SNR is generally below 10%, but it decreases for elevated material temperatures due to the power function rise of the thermal radiation as a function of T in contrast to the linear increase of lamp energy density necessary to generate this increase in T.

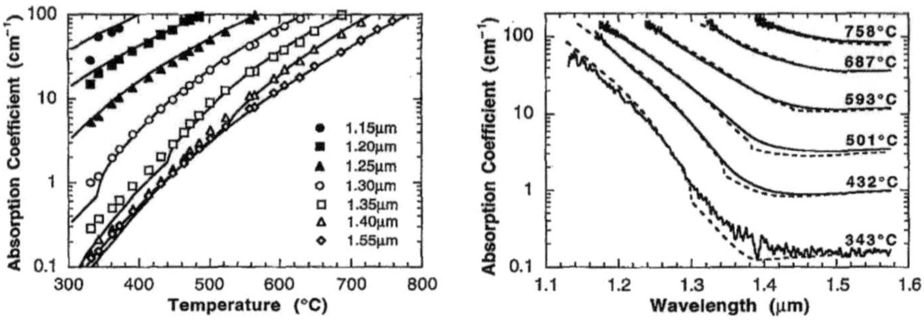

Fig. 3.7 IR absorption coefficient for Si as a function of wavelength and temperature for 10 mm square samples and a thickness of 390 μm. Reprinted from [3], with the permission of AIP Publishing

However, *background* is not solely an issue of temperature measurement during FLA itself but also during preheating. The opacity of the wafer w.r.t. preheating by the halogen lamps is a function of temperature and wavelength. Background occurs if the transmitted lamp power outsizes the noise equivalent power of the pyrometer. Hereby, the penetration depth $1/\alpha$, where α is the absorption coefficient, is a useful parameter. Figure 3.7 suggests that at a wavelength of 1450 nm it decreases beneath typical values for wafer thickness for a temperature larger than at least 600 °C which is explained by the IR band gap shift for elevated temperatures [3]. To illustrate this very important issue of temperature measurement of Si and other semiconductors let us assume a preheat temperature of 600 °C and a wafer thickness which corresponds exactly to the penetration depth at 1450 nm at 600 °C. Further, let us presume for ease thermal equilibrium, even if typical preheating times of a few minutes may not allow for this assumption. Thus, the energy density which is emitted by the halogen lamps, equals the one emitted by the wafer due to its temperature of 600 °C. Now, following this idea, any sensor at the surface of the wafer opposing the preheating on its back would not only see the thermal radiation of the wafer, but also part of the radiation of the halogen lamps behind the wafer due to the finite thickness of the latter. This means, the pyrometer would detect $e^{\alpha d} \times E_{1450\text{nm}}$ (600 °C), where $E_{1450\text{nm}}$ refers to the energy density at 600 °C. In total, this equals an alleged temperature reading of about 625 °C for $\alpha = -1/d$. Due to the exponential relationship between depth and remaining energy density this value drops rapidly with increasing wafer thickness. However, only if the wafer thickness is doubled the temperature can be cut to a deviation below 10 K. Furthermore, the penetration depth decreases with temperature according to Fig. 3.7.

This approach according to [11], however, uses at least two water windows, one to diminish the background of the flash lamps and a second to achieve the same result with the halogen lamp background radiation [20, 21]. Moreover, the flash lamps are bathed in water (Fig. 2.7) which serves two purposes at the same time: cooling and filtering. It is not told how thick the thus created water coat around the lamps actually is, but the penetration depth for radiation at a wavelength of 1450 nm in water is about 0.45 mm [12]. To give an estimate what this means for the thickness of the water

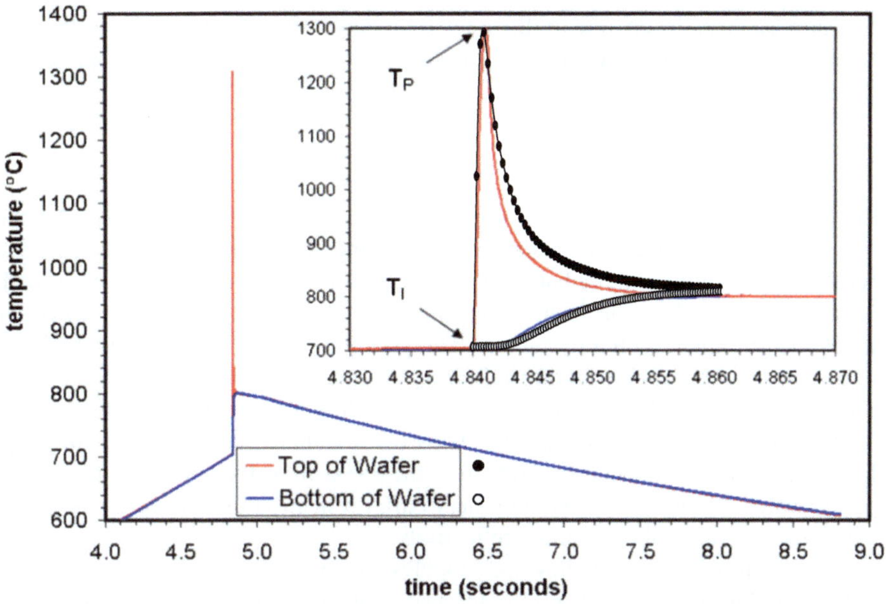

Fig. 3.8 Wafer top- and bottom-side temperatures during FLA (lines) [13] and simulation of wafer top- and bottom-side temperatures (dots) during FLA at 32.5 J/cm^2 for a pulse length of 1 ms at a preheat temperature of 700 °C and a wafer thickness of 775 μm. Deviations may originate from the unknown specific flash lamp spectrum and pulse shape of the authors [13] or surface modifications of the substrate

coat let us first infer the energy density of the flash lamps from [13, 21]. Figure 3.8 shows the wafer top and bottom side temperatures. The overlay in the inset, however, has been created assuming an incident flash energy density of 32.5 J/cm^2 at a pulse length of 1 ms, a preheat temperature of 700 °C and a wafer thickness of 775 μm to create matching conditions to the paper. As both figures are in good agreement, the assumption that the energy density is about 38 J/cm^2 is taken to be true. For an absolute temperature of 1300 °C a measurement accuracy of e.g. 5 K means that no more than $(10 * F)‰$ of the lamp radiation should be able to reach the pyrometer detector, where F characterizes the unknown multiple of the total energy density w.r.t. its share of the flash lamp spectrum around 1450 nm. To keep to the guidelines of the roadmap and taken the wafer surface only reflects 30% of lamp radiation into the detector this radiation needs to be diluted by more than $e^{-3.5}$ for $F = 1000$ which corresponds to a water coat with a thickness of at least 1.5 mm.

In the year 2010, an advancement of the above described method has been presented by a Japanese team of researchers. Based on background-corrected temperature measurement via the water filter approach they introduced online emissivity-compensation by means of polaradiometry. The method is related to ellipsometry, but using a black body source instead of a laser. The heart of this approach is to gain the reflectivity of the sample by analysing the reflection of the black body source on the wafer surface. To be able to distinguish the latter from the unpolarised thermal radiation, the *s*- and *p*-polarisation of the source is individually measured [5, 22].

3.1.2.3 Ripple Pyrometry

Ripple Pyrometry goes back to an invention by Charles Schietinger and Bruce Adams from the 90s on temperature measurement during RTA. Then they introduced a method that modulates the radiation of halogen lamps which cause the background. Thus this radiation becomes distinguishable from the thermal radiation of the wafer during measurement with a pyrometer. The sketch in Fig. 3.9 shows how the lamp radiation (orange arrow) is modulated (orange dotted arrow) and is detected recombined with the thermal radiation of the wafer (red dotted arrow) by a pyrometer. Further a second pyrometer is installed to view the modulated lamp radiation directly. However, its modulation depth differs from the one seen by the first pyrometer viewing the wafer. In fact the modulation depth after reflection off the wafer surface is reduced to the value of the wafer's reflectivity.

Knowing its reflectivity R, however, enables one to deduce the share of modulated background lamp radiation $R * I_L$ in the wafer signal I_W and thus to determine the sample's temperature by applying Planck's Law to $E_W = I_W - R * I_L$ (Fig. 3.10). Considering the material reflectivity E_W reveals a constant value of thermal emission from the wafer. However, while for RTA a typical SNR of 1:2 applies, during FLA ratios of 1:10 have to be considered as shown in Table 3.1 [23].

Recently, this approach had been adopted to perform temperature measurement during FLA. Modulation during RTA is implemented by using the intrinsic ripple of the electrical socket-driven halogen lamps, which stems from the 50 Hz alternating current. However, due to the operation of flash lamps via the discharge of a capacitor (cf. Chap. 2) the amplitude modulation of the power supply is missing and thus another solution had to be found. A voice coil-driven modulator was used instead

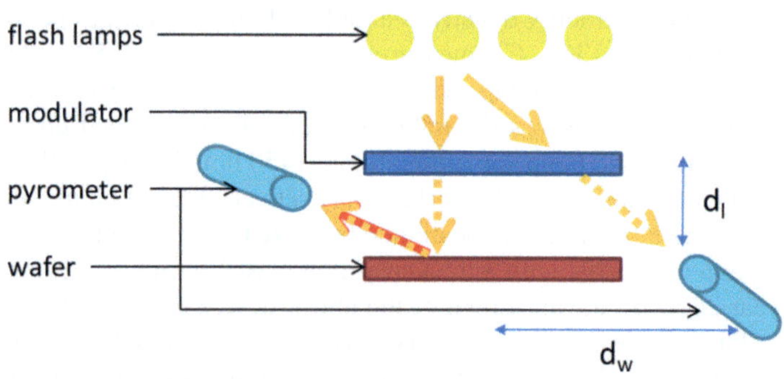

Fig. 3.9 Illustration of a set-up for temperature measurement using ripple pyrometry (modulator-approach). Two pyrometers/detectors are required, one in direct sight with the lamps (bottom right), the other viewing the wafer (left)

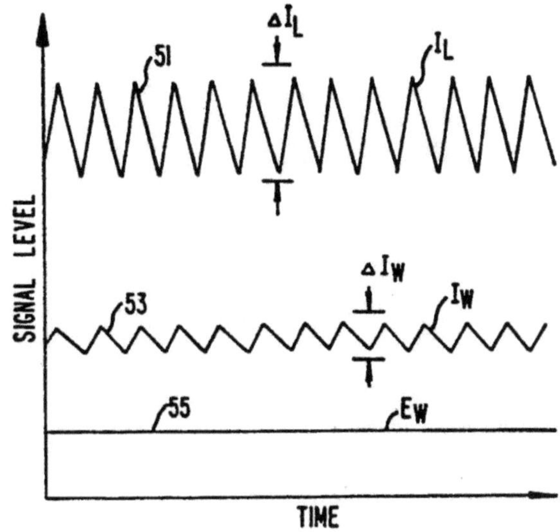

Fig. 3.10 Relative light signal intensities for the flash lamp radiation (I_L), the apparent thermal radiation of the wafer surface (I_W) and the true thermal radiation of the wafer (E_W) [23]

which imposes an amplitude modulation on the flash independently of lamp operation [5, 24].

Figure 3.11 gives the general idea of the set-up for amplitude modulation of the flash lamps. A voice coil similar to those known from loudspeakers pushes periodically a modulator beneath the flash lamps and some kind of mechanical spring keeps the system in permanent oscillation at a minimum of energy input. Just as described before an amplitude modulation is created on the flash pulse and its magnitude compared to the signal that is detected by the pyrometer viewing the wafer. The suitability for flash pulses down to 1 ms could be shown. Moreover, it is possible to increase the effective modulation frequency by the suitable choice of modulator displacement and grating constant of the modulator. However, this method can only work for temperature measurement if the chamber is designed such that no flash light can hit the wafer without being modulated. For the intrinsic modulation of halogen lamps this was impossible by its nature. For an external modulator additional precautions must be already met when designing the process chamber. Also the position of the two pyrometers cannot be chosen randomly. As the determination of the sample's reflectivity R from the amplitude modulation depth of the lamp radiation works best for plane-wave illumination, the detectors are best placed just above the wafer surface, i.e. furthest away from the modulator and the lamps and yet in closest mutual vicinity. Figure 3.12 shows the as-measured reflectivity R' as a function of detector position if R was assumed to be *one*. It can be clearly seen that only for maximum distance from the mechanical oscillator R' and R match. The horizontal position of the detectors also plays an important role. In this case standard chamber walls, which are highly reflective for minimum intensity losses, are assumed which optically enlarge the chamber thus leading to enhanced as-measured reflectivity. Further, through the

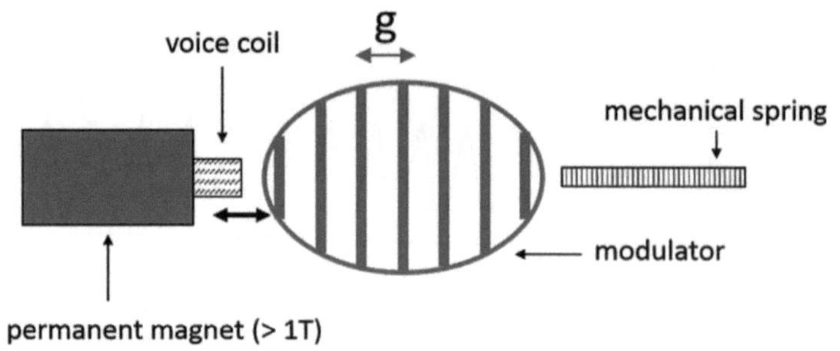

Fig. 3.11 Schematic drawing of a modulator using a voice coil [5]

Fig. 3.12 As-measured reflectivity for increasing vertical distance d_l of the lamp detector w.r.t. the grid of the mechanical oscillator. The values in the legend describe the horizontal distance d_w from the centre of the mechanical oscillator in units of the grating constant g, while the mutual position of the two detectors is kept constant (Figs. 3.9 and 3.11) [5]

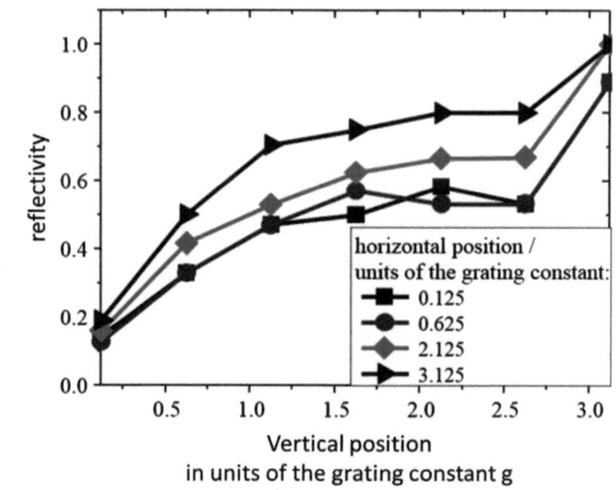

ability of this approach to determine R online this might be a promising way for emissivity-independent pyrometry [5, 24].

Ripple pyrometry lives on the partial shading of the flash light. During RTA the processing times are so long that the thermal mass of the wafer does not follow the modulation. Because of it the modulation detected by the wafer pyrometer would not be solely a consequence of lamp light modulation. Due to the small time regimes during FLA the mass of the heated wafer, however, is a lot smaller than during RTA for flash pulses which are short enough such that the wafer is not completely heated, of course. Yet, it could be shown that no significant modulation of the thermal radiation of the wafer occurs. Further details may be found in [5].

3.1.2.4 Temperature Measurement of Patterned Wafers

A device-loaded wafer shows an inhomogeneity w.r.t. temperature distribution yet alone due to the various emissivities of the transistor constituents. Vice versa light couples differently to them and thus creating true temperature variations across metal and semiconductor surfaces, which has been found to be in the order of 100 K [25]. Emissivity was studied by thermal wave measurement for three different cases: Si trench structure, shallow trench isolation and transistor structure. Thermal Wave Imaging comprises the generation of a highly damped plane wave by means of periodical heating of the sample, preferably by a laser. The thus created temperature field on its surface is subsequently observed by IR imaging which is not a temperature measurement yet. The thermal wave signal depends on the residual ion-implant damage and is thus an indirect measure for the local temperature, provided that the correlation between temperature, remaining damage and the thermal wave signal is known.

A sketch for illustration of the geometric conditions as used in [25] can be found in Fig. 3.13a. Relative emissivity increases with increasing trench depth due to the growing black body behaviour of the trench (not shown) and Fig. 3.13b gives the influence of decreasing pattern density which are referred to in the paper with A–F, respectively. For all three cases the relative emissivity rises for denser patterns. The weak gain for B and D is likely due to thermal diffusion from neighbouring patterns. Rabus et al. [26] did similar investigations in furnace- and lamp-driven RTP systems where they found four times larger temperature deviation on a lamp annealed patterned wafer. Again temperature measurement has been carried out by comparing the results to test processes.

The main challenge to temperature measurement on patterned wafers is array-like emissivity-compensation. So far thermal imaging may support the reproducibility of the FLA process while considering the constraints which are discussed in this chapter.

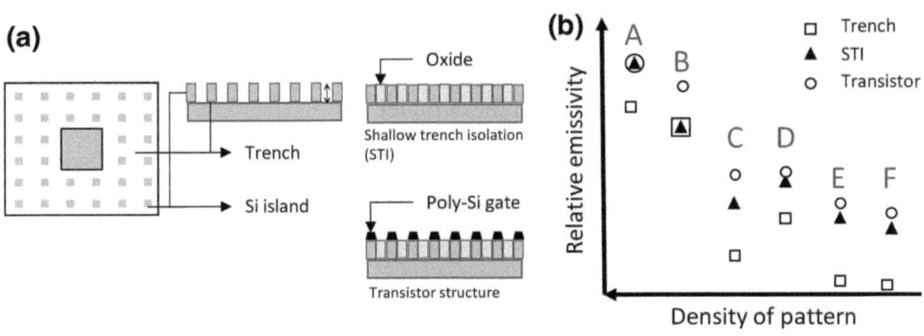

Fig. 3.13 Model of a patterned wafer for emissivity testing (**a**) and emissivity as a function of decreasing pattern density (A–F) on the wafer (**b**) according to [25]

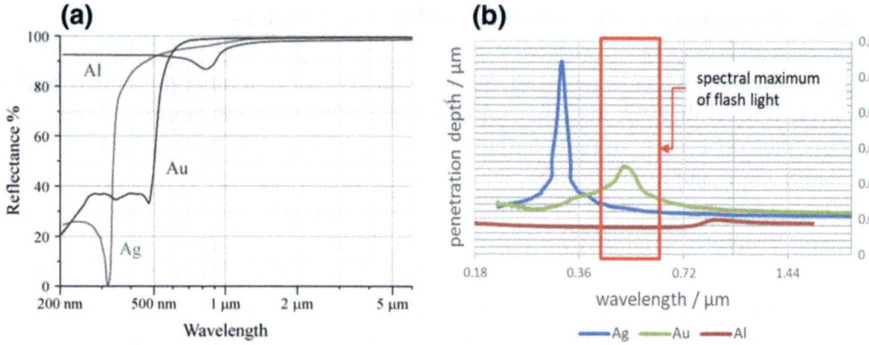

Fig. 3.14 Reflectivity (**a**) [31] and penetration depth of silver (Ag), gold (Au) and aluminium (Al) (**b**). Data (for zero transmissivity at RT) is taken from [32] (Ag, Au) and [33] (Al)

There are also various simulation approaches to investigate temperature evolution during FLA. However, as these rely even more on assumptions on material properties, solutions which comprise pure measurement are in favour by practitioners in industry. More detailed information may be found in Sect. 3.2.

3.1.3 Temperature Measurement During Annealing of Non-semiconductors

The methods introduced above are the only practical methods known for temperature measurement during FLA. They were used while processing semiconductors because they are the central and most established theme of thermal annealing. Nowadays other materials gain increasingly the researchers' interest such as thin foils, transparent conductive oxides, metallic films and glass substrates [27]. Nevertheless, experimental temperature measurement is not focused on. The following section shall therefore draw attention to the physical fundamentals to be considered when attempting contactless temperature measurement on these materials.

3.1.3.1 Temperature Measurement of Metals

In contrast to semiconductors metals possess a very high reflectivity already in the visible spectrum (Fig. 3.14a), namely at the maximum of the flash lamp emission. Therefore, only little energy can be absorbed to start with and the free electron gas raises the absorption coefficient by several orders of magnitude (dependent on the respective spectral regime) compared to many semiconductors. This results in a shallow penetration depth of the flash of less than 100 nm (Fig. 3.14b) and makes them unusual materials for FLA. However, this circumstance becomes advantageous if metallic films need to be put on temperature-sensitive substrates like plastic (foils) or cardboard for optoelectronic applications or package labelling [27–29]. For other, already established processes like *flash-enhanced atomic layer deposition* creates a significant reduction in processing time and thus saves costs [30].

Though the emissivity above a wavelength of 1 μm must be small when considering Kirchhoff's Law, conservation of radiation and Fig. 3.14a, it decreases steadily in contrast to the abrupt drop at the bandgap of a semiconductor. Still, it remains favourable for contactless temperature measurement to choose a detection range which is a balance between shorter wavelengths with higher emissivity and longer wavelengths with lower flash lamp background. Depending on the melting point of the metal, its thermal conductivity and the temperature threshold of the substrate FLA on metallic thin films is often a low temperature process well below 1000 °C. Although this further decreases the SNR between thermal and flash lamp radiation the choice of a high-emissivity regime needs to be outweighed to the gain of SNR due to moving well outside the flash maximum (Fig. 3.5).

3.1.3.2 Temperature Measurement of Thin Foils

The reduction of substrate thickness to achieve a maximum in flexibility for various applications and a reduction in cost and time due to the ability of roll-to-roll (R2R) processing and due to less material consumption has led to a wide use of thin foils as substrates with a thickness of about 100 μm. These foils, which are often made of glass or Polyethylene terephthalate (PET), a thermoplastic polymer, are temperature-sensitive due to their low melting or softening point. Yet, if they serve as a substrate, the top layer might have to undergo thermal annealing as it is the case for transparent conducting oxides (TCOs), for example. FLA as a surface annealing process is thus very well suited for this purpose [34].

TCOs possess a high transmissivity in the visible range, already before thermal annealing, which complicates the coupling of flash lamp radiation. Thermal annealing, however, at elevated temperatures up to 1000 °C is necessary to reduce the sheet resistance in order to make them highly conductive for the use as displays or contact for flexible PV [34]. This makes contactless temperature measurement difficult as the large transmissivity lowers the emittance, while the background of the flash is increased. Figure 3.15 reveals that moving into the IR regime raises its reflectance which yet again means a lower emittance. However, above 1500 nm at the plasma frequency of the specific indium tin oxide (ITO)—dependent on its properties—a maximum in absorbance can be found. This leads to both transmissivity and reflectivity reaching an optimal value for contactless temperature measurement because the first is about to decrease to a minimum while the latter has not yet come to its maximum [35].

Freestanding thin foils without substrates are also known. Polycrystalline Si foils of 8 μm have been made by spray-coating of Si nanoparticle dispersions by [36]. The sintering of these nanoparticles and the related phase change, however, introduces

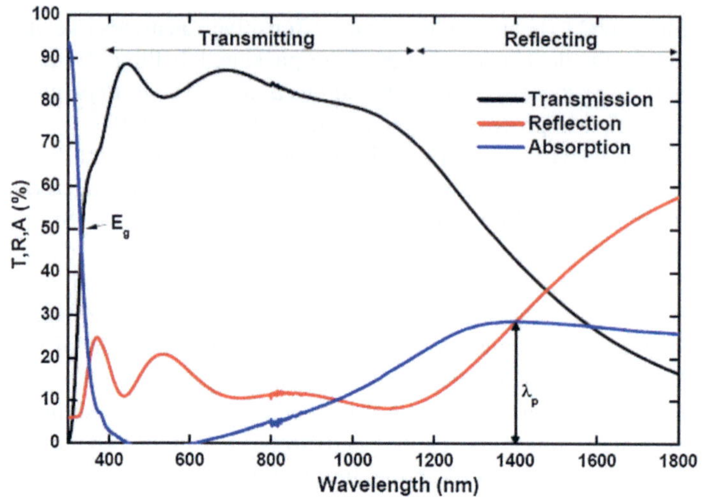

Fig. 3.15 Transmission, reflection and absorption spectral curves of a typical ITO, thickness unknown. The absolute values shall not matter here. λ_p refers to the plasma frequency of the ITO, where the absorption is largest while transmission and reflection are just about to in- and decrease, respectively [35]

additional challenges as the emissivity change in this case is comparatively large. Moreover, it may not only change each time, it may also vary laterally.

3.1.4 True Temperature Measurement

As a final remark attention shall be drawn to what is actually measured when using contactless means to determine the temperature of a sample. FLA lives on the fact that a thermal gradient is created through the thickness of a substrate whose colder back serves as a heating sink for the flash-heated surface. In the case that the sample is not heated up uniformly—neither laterally nor transiently—it needs to be studied which temperature is actually measured or in other words from which depth of the material does the radiation stem? There is very little literature on the absorption depth as a function of temperature for most semiconductors (although these considerations are also valid for other semi-transparent materials). However, for Si substrates data is available (Fig. 3.16a) which shall serve as a model material to demonstrate the effect of absorption depth on true temperature measurement during FLA. From Fig. 3.16a one can deduce the maximum depth from which thermal radiation can stem during contactless temperature measurement for a given diagnostic wavelength. It can be therefore concluded that a wafer with a thickness of 520 μm at RT is transparent to all radiation detectors at a wavelength above 1000 nm. Let us take this thought further to elevated temperatures, at which data on the penetration depth is also available for

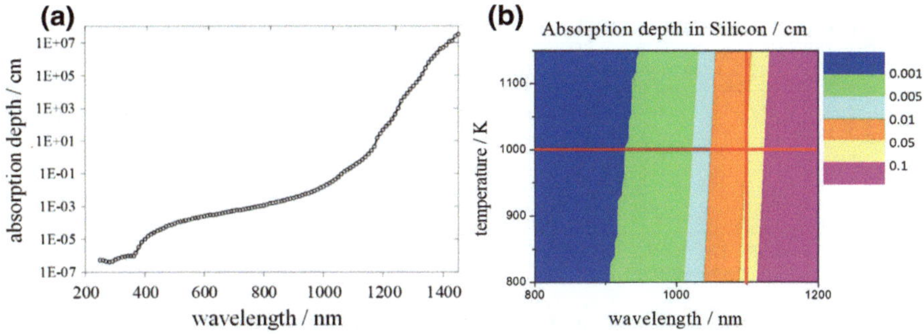

Fig. 3.16 Absorption depth for Si at RT. Data taken from [38] (**a**) and penetration depth in Si as a function of wavelength and temperature (**b**); wavelength dependence taken from [38], temperature dependence taken from [37]. The red lines are inserted for guidance to the eye

Si. Sun et al. [37] found that the following empirical equation holds for temperatures between 298 and 1183 K and for wavelengths between 1100 and 250 nm:

$$\alpha(\lambda, T) = \alpha_0(\lambda)\exp\left(\frac{T}{T_0}\right), \quad \text{where } T_0 = 703\ \text{K} \tag{3.3}$$

According to (3.3) Fig. 3.16b shows the penetration depth in Si as a function of both wavelength and temperature and reveals that so-called *bandgap pyrometry*, where the diagnostic wavelength is just above the bandgap of Si at about 1100 nm (as it is a function of temperature itself), are always faced with the circumstance that wafers with typical thicknesses below 1 mm remain transparent to the detector if the preheat temperature is not chosen well above 1000 K. Backside preheat lamps may interfere with the thermal radiation of the wafer and falsify the pyrometer reading though the emissivity has taken on a stable value (Fig. 3.3). This contradicts—without further precautions as described in Sects. 3.1.2.1 and 3.1.2.2—the annealer's desire to record the temperature profile as early as possible in the process ideally starting at RT.

Also, emissivity is a function of wafer thickness for transparent substrates as shown in Fig. 3.17a for RT. Therefore, for wavelengths smaller than the penetration depth of the particular thickness no difference in the emissivity curve can be seen. However, if the wavelength which corresponds to the penetration depth is exceeded, the dependence of the emissivity on wafer thickness becomes obvious. For elevated temperatures the bandgap shifts further towards the IR part of the spectrum. This enhances absorption in the near infrared (NIR) and likewise reduces the penetration depth of NIR radiation in Si which is why the emissivity in Fig. 3.17b is independent of thickness up to appr. 850 nm while in Fig. 3.17a the deviation of the curves begins already at around 650 nm.

Now, variation of the sample emissivity as a result of wafer transparency and thus penetration depth in Si is only one consequence. The effect on the temperature reading with a pyrometer is another question to be answered even for opaque

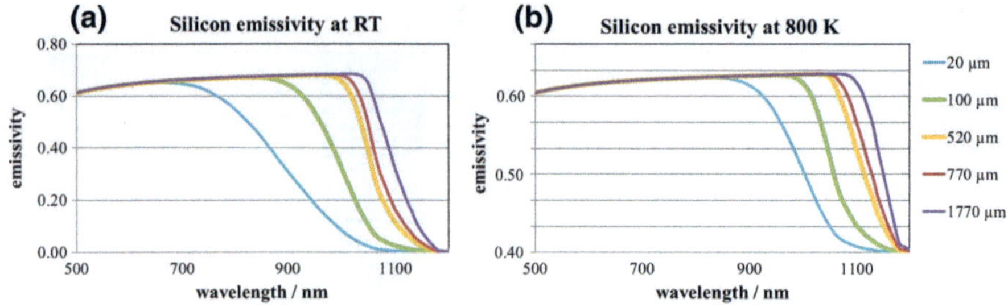

Fig. 3.17 Emissivity of a Si wafer as a function of wavelength and wafer thickness at RT (**a**) and at 1000 K (**b**) [37, 38]

wafers. Radiation from the bulk may—subject to its temperature and wavelength—have enough energy to travel to the wafer surface. This additional background that hits the pyrometer will distort the measurement. What does this mean? To start with let us assume a homogenously heated wafer at 1100 K and a diagnostic wavelength of 1050 nm. From [37] we know that these values account for a penetration depth of 128 μm which means that *surface temperature measurement* in this case refers to the measurement of the total thermal radiation across a depth of 128 μm. Strictly speaking, a wafer of such a thickness will only transmit 30% of background radiation, but for the sake of a better understanding let us ignore this for a minute. Through gauging the pyrometer this does not influence the measurement as long as the penetration depth is considerably smaller than the wafer thickness (Fig. 3.17).

The transient temperature distribution in a wafer that has undergone millisecond processing is not homogeneous though, but a thermal gradient is created inside the material. Subject to the penetration depth at the diagnostic wavelength the gradient that needs to be considered may take on different sizes. The larger the thermal gradient to be taken into account the less the temperature reading will deviate from the actual temperature at the "surface". Although Fig. 3.16b tells us that the penetration depth increases for decreasing wafer temperature the power function relationship between thermal radiation and temperature makes up for it. To give an example let us again assume a diagnostic wavelength of 1050 nm which is not too close to the flash lamp spectrum yet not too far off such that a wafer with a thickness of 520 μm will appear opaque at a preheat temperature of 800 K. To simplify the computation let me take the penetration depth at this preheat temperature which is known from (3.3) to be 197 μm. We know from experiments that for a flash pulse of say 3 ms and a surface temperature of 1500 K the temperature at a depth of almost 200 μm is about 1100 K [5]. At this temperature [37] predicts an absorption coefficient of roughly 78 cm^{-1}. Now, doing some simple calculations according to Lambert-Beer's Law this gives about 20% of thermal radiation (for a temperature of 1100 K) that reaches the surfaces after travelling 197 μm towards the surface and that can thus falsify the temperature measurement. The deviation comes out to be less than 3 K as expected for these elevated temperatures. However, it contributes to the other uncertainties discussed.

3.2 Temperature Simulations

3.2.1 Mathematical Background

Simulation aims to determine the thermal cycle a sample undergoes during FLA, expressed as the temperature field $T(t, \vec{r})$. It starts with the normalized flash lamp spectrum $f(E, t)$ which depends on time t and photon energy E. Normalization means that the integral over E and t equals the energy density E_D. This light interacts with a material system which is characterized by a complex refractive index $\tilde{n} = n + ik$ depending on E, on the location $\vec{r}$ (which in turn depends on geometry and material composition) and on the temperature T. As T varies with time any temperature dependence is a time dependence as well. The interaction results in an absorption profile $a(E, t, \vec{r})$, and integration over E provides a source term $S_i(t, \vec{r})$ describing the input of energy to the system in form of a volumetric power density. The temperature development within the system is described by the heat equation in a general form:

$$\frac{\partial}{\partial t}(\rho c_P T) = \nabla(\lambda_T \nabla T) + S_i - S_O \tag{3.4}$$

where ρ is the mass density, c_P is the heat capacity, λ_T is the thermal conductivity, and S_O is the outflow of energy by dissipative processes, namely convective heat transfer and thermal radiation. The change of temperature starts a new iteration with a recalculation of the optical constants, the absorption profile and so on. The approach is sketched in Fig. 3.18a, but the calculation is rather complex and time-consuming, although it already contains a few approximations. One of these approximations is the neglect of flash lamp radiation patterns which in fact cause a distribution of incident angles and thus a distribution of penetration depths.

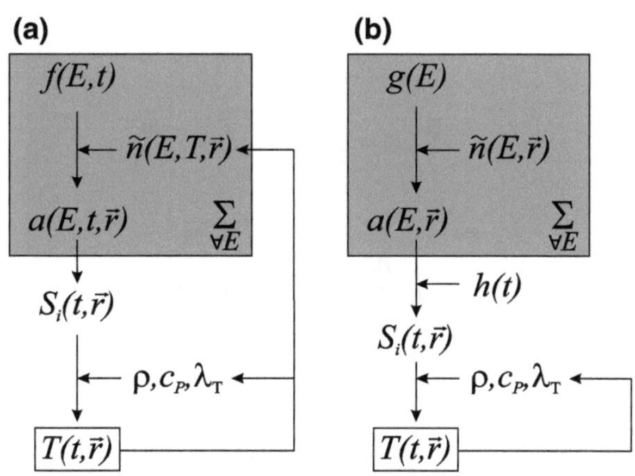

Fig. 3.18 Simulation scheme where optics and the heat equation are coupled (**a**) and decoupled using the approximations of a time-independent spectrum shape and temperature-independent optical constants (**b**)

However, the situation can be strongly simplified using the following approximations. In most cases the flash lamp spectrum changes only slightly with time, and the time-integrated spectrum is a good approximation. Mathematically, the spectrum $f(E, t)$ can be expressed as

$$f(E,t) = g(E)h(t) \qquad \int g(E)\mathrm{d}E = 1 \qquad \int h(t)\mathrm{d}t = E_\mathrm{D} \tag{3.5}$$

where $g(E)$ is the time-integrated flash lamp spectrum and $h(t)$ is the pulse shape in time. If complex material changes like diffusion or a phase transition during annealing are not considered, the optical constants usually do not change very much with temperature. This decouples the heat equation from optics and allows the calculation of the absorption profile in advance (Fig. 3.18b). Another common approximation is the neglect of lateral temperature variations which replaces the vector $\vec{r}$ by a one-dimensional depth coordinate x. Finally, a recent study investigated the effect of different boundary approximations to the solution of the heat equation in case of FLA [39].

Optical modelling is usually performed by the wave transfer matrix method [40] and aims to determine the absorption profile and the absorbed energy density A. The latter is generally given by

$$A = (1 - R_\mathrm{S})\Gamma E_\mathrm{D} \tag{3.6}$$

where R_S is the reflectance of a sample averaged over the flash lamp spectrum and Γ is a correction factor which considers the chamber geometry and chamber wall reflections (see next section). The wave transfer matrix method assumes that an incoming wave with the intensity 1 strikes the surface of a thin film system, and the interaction with the system causes a reflected and transmitted wave with the complex wave amplitude E_R and E_T, respectively (Fig. 3.19). The response of the system is described by

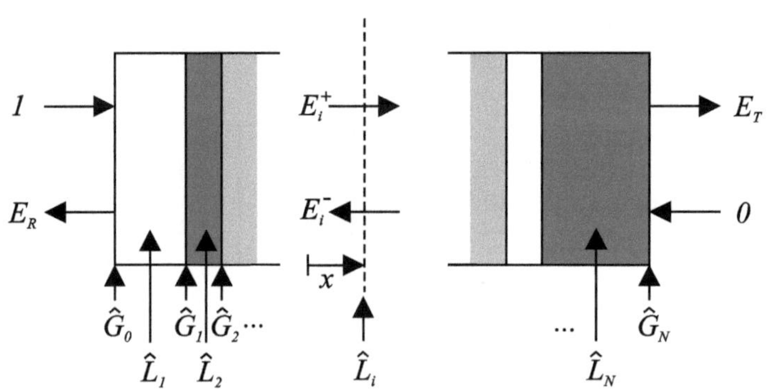

Fig. 3.19 Calculation scheme for the electric field inside the sample

$$\begin{pmatrix} 1 \\ E_{\mathrm{R}} \end{pmatrix} = \widehat{M} \times \begin{pmatrix} E_{\mathrm{T}} \\ 0 \end{pmatrix} \quad \widehat{M} = \begin{pmatrix} m_{00} & m_{01} \\ m_{10} & m_{11} \end{pmatrix} \tag{3.7}$$

where E_{R} and E_{T} can be extracted by

$$E_{\mathrm{T}} = \frac{1}{m_{00}} \qquad E_{\mathrm{R}} = \frac{m_{10}}{m_{00}} \tag{3.8}$$

The matrix $\widehat{M}$ is obtained by matrix multiplication in which each interface and each layer of the thin film system is represented by an interface matrix $\widehat{G}$ and a layer matrix $\widehat{L}$, respectively. Assuming the system has N layers, matrix multiplication is performed as

$$\widehat{M} = \widehat{G}_0 \times \widehat{L}_1 \times \widehat{G}_1 \times \widehat{L}_2 \times \cdots \times \widehat{G}_{N-1} \times \widehat{L}_N \times \widehat{G}_N \tag{3.9}$$

$$\widehat{G}_i = \frac{1}{t_{i,i+1}} \begin{pmatrix} 1 & r_{i,i+1} \\ r_{i,i+1} & 1 \end{pmatrix} \tag{3.10}$$

$$\widehat{L}_i = \begin{pmatrix} \Delta & 0 \\ 0 & \Delta^{-1} \end{pmatrix} \quad \Delta = \exp(-i\varphi) \quad \varphi = \frac{2\pi}{\lambda}\tilde{n}_i d_i \tag{3.11}$$

where $r_{i,i+1}$ and $t_{i,i+1}$ are the reflection and transmission coefficients at the interface between layer i and $i+1$ according to the Fresnel equations, $\tilde{n}_i$ is the complex refractive index and d_i is the thickness of layer i. The indices *0* and $N+1$ denote the half space at the front and backside, respectively. The reflectance and transmittance are given by

$$R = \|E_{\mathrm{R}}\|^2 \qquad T = \|E_{\mathrm{T}}\|^2 \frac{Re\{\tilde{n}_{N+1}\cos\theta_{N+1}\}}{Re\{\tilde{n}_0\cos\theta_0\}} \tag{3.12}$$

where $\tilde{n}_0/\tilde{n}_{N+1}$ is the complex refractive index of the front/back half space and θ_0/θ_{N+1} is the entrance/exit angle. In case of FLA the second factor in the transmittance formula (3.12) becomes 1 because the front and back medium is usually the same. However, the factor has to be considered when calculating the electric field inside the system. Please also note that $r_{i,i+1}$ und $t_{i,i+1}$ are different for *s*- and *p*-polarization in the case that incidence of light is not perpendicular.

The amplitude of the electric field inside the layer i and in a distance x_i from the interface $(i-1, i)$ is given by solving the equation

$$\begin{pmatrix} 1 \\ E_{\mathrm{R}} \end{pmatrix} = \widehat{M}_i \times \begin{pmatrix} E_i^+ \\ E_i^- \end{pmatrix} \quad \widehat{M}_i = \begin{pmatrix} m_{00} & m_{01} \\ m_{10} & m_{11} \end{pmatrix} \tag{3.13}$$

$$E_i^+ = \frac{m_{11} - m_{01}E_{\mathrm{R}}}{m_{00}m_{11} - m_{01}m_{10}} \qquad E_i^- = \frac{1 - m_{00}E_i^+}{m_{01}} \tag{3.14}$$

Thereby, $\widehat{M}_i$ is obtained by a matrix multiplication from the first interface up to layer i:

$$\widehat{M}_i = \widehat{G}_0 \times \widehat{L}_1 \times \widehat{G}_1 \times \cdots \times \widehat{G}_{i-1} \times \widehat{L}_i^* \tag{3.15}$$

where $\widehat{L}_i^*$ is the layer matrix i according to (3.11) with d_i replaced by x_i. E_i^+ and E_i^- interfere with each other, and the power density Π of the electromagnetic wave in layer i at location x_i for the absorption a at this point scales with

$$\Pi \propto \Pi_0 = \left\| E_i^+ + E_i^- \right\|^2 \frac{Re\{\tilde{n}_i \cos\theta_i\}}{Re\{\tilde{n}_0 \cos\theta_0\}} \quad a = \frac{\Pi_0 k}{\lambda} \tag{3.16}$$

Please note that Π_0 is a dimensionless quantity and that the final form of the absorption profile for a given photon energy (or wavelength) is obtained by normalizing it to the absorbed energy density A:

$$\int_{x=0}^{D} a(x)\mathrm{d}x = \frac{A}{E_\mathrm{D}} \tag{3.17}$$

where D is the total thickness of the layer stack. Attention should be paid for thick, transparent substrates with an unpolished backside as the previous equations do not consider diffuse scattering. In this case it might be better to treat the substrate as an infinite half space, provided that the rough backside prevents any back reflection.

There are further hidden pitfalls in the implementation of the wave transfer matrix method. The most obvious one is the case of a thick and opaque substrate which sets the value of E_T to zero and triggers a zero division in (3.9). This requires the definition of a useful stop criterion which finally implies the introduction of an auxiliary layer stack in which the fully absorbing layer (and all layers behind) are replaced by a thinned absorbing layer in such a way that the transmittance is close to zero, but large enough to avoid zero division and the build-up of rounding errors. Furthermore, the multiplication of complex matrices is not necessarily commutative, and part of the literature defines an inverse order of multiplication in (3.7) and/or (3.9) with corresponding modifications of $\widehat{G}$ and $\widehat{L}$ in (3.10) and (3.11). Therefore, any implementation should be checked by simple examples where the analytic solution is known.

Real systems are usually composed of thin and thick layers which differ from each other by the question if interference effects are considered or not. Without going too much into detail, the general strategy, taking the individual case shown in Fig. 3.20, is the following. At first, the sequence of *thin* films is regarded as a subsystem and is calculated first using (3.6–3.16). For the present example the matrix $\widehat{M}_\mathrm{sub}$ is calculated by

$$\widehat{M}_\mathrm{sub} = \widehat{G}_0 \times \widehat{L}_1 \times \widehat{G}_1 \times \widehat{L}_2 \times \widehat{G}_2 \tag{3.18}$$

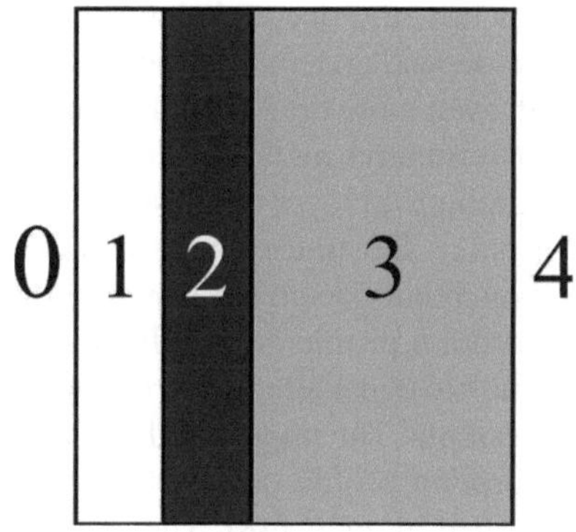

Fig. 3.20 Layer system made of two thin layers (1, 2) and a thick substrate (3)

From this, the reflectance R_{sub} and the transmittance T_{sub} is calculated, whereby layer 3 serves as the back half space and is *different* from the front half space. A system of thick layers can be calculated in an analogous way, but the complex wave amplitudes are now replaced by real intensities. Thus, the analytic expression of $\widehat{G}$ and $\widehat{L}$ changes to

$$\widehat{G}_i = \frac{1}{T_{i,i+1}} \begin{pmatrix} 1 & -R_{i,i+1} \\ R_{i,i+1} & 1 - 2R_{i,i+1} \end{pmatrix} \tag{3.19}$$

$$\widehat{L}_i = \begin{pmatrix} \Delta & 0 \\ 0 & \Delta^{-1} \end{pmatrix} \quad \Delta = \exp(-\alpha_i d_i) \quad \alpha_i = \frac{4\pi k_i}{\lambda} \tag{3.20}$$

where $R_{i,i+1}$ und $T_{i,i+1}$ are the reflectance and transmittance at the interface between layer i and $i + 1$, α_i is the absorption coefficient and k_i is the imaginary part of the refraction index of layer i. The thin film subsystem is now replaced by a virtual interface $\widehat{\Gamma}$, and the total reflectance and transmittance of the whole system is calculated by $\widehat{M}$ according to

$$\widehat{M} = \widehat{\Gamma} \times \widehat{L}_3 \times \widehat{G}_3 \quad \widehat{\Gamma} = \frac{1}{T_{\text{sub}}} \begin{pmatrix} 1 & -R_{\text{sub}} \\ R_{\text{sub}} & 1 - 2R_{\text{sub}} \end{pmatrix} \tag{3.21}$$

where $\widehat{G}_3$ und $\widehat{L}_3$ are the matrices according to (3.19) and (3.20), respectively.

Finally, a simulation can only be as good as the input parameters. Whereas the temperature-dependent material constants like λ_{T} or c_p of Si or Ge are well known, temperature-dependent data are often missing for new or complex materials. In some cases the properties strongly depend on the processing history, e.g. on the details of deposition. In addition, any deviations from the ideal layer geometry like diffuse interfaces, lateral inhomogeneities, or material gradients make the simulation more difficult. The implementation of the wave transfer matrix method in combination with the heat equation is discussed in more detail in [19, 41].

In fact, the decoupled approach is widely applied in literature to calculate the time-dependent temperature profiles during laser annealing (LA) and FLA, and partly with even more drastic approximations than those illustrated in Fig. 3.18b. The wave transfer matrix method is usually used if interference matters, e.g. in case of a thin amorphous Si layer on glass which has been crystallized by FLA [42, 43]. However, for bulky substrates like a Si wafer it is sufficient to assume an absorption profile exponentially decaying from the surface into the depth [44–47]. As the depth of the absorption profile is usually small compared to substrate thickness, it can be even enough to consider the energy input of the flash lamp as a heat source at the surface of the sample. The impact of the two different heat transport models, namely an explicit absorption profile or a surface heat source, was investigated in case of a Si wafer annealed by FLA [48]. The study revealed that the temperature difference between the two models was always below 10 K with a slight overestimation of the surface temperature in case of a surface heat source. This basic assumption was also used in FLA simulations of thin amorphous Si on crystalline Si [49] and LA simulations of thin GaMnAs layers on GaAs [50].

However, the energy flow through the boundary is a power density per area, whereas (3.4) requires a volumetric power density. To consider this, a suitable auxiliary function $f(x)$ like the Dirac delta function has to be introduced whose integral over space is normalized to 1 and whose time dependence is given by $h(t)$ of (3.5). The same applies to dissipation processes which are usually also given as a power density per area. Finally, phase changes like melting were considered in order to simulate melt depths and crystallization velocities [19, 51, 52]. However, phase changes significantly alter the optical properties, and in such cases the decoupled approach is only reasonable on the rising edge of the flash (or laser) pulse up to the point where the phase change occurs for the first time. In some cases commercial FLA tools are already equipped with a temperature simulation tool [53].

3.2.2 *Chamber Wall Reflection*

A fundamental question for the design of an FLA chamber is if their walls should be dark or reflective. In a dark chamber the walls absorb the major part of the flash light and the sample absorbs, in a first approximation, only the light which is emitted in the direct line of sight to it. A dark chamber simplifies optical temperature diagnostics as disturbances by multiple reflections of the flash light will be minimized. For the same reason, the determination of the absorbed energy density A is less faulty. In fact, $\Gamma = 1$ holds for a dark chamber in (3.6). However, the crucial disadvantage is that the major part of the flash light is not used to anneal the sample but to heat up the sample walls. In order to ensure a good homogeneity of the flash light, the size of the lamp field is usually larger than the sample size which, in turn, leads to an additional loss of flash light. Such a low efficiency is not acceptable for most applications, and consequently, dark FLA chambers are nowadays used in few individual cases only.

If chamber walls with a high reflectivity or reflectance are considered the situation becomes more complex. The reflectance R_C of the chamber wall and R_S of the sample are barely independent of wavelength, and the spectrum after multiple reflections may significantly differ from the original one. This especially applies to the ultraviolet (UV) part as in many cases the reflectance is smaller in the UV than in the visible and near IR spectral region. So it was found that the reflectance of a standard Al reflector (which in fact is a technical Al alloy) drops down to 20% in the blue and UV spectral region [42]. If this is critically, the reflector can be coated by a suitable thin film to enhance reflectance. In the following, an average reflectance R according to

$$R = \int R(E)g(E)\mathrm{d}E / \int g(E)\mathrm{d}E \tag{3.22}$$

is used where $R(E)$ is the reflectance at the photon energy E and $g(E)$ is the corresponding intensity of the spectrum. Depending on wavelength and operation conditions, the flash lamps themselves are transparent, semi-transparent or opaque with the corresponding effect on the reflected light. In addition, the chamber geometry can introduce, if not properly designed, inhomogeneities in the distribution of E_D.

The next, more advanced approximation is the chamber geometry as shown in Fig. 3.21 in which the sample is homogeneous, opaque and covers the lower half space. In addition, the lamps are assumed to be transparent and the chamber is approximated by a large reflector covering the upper half space. In addition, the angular dependence of reflectance is neglected. In real cases a part of the reflected light is reabsorbed in the flash lamp. Fortunately, this light is not lost but additional heats the flash lamp: the temperature of the plasma is higher, its conductivity is lower, and the peak current density is higher for given operation conditions. In optimum cases this effect can give a contribution in the order of 10% [54]. Applying simple ray optics, the intensity of a light beam reflected i times between the upper and lower part is proportional to $(R_S R_C)^i$, and summed up over i the amount of light absorbed in the sample is given by

$$\mathrm{A} \propto \frac{(1 - R_S)(1 + R_C)}{2(1 - R_S R_C)} \qquad \Gamma = \frac{(1 + R_C)}{2(1 - R_S R_C)} \tag{3.23}$$

The graph of Fig. 3.21 displays the absorbed fraction of light A, normalized to 1, as a function of R_S and R_C, provided that E_D is a true measurement. Although (3.23) is a rough guess it visualizes the main tendencies. For $R_C = 0$ (dark chamber conditions) the sample absorption is 0.5 at its best. However, without anti-reflection coatings sample reflectance is often in the range of 0.2–0.5. As an example, the reflectance of a Si and Ge wafer averaged over a Xe flash lamp spectrum as shown in Fig. 2.23a is 0.4 and 0.48, respectively. With increasing R_C the absorbed fraction of light also increases and becomes more and more independent from the sample reflectance. This is an advantage especially in case of thin film applications where the reflectance strongly depends on the layer thickness and the optical constants of

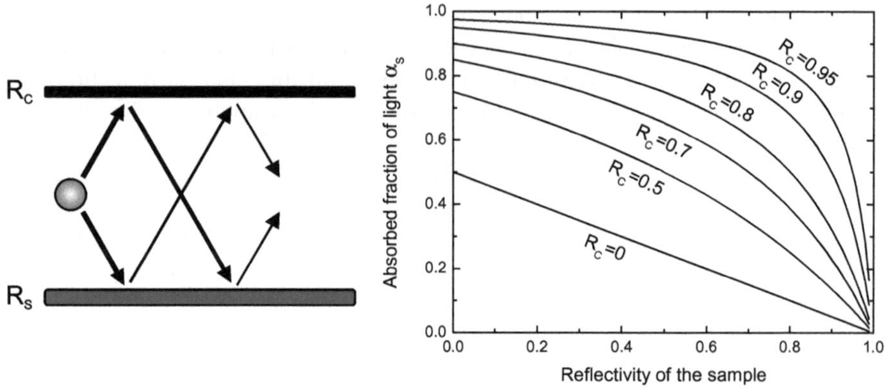

Fig. 3.21 Basic chamber geometry (left) and the normalized, absorbed fraction of light *A* as a function of sample and chamber wall reflectivity (right)

the films. Nearly the same graph as in Fig. 3.21 can be found in [55], but for the case of RTA.

The Γ value in (3.23) is the correction for the case that E_D is the *true* energy density unaffected by reflections in the chamber. However, the more practical case is that E_D was measured in the same chamber. It is further assumed that the measurement was done by a suitable, small area detector whose reflectance is close to zero or whose calibration data sheet considers its reflectance. Furthermore, the detector should be mounted on a plate whose reflectance is R_D. Under such conditions, (3.23) changes to

$$\mathrm{E_D} \propto \frac{(1 + R_C)}{2(1 - R_D R_C)} \tag{3.24}$$

The combination of (3.23) and (3.24) results in

$$\Gamma = \frac{(1 - R_D R_C)}{(1 - R_S R_C)} \tag{3.25}$$

where Γ is now the correction factor between the absorbed energy density *A* and the *measured* energy density E_D. Depending if the reflectance of the sample is smaller or larger than R_D, Γ is smaller or larger than 1.

A similar situation is FLA of a small sample with the reflectance R_S placed on a larger substrate as shown in Fig. 3.22. In this case the chamber has to be divided at least in an upper and lower part characterized by an average reflectance R_U and R_L, respectively (please note that R_S has now a different meaning than in Fig. 3.21). The area of the sample is assumed to be sufficiently small compared to the area of the substrate. Calculation is analogous to the previous case resulting in

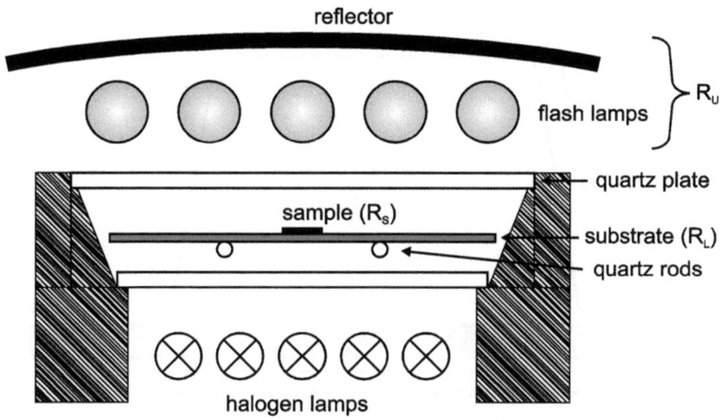

Fig. 3.22 Basic scheme of the chamber geometry with a small sample and a substrate. The correction factor Γ varies with different substrates. After [56]

$$\Gamma = \frac{(1 - R_D R_U)}{(1 - R_L R_U)} \tag{3.26}$$

This is very similar to (3.25), only the meaning of R_L and R_U is different.

The correction factor Γ can be determined by performing a calibration experiment in which the achieved maximum surface temperature during FLA is known. One possibility is to flash samples with increasing E_D up to the point where first traces of surface melting appear in the middle of a sample. It has to be considered that first melt seeds will already occur at lower energy densities at the corners and the edges which is due to the additional absorption of light at the side faces (next section). If (3.6) and (3.26) are both applied for the measurement of E_D and A, the latter one can be expressed as

$$A = E_D \frac{(1 - R_S)(1 - R_D R_U)}{(1 - R_L R_U)} \tag{3.27}$$

It is assumed that all values are known except R_U: R_L and R_D can be measured or are known (e.g. in case of a Si wafer as substrate), R_S is the sample reflectance, and A is the absorbed energy density. In such a case R_U can be estimated by

$$R_U = \frac{1 - R_S - A/E_D}{R_D(1 - R_S) - A R_L / E_D} \tag{3.28}$$

If R_U is once known, R_L of an *unknown* configuration can be obtained according to:

$$R_L = \frac{1}{R_U}\left[1 - \frac{E_D}{A}(1 - R_S)(1 - R_D R_U)\right] \tag{3.29}$$

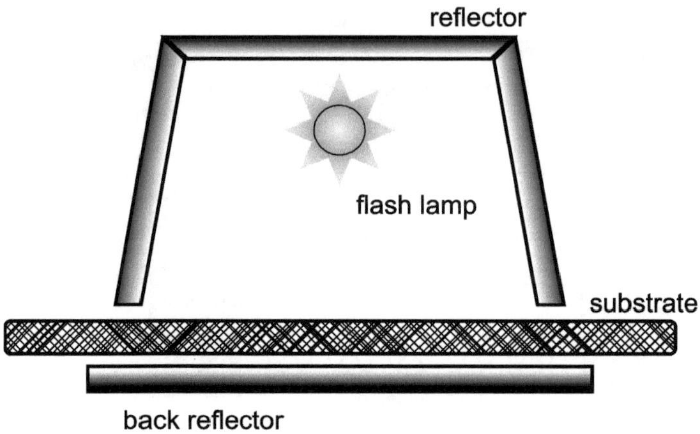

Fig. 3.23 FLA system for transparent substrates as used in [57]

As an example, this method was used to estimate Γ of three different configurations: a Si substrate, a quarter of a Si wafer and a quartz substrate. As the Si substrate is opaque to the flash light and its reflectance is well known, the measured and the simulated value A were used to get R_U and Γ of this configuration according to (3.28) and (3.26), respectively. Based on this, R_L and Γ of the other two configurations can be calculated. R_L and Γ are expected to decrease in the order of the Si substrate, the quarter wafer and the quartz substrate, and in fact, values of 1.29, 1.12 and 0.93 were measured for Γ. It also has to be mentioned that this method gives only a rough guess. For a more precise modelling, diffuse reflection has to be taken into account, and the use of raytracer software can give a more correct description of the chamber geometry. More details about this example were published elsewhere [56].

Finally, highly reflective chamber walls are even more important if the sample is only weakly absorbing, e.g. in case of TCO layers on glass. In order to have a maximum harvesting of the energy reflectors should also be mounted at side and bottom walls. An example for such an arrangement can be found in [57] where the sample, made of 100 nm ITO on a 500 μm thick glass carrier, is confined between a top and bottom reflector (Fig. 3.23). By using this arrangement, the authors were able to increase the absorbed light by a factor of 2.8 when they introduced the additional back reflector made of polished aluminium.

3.2.3 *Temperature Profiles*

Although the temperature profiles during FLA are complex, there are some general trends which should be discussed in the following. Thereby, the absorption profile was determined with the help of an in-house software using the wave transfer matrix method (Sect. 3.2.1). However, the calculation of reflectance and transmittance of a thin film system is the standard task of any ellipsometer software, and some of them are probably smart enough to allow the calculation of an intensity profile of an incoming wave. The temperature simulations using these absorption profiles as source term were performed by means of the COMSOL Multiphysics© software, but any software which allows the solving of partial differential equations should also be sufficient. Potential errors may lie in an inappropriate mesh size: meshes too coarse can give large deviations, and meshes too fine may provoke rounding errors. If not otherwise stated, the following assumptions have been made:

(i) The time-integrated Xe spectrum like in Fig. 2.23 has been used. Any dependence on pulse length and intensity has been assumed to be insignificant.
(ii) The temperature dependence of optical constants is negligible. This allows, together with the previous assumption, the decoupling of the optical and thermodynamic simulation according to Fig. 3.18b.
(iii) The angle dependence of the incoming light has been ignored, i.e. the simulation assumes perpendicular incidence.
(iv) Potential deviations of material constants in thin films compared to the bulk values, especially in case of thermal conductivity, have not been considered.
(v) The thermal resistance of interfaces and phase changes were not considered.

To be independent from a specific chamber configuration, the following simulations solely depend on the absorbed energy density A. The first issue to discuss is the general temperature dependence as illustrated in Fig. 3.24. In this example a 525 μm thick Si wafer is exposed to a 1.3 ms flash (Fig. 3.24a) with $A = 25\ \mathrm{J\,cm^{-2}}$. At first, the temperature at the surface strongly rises with the rising edge of the light pulse, but slightly delayed (Fig. 3.24b). In fact, temperature reaches a maximum of nearly 750 K at 1.1 ms which is on the falling edge of the light pulse. This maximum marks the time when the heating of the absorption zone by the light pulse is balanced by thermal conduction towards the substrate. Afterwards, temperature drops down to an equilibrium temperature T_{equ} of about 550 K after 6–8 ms. In a very first approximation T_{equ} scales with

$$T_{\mathrm{equ}} \approx T_0 + \frac{E_{\mathrm{D}}}{c_p\, d\rho} \tag{3.30}$$

where T_0 is the starting temperature (either RT or a certain preheating temperature), c_p is the specific heat capacity of the substrate, d is the substrate thickness, and ρ is its mass density. This approximation assumes that the temperature dependence of c_p is negligible and that heat dissipation by thermal radiation or convective heat transfer takes place on a time scale much longer than the pulse length. Both assumptions can be met if the temperature jump is not too large and the thermal conductivity of the substrate is not too small. The decrease of the surface temperature depends on several factors: the edge steepness of the pulse, the thermal conductivity of the material and the thickness of the substrate. In contrast to the surface, the backside temperature monotonically increases from RT to the equilibrium temperature, and temperature profiles inside the wafer are in between (Fig. 3.24b). The temperature

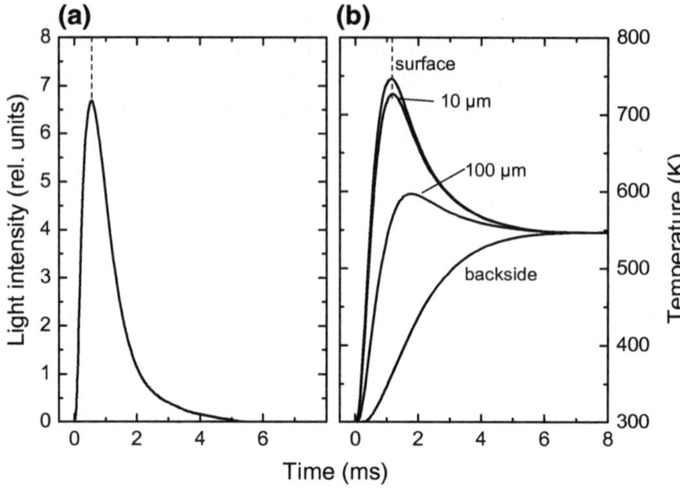

Fig. 3.24 Time profile of a 1.3 ms light pulse as measured by a photodiode (**a**) and the corresponding temperature profiles at the surface, in a depth of 10 and 100 μm, and at the backside of a Si wafer (**b**). The simulation assumes $A = 25$ J cm^{-2}

curves also reveal that the temperature difference within the first μm is small, at least for materials with a similar thermal conductivity as Si.

Figure 3.25a displays the surface temperature of a Si wafer as a function of time if exposed to flash pulses with different lengths, namely 0.4, 1.3, 3, 12 and 20 ms. The absorbed energy density was chosen in such a way that the surface peak temperature is always around 1000 K. The edge steepness of the temperature profile is the larger the shorter the pulse, and the temperature difference between front and backside also increases with decreasing pulse length. It can be easily assumed that thermal stress induced by the flash pulse is following this trend. In order to achieve a surface temperature of 1000 K, A increases with increasing pulse time in the order 18.2, 34, 39.2, 67.2, and 74 J cm^{-2}, accompanied by a corresponding increase of T_{equ} as the absorbed energy available for the heating of the whole sample increases. Reversely, if A is fixed, the surface temperature increases if the pulse becomes shorter, whereas the equilibrium temperature remains nearly unchanged (Fig. 3.28b).

As seen in (3.30), the substrate thickness has a significant influence on the equilibrium temperature. This dependence is illustrated in Fig. 3.26 for a 1.3 ms flash with $A = 25$ J cm^{-2} applied to Si wafers of different thickness. For thick wafers, the surface temperature passes through a maximum before it converges to T_{equ} for longer times. Above a certain thickness (here at about 525 μm) this maximum becomes independent of the wafer thickness. However, for thinner wafers, namely for thicknesses $\leq$175 μm, the surface temperature monotonically increases up to T_{equ}. In fact, the thermal diffusion length is about 160 μm, if using the material properties of Si at 800 K. This change in temperature behaviour marks the transition from the thermal flux regime to isothermal annealing (Sect. 1.3). After all, the temperature

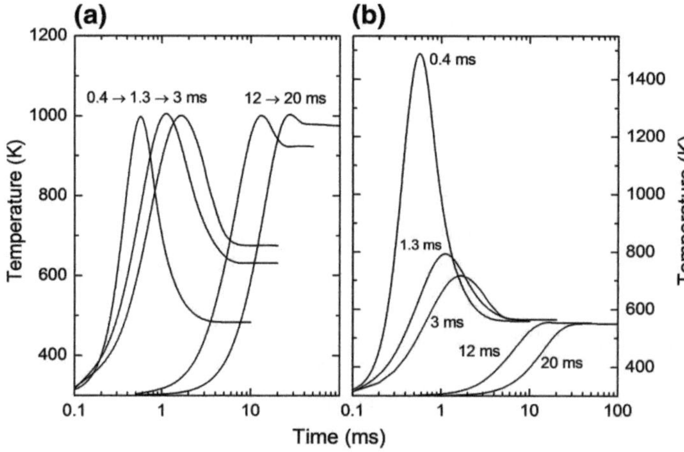

Fig. 3.25 Surface temperature during flash pulses of different length when the surface temperature is fixed (**a**) or with a constant absorbed energy density A of 25 J cm^{-2}

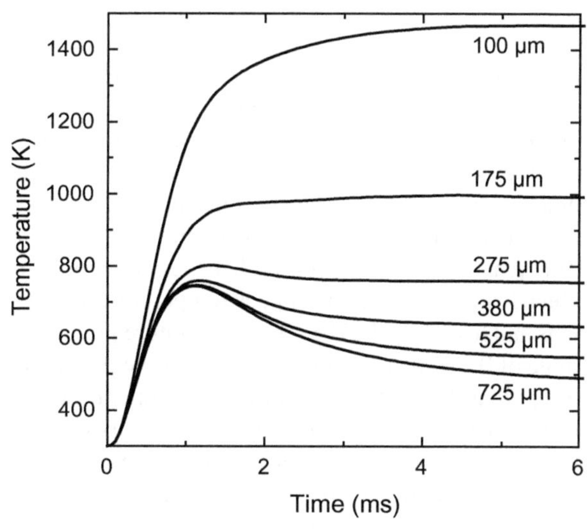

Fig. 3.26 Surface temperature during a 1.3 ms flash pulse with $A =$ 25 J cm^{-2} applied to Si wafers with different thickness and a start temperature of $T_0 = 300$ K. The equilibrium temperature T_{equ} is reached after 5–6 ms

rise T_{equ}–T_0 shows a nearly inverse dependence on the wafer thickness as expected by (3.30).

The specific profile strongly depends on the thermal conductivity of the substrate. To illustrate this, a model system with a 300 nm thick Cr layer on a 525 μm thick substrate made of Cu, Si or quartz glass was simulated (Fig. 3.27). The Cr layer thickness is sufficient to absorb nearly all light which is not reflected; thus all 3 systems are equal from an optical point of view. The simulation assumes that these systems are exposed to a 1.3 ms flash pulse with $A = 10.2$ J cm^{-2}. In case of quartz glass as a substrate the thermal conductivity is so low (~1.4 W m^{-1} K^{-1}

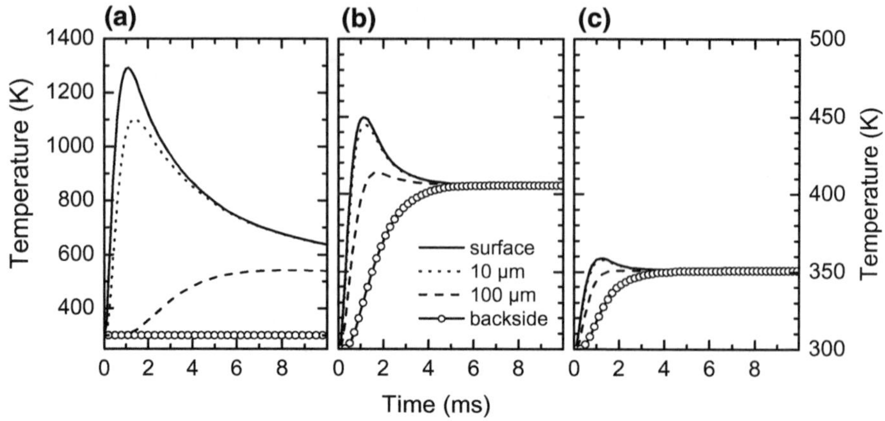

Fig. 3.27 Temperature as a function of time at the surface (solid line), in a depth of 10 μm (dot line), of 100 μm (dashed line) and on the backside (open circles). The systems is a 300 nm thick Cr layer on top of a 525 μm thick substrate made of quartz glass (**a**), Si (**b**) and Cu (**c**). The systems are exposed to a 1.3 ms flash pulse with $A = 10.2$ J cm^{-2}

at RT) that the peak temperature of the surface reaches a value of nearly 1300 K (Fig. 3.27a). In contrast, the backside remains at a much lower temperature, and temperature equalization is not reached before 100 ms. The temperature gradients which evolve in the material strongly depend on thermal conductivity: whereas the temperature difference between the upper and lower side of the Cr layer is well below 1 K, temperature gradients in the quartz glass substrate reach values of several 10 K μm^{-1}. Si has a much higher thermal conductivity (~156 W m^{-1} K^{-1} at RT) which leads to a much lower peak temperature and a temperature equalization time of 5 ms and less (Fig. 3.27b). The temperature gradients are now much lower and in the range of 0.5 K μm^{-1}. For Cu as a substrate the situation is similar to Si (Fig. 3.27c) except that the peak and equilibrium temperature are even smaller due to the larger heat capacity of Cu (the specific heat capacity of Cu is only half of that of Si, but the mass density is nearly 4 times larger). Consequently, the temperature gradients are even lower and in the order of 0.1 K μm^{-1}.

As a consequence, substrates with a low thermal conductivity like glass or ceramic targets can shift FLA in the direction of LA where only a surface region is really heated. In contrast, a high thermal conductivity as typical for metals shifts FLA towards RTA. However, the case of quartz glass also shows that it needs several μm or even tens of μm of buffer layer thickness to effectively shield a temperature-sensible substrate from the high surface temperature. To model this, a 1 μm buffer layer made of thermal SiO_2 was introduced between the Cr layer and the Si substrate to reduce thermal conduction to the back. As seen in Fig. 3.28a, the upper side of the buffer layer is at a significant higher temperature than the lower side. The difference is approximately 40 K at the surface temperature maximum. If SiO_2 is now replaced by an hypothetic material which has only 20% of the thermal conductivity of ther-

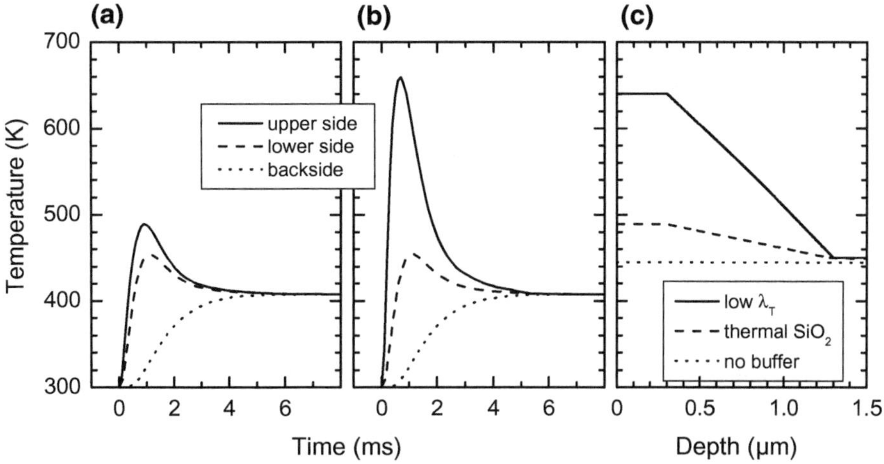

Fig. 3.28 Temperature profile during a 1.3 ms flash pulse $A = 10.2$ J cm^{-2} applied to Si wafers with a 1 μm thick SiO_2 layer and a 300 nm Cr layer on top (**a**) and to the same system with a 1 μm thick material having only 20% of the thermal conductivity of thermally grown SiO_2 (**b**). The temperature gradients after 0.9 ms for systems (**a**), (**b**) and a system without a buffer layer (Fig. 3.27b) are given in (**c**)

mal SiO_2, this difference can be enhanced to 200 K allowing a much higher surface temperature (Fig. 3.28b). The corresponding temperature gradients after 0.9 ms are shown in Fig. 3.28c. Whereas the temperature gradient in Cr and Si is very small, values of 40 and 200 K μm^{-1} can be achieved for thermal SiO_2 and the low thermal conductivity material, respectively. Although these specific values depend on the individual layer system and the flash pulse, it becomes clear that an effective temperature shielding requires buffer layers with both a thickness in the μm range *and* a low thermal conductivity.

All the details discussed before can be summarized in a couple of general trends describing the impact of certain FLA or material parameters to the temperature profile [58]:

(i) Energy density. The higher the energy density is, the higher the temperature in general. However, this relationship is only linear in a rough approximation because of the temperature dependence of heat capacity and thermal conductivity.
(ii) Pulse length. If the pulse becomes shorter for a given input energy, the maximum temperature at surface increases whereas the equilibrium temperature remains roughly the same. In addition, temperature gradients also rise. If the pulse becomes shorter for a given surface temperature maximum, the required input energy and the equilibrium temperature decreases.
(iii) Sample thickness. If the thickness is smaller than the thermal diffusion length, the temperature increase is inversely proportional to the thickness. If the thick-

ness becomes larger, the maximum temperature at the surface continues to decreases at first, but goes into saturation for even further thicknesses.

(iv) Optical constants of the sample. At first, they determine the reflectance and thus the amount of light which can be absorbed at all. At second, the absorption coefficient determines the penetration depth and the width of the absorption profile. In most cases the impact of reflectance on the temperature profiles is dominant.

(v) Heat capacity. The higher the heat capacity of the substrate is, the lower the temperature in general. However, the dependence is complex, and phenomena like phase changes, overheating and undercooling may significantly influence the temperature profiles.

(vi) Thermal conductivity. A low thermal conductivity increases the maximum temperature at the surface, increases the temperature gradients, and delays the equalization of temperature between the front and backside. A high thermal conductivity will induce the opposite behaviour.

3.2.4 Heat Dissipation

Whereas the heating rate of the surface region mainly depends on the rising edge of the flash pulse, several processes have an influence on the cooling rate: the falling edge of the flash pulse, thermal conduction towards the backside, thermal radiation and convective heat transfer by the annealing ambient. Considering the ms or even μs time scale of FLA, thermal radiation and convective heat transfer will play only a minor role. This is illustrated in Fig. 3.29a showing the profile of the surface temperature of a Si wafer after a 3 ms pulse with $A = 58$ J cm^{-2}. The surface temperature reaches a maximum at about 3 ms, followed by a rapid cooling phase and a plateau at the equilibrium temperature. Only on a time scale of 100 ms and longer thermal radiation and convective heat transfer are able to lower the temperature. The graph for convective heat transfer in Fig. 3.29a assumes a typical heat transfer coefficient of 20 W m^{-1} K^{-1}, but even higher heat transfer coefficients as known for forced gas cooling will not significantly shorten the plateau of equilibrium temperature; they only accelerate cooling on the long time scale. In conclusion, thermal radiation and convective heat transfer only matter in specific cases like that of a glass substrate with a very low thermal conductivity. Another specific case is multiple flash operation with repetition rates of several Hz and more in which the cooling rate can influence the elevated background temperature of the sample. A simple example is given in Fig. 3.29b in which 10 pulses with a length of 1.2 ms, an energy density of $E_D =$ 35 J cm^{-2} and a repetition rate of 0.83 Hz are applied to a 1.3 μm thick NbSn film on a 750 μm thick Cu substrate. Because of the large thermal conductivity of Cu, temperature gradients within the layer stack can be neglected. As seen, each flash is able to increase the temperature by a value in the order of 100 K, and for the first pulses this temperature jumps really accumulate. However, if T_{equ} after the flash comes close to 1000 K, the temperature will significantly fall within the pause time

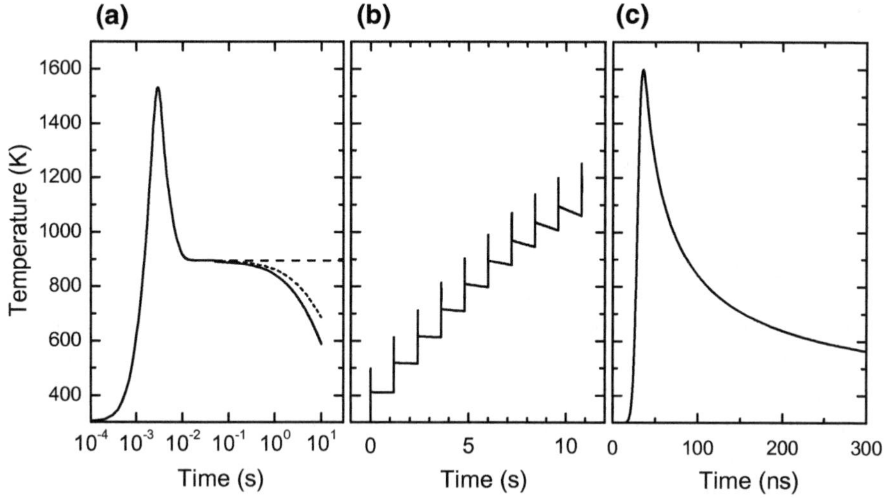

Fig. 3.29 Surface temperature of a 525 μm thick Si wafer during a 3 ms flash pulse with $A = 58\ \mathrm{J\ cm^{-2}}$ (**a**) in case of no heat dissipation (dashed line), thermal radiation (dot line) and thermal radiation plus convective heat transfer (solid line). **b** Surface temperature of a NbSn thin film on Cu under multiple FLA (**b**) and of Si during a hypothetical 10 ns pulse (**c**). After [58]

between two flashes due to thermal radiation. Convective heat transfer was neglected as in this case FLA in vacuum was simulated.

The fast cooling rate after the temperature maximum is governed by thermal conduction and the falling edge of the light pulse. Clearly, as long as significant amounts of light are absorbed in the sample its temperature rises, for which reason the decrease of temperature starts on falling edge of the light pulse. Thus, the edge steepness of the falling edge of the light pulse gives an upper limit for the cooling rate. But on the other hand, the cooling rate will also not be much smaller than that. This is illustrated in Fig. 3.29c which displays the temperature profile during a hypothetical light pulse with a pulse length of 10 ns. Although the thermal conductivity is the same as in the case of the 3 ms pulse, the decay time is only a few times longer than the pulse length. However, in case of 10 ns it may take much longer to reach the equilibrium temperature. The main reason for the fast cooling rate is the product $\lambda_T \nabla T$ in (3.4). If the thermal conductivity increases, the temperature gradient that evolves decreases in accordance with this, and the product of both is nearly constant. For a long pulse a high amount of energy is needed to achieve a certain surface temperature, and a given heat flow will need a correspondingly long time to transport the heat away. Consequently, a short pulse needs only a small amount of energy which can be dissipated in a short time. However, the relationship between pulse time and the energy density which is required to achieve a certain peak temperature is not linear. Temperature profiles in Si under ns LA can also be found in [59].

3.3 Homogeneity

3.3.1 *The Homogeneity of Irradiation*

The homogeneity of a FLA process is mainly related to the lateral temperature distribution on a sample which is determined by two factors: (i) the homogeneity of the light which hits the surface and (ii) the local variation of material parameters if the sample or wafer is patterned. A homogeneous temperature distribution and thus a homogeneous irradiation of the sample are of major importance as local variations of the energy density will manifest in adequate variations of absorption and temperature. The thermal diffusion length is usually too short to level out these differences. To illustrate the problem, Fig. 3.30 displays a drastic, but virtual example already discussed in [58] in which two flash lamps with a bore diameter of 1 cm are separated by 4 cm and located 2 cm above a Si wafer (Fig. 3.30a). The corresponding pattern of light intensity on the surface as shown in Fig. 3.30b is a superposition of the analytical solution of a single lamp, i.e. the flash lamps are regarded to be transparent. This pattern shows a difference of 11% between the minimum and the maximum, and this difference is imprinted on the temperature distribution at the sample surface. Figure 3.30c displays the surface temperature of a Si wafer at different times if exposed to a 20 ms flash pulse with an energy density of 100 J cm^{-2} without preheating. For 23 ms, temperature still differs by 11%, and this difference equalizes only slowly on a much longer time scale. More importantly, the position directly under the lamp has seen a maximum temperature which is ca. 120 K higher than that at the mid position, and this difference cannot be reversed by any equalization process.

To ensure a certain homogeneity of the flash light on a sample surface the arrangement of flash lamps and reflectors have to be optimised. The following distributions are the result of a simulation with an in-house software which combines a Monte-Carlo approach (random photon emission) with a raytracer. The distributions have the

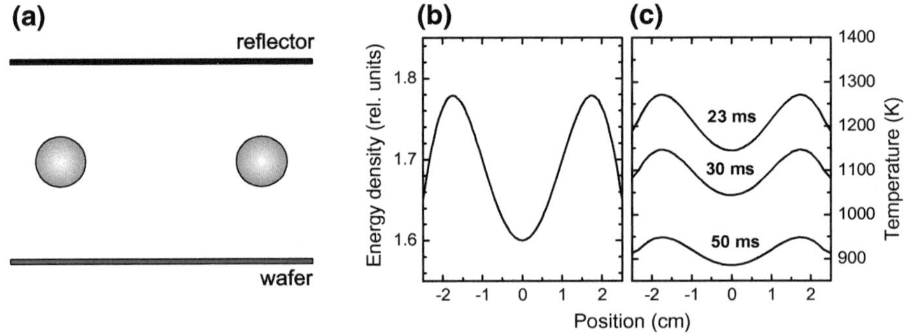

Fig. 3.30 Simulation of the lamp intensity (**b**) and the resulting distribution of the surface temperature of a Si wafer (**c**) for the flash lamp arrangement shown in (**a**). After [58]

unit ppm/cm^2 which is in fact the number of photons absorbed in a surface element divided by the total amount of emitted photons and the area of the surface element. The typical size of such an element is in the mm^2 range. The lamps are assumed to be a Lambertian radiator with constant intensity and to be opaque. Figure 3.31a displays the result for an arrangement of 3 lamps with a bore diameter of 20 mm, a distance of 80 mm from the target wafer and a distance of 60 mm from each other as shown in Fig. 3.31b. Although the approximations used in this simulation are rough, one conclusion can be immediately drawn: the area of reasonable homogeneity, probably the central region with an extension of 5×10 cm^2, is significantly smaller than the area covered by the lamps. Thus, for a given area which should be irradiated homogeneously, the lamp field has to extend beyond each edge of this area. This extension depends on the distance between lamp plain and the substrate surface, but should amount at least 20–30%.

As seen in Fig. 3.31c, homogeneity also depends on the distance between the lamps, here measured between the central axes of the cylindrical lamps. Even with an optimum distance (ca. 6 cm in this example) it becomes rather difficult to achieve an homogeneity better than ca. 3%, not to forget the imperfections of the real world compared to the ideal case of an simulation. There are a couple of strategies to address this issue. Usually, engineers use slightly parabolic reflectors instead of plane reflectors, or individual parabolic reflectors for each lamp. The lamp does not necessarily need to be placed in the focal point of the reflector as this position ensures

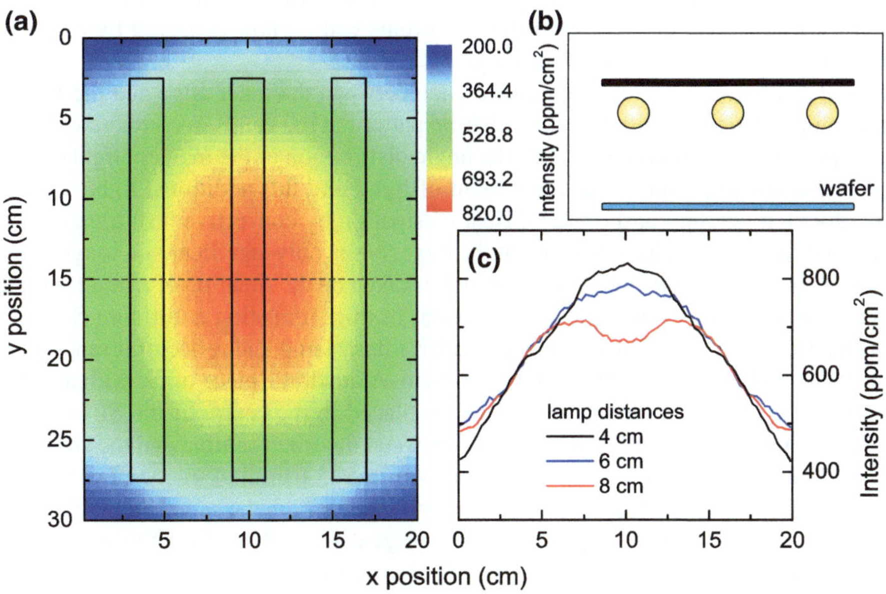

Fig. 3.31 Simulation of the light intensity distribution on a Si target if illuminated by 3 flash lamps (**a**). The lamps and the reflector arrangement is sketched in (**b**). Light intensity distribution for $y =$ 15 cm if the distance between the lamps is varied (**c**). More details can be found in the text

a parallel radiation path but not a constant intensity of the cross section. Considering that the real flash lamp is an extended and not a point source, a position more close to the reflector is mostly more advantageous for homogeneity than a position close to the target. This tendency is in accordance with simulations and measurements performed by Kim et al. [42] for a single lamp and a parabolic reflector. In contrast to the case of a lamp field, the use of a single lamp with a parabolic reflector usually results in a narrow stripe of constant light intensity, and the width of this stripe can be as small as 10 mm [60].

Another possibility is to arrange lamps in two planes which are shifted relative to each other in the plane, similar to Fig. 2.8. Using this arrangement, Usami et al. already achieved an uniformity of the light irradiation of 5% in a $40 \times 40\ \text{mm}^2$ area in 1986 [61]. Uniformity in this context is defined as the maximum deviation of light intensity from an average value. Later on, a temperature uniformity of 1.5% in the central 2-in. area of a 4 in. wafer, and probably the double value for the whole wafer except the edge zone, was realized [62]. Interestingly, the temperature distribution along the lamp axes did not significantly differ from that perpendicular to the lamp axes. If there is the technical possibility to control lamp currents individually, a controlled variation of the lamp currents can improve homogeneity. This was done in case of flash-RTA by using a so called lamp correction table [63], and the authors were able to reduce the centre-to-edge variation of 300 mm Si wafers from 3 to 4% to about 1%. In this publication uniformity is defined as the 1σ value of a distribution of sheet resistance values. A basic and simple approach to this is the use of 3–4 individual lamp circuits for individual zones of the target, each of it with its own pulse forming network. Different lamp currents can be now achieved by different charging voltages at the same pulse length.

Finally, there is a temperature difference between the outer edge and the rest of the sample which is caused by the radiation pattern of the lamps in combination with the specific reflector arrangement. The angle distribution of the incident light causes the sidewalls to absorb a certain amount of light, and this additional light leads to a higher temperature at the edges. Heat dissipation by convective heat transfer and thermal radiation is also enhanced at the edge, but is only significant on longer time scales. To go into more detail, a closer look is taken on the situation of a small sample lying on a target as it is typical in research. In this simulation a 100 mm Si wafer with a thickness of 525 µm is placed under a flash lamp and reflector arrangement as shown in Fig. 3.31b with a distance of 6 cm. A quadratic piece of Ge with an edge length of 2 cm and a thickness of 1 mm is placed in the centre of the Si wafer. In such a configuration the thermal contact between the small sample and the substrate is not well defined. In a real experiment with a 525 µm thick piece of Si simply lying on a 525 µm thick Si wafer it was found that for both the sample and the wafer surface melting occurs nearly at the same energy density. Therefore, the assumption of no thermal contact appears to be a reasonable approximation. In simulation, this is modelled by a significant air gap. Furthermore, the raytracer simulation (not shown) reveals that the energy density at the sidewalls of the Ge sample is approximately 21% of that at the front surface.

The 3-dimensional temperature distribution at $t = 24$ ms of this system during FLA with a 20 ms pulse and $E_D = 70$ J cm^{-2} is shown in Fig. 3.32. It can be seen that the corners and edges are hotter than the rest of the surface which is caused by the additional absorption of the sidewalls. However, the Ge piece is, on average, significantly cooler than the Si surface. This is a combined action of the larger reflectance of Ge compared to that of Si and the larger thickness of 1 mm. In this simulation the Ge piece has no thermal contact with the Si wafer, but in reality it might be somewhere between no contact and full thermal contact. In this case the substrate is an additional heat sink leading to a somewhat lower Ge temperature compared to simulation. Nevertheless, consideration should be given to the fact that the annealing temperature of a small sample on a substrate not only depends on the sample but also on the thickness and thermal conductivity of the substrate as well as on the thermal contact between them.

The situation at the corners of the Ge sample is more complex. According to the enlarged section in Fig. 3.32 the upper corner is the hottest point due to the combined absorption at the front side and the sidewalls. However, the lower part of a sidewall is only heated by sidewall absorption and thus at much lower temperature (but not as low as the temperature of the backside). Thus, during FLA the *upper* edges and

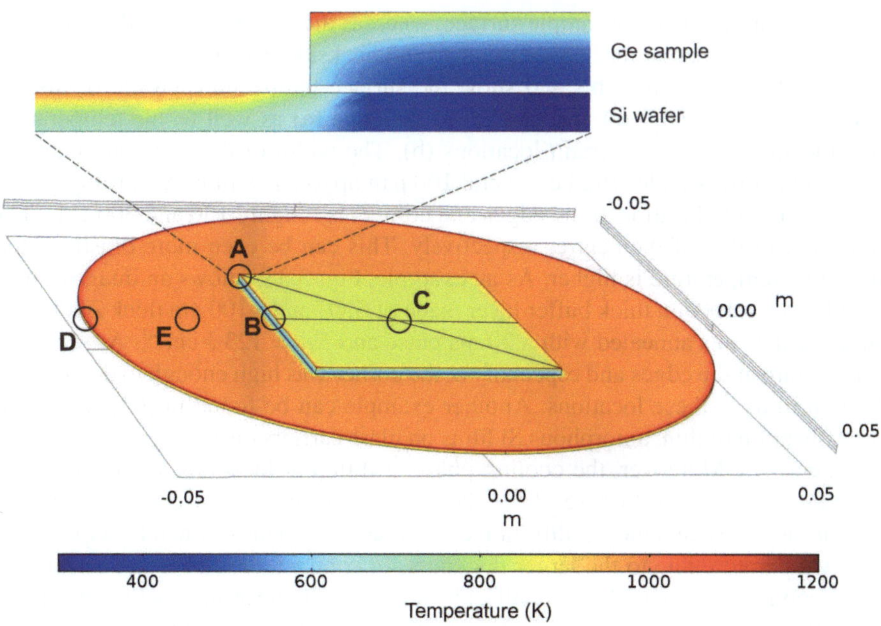

Fig. 3.32 Temperature distribution at $t = 24$ ms for a Ge sample lying on a Si wafer and annealed with a 20 ms pulse and $E_D = 70$ J cm^{-2}. The enlarged section displays the two-dimensional temperature distribution at the corner of the Ge piece lying in the section plane along the diagonal of the Ge piece. The circles mark surface locations whose temperature profile in time is given in Fig. 3.33b. The simulation was performed by the COMSOL Multiphysics© software

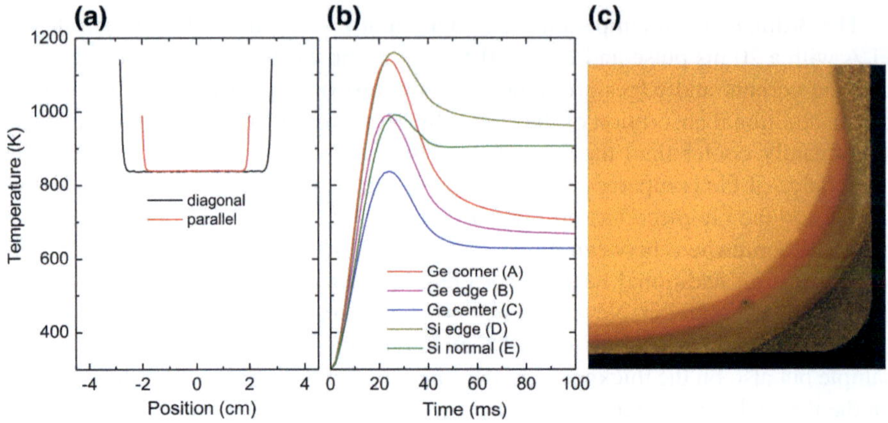

Fig. 3.33 **a** Surface temperature of the Ge sample at $t = 24$ ms along the diagonal (black) and a cut line parallel to the edge (red) as indicated as a thin black line in Fig. 3.32. **b** Surface temperature profile during FLA at the locations marked in Fig. 3.32. **c** Microscope image of a sample exposed to a 20 ms flash with $E_D = 123$ J cm^{-2} in an oxygen ambient [64]. The surface layer made of ZnO has decomposed at the corner due to the extraordinary high temperature during FLA

corners are hotter than the sample surface, whereas the *lower* edges and corners are cooler with a temperature between that of the backside and that of the front side. Figure 3.33a, b quantify this and show the surface temperature profile of the Ge sample along the diagonal and parallel to the edge (a) as well as the temperature as a function of time at certain locations (b). The width of this hot zone is in the order of the diffusion length, i.e. several 100 μm up to a few mm. According to this the temperature elevation at the edges and corners is roughly 150 and 300 K higher compared to that of the centre, respectively. This can be even more drastic if the annealing temperature is higher. As an example, Fig. 3.33c shows an image of a Si sample with a 100 nm thick buffer layer made of SiO_2 and a 100 nm thick ZnO layer on top which was annealed with a 20 ms pulse and $E_D = 123$ J cm^{-2}. As seen, the temperature at the edges and especially at the corner was high enough to decompose the thin films at these locations. Another example can be found in [65] where the crystallization of thin, amorphous Si films on glass carriers during 5 ms FLA started at the corners. Moreover, the cooling phase at different locations at the surface is different (Fig. 3.33b). In case of Ge the temperature maximum is reached earlier than in case of Si, and the equilibrium temperature at 100 ms is much lower than the peak temperature due to the larger thickness. In contrast, the Si surface undergoes little cooling in the first 20 ms. Finally, all equilibrium temperatures will converge to each other, mainly by lateral thermal conduction or heat transport via the air gap, but this is a matter of seconds or even minutes. If the edge effect should be avoided, the sample has to be closely encircled by fitting dummy samples of the same thickness like in a puzzle. In case of wafers it is a common idea to cut out a fitting dummy ring from another wafer with the same thickness but larger diameter.

3.3.2 *Pattern Effects*

Even if the light irradiation is completely homogeneous, local variations of the reflectance and absorption will lead to pattern effects in temperature [66]. Typical thermal diffusion lengths for ms-range annealing are roughly in the order of 100 μm [67], i.e. structures being much smaller than that should have, at a first glance, nearly the same temperature. Unfortunately, this statement only describes the situation *after* the temperature has been equalized, even if this happens within 1 ms or shorter.

If the wafer is optically inhomogeneous on a large scale, e.g. if the left and right side of the wafer is bright and dark, respectively, the temperature differences will induce corresponding differences in the thermal stress distribution and the wafer warpage. Fortunately, this is a rare case. However, optical inhomogeneity on a medium scale of mm and cm are much more common, e.g. between logic areas and dense memory areas on a chip [68]. So it was found that the absorbance of flash light varies between 58 and 78% depending on the local area ratio between blank Si and Si capped with SiO_2 [69]. Furthermore, the reflectance was found to vary with the pattern pitch with a minimum for a pitch of 200 nm [68]. Generally, dense patterns tend to higher absorption than sparse ones, and the resulting temperature difference can be as large as 130 K [70]. A similar trend was found for the density of trenches [71].

Inhomogeneity on a small scale of only a few μm and below ensures that small local areas experience different peak temperatures before thermal conduction can level out these differences. So it was found in case of trench isolation that the absorption and thus the maximum temperature increase with the trench depth [25]. The reason for this is that deeper trenches absorb more light due to the higher inner surface area (or due to multiple reflections in the picture of ray optics), but the effect is reduced if the flash pulse length becomes longer. Another example is the fabrication of fully silicided gate metal-insulator-semiconductor field-effect transistors (MISFETs) [73]. If uncapped, the NiSi Source/Drain areas are directly exposed to flash light leading to unintentionally high temperatures and electrical degradation. In addition, the direct exposure to flash light has been reported to increase the density of traps at the SiON/Si interface of metal-oxide-semiconductor field-effect transistors (MOSFETs) which in turn causes higher gate current fluctuations [72]. In this case additional interface traps are created at the gate edge of the MOSFET by the laterally irradiated UV light of the flash. The impact, with respect to the channel length, is obviously more severe for short channel than for long channel transistors (Fig. 3.34). This degradation can be minimized by using low FLA energies combined with high preheating temperatures [72] or by performing a post-FLA recovering anneal [74].

Inhomogeneity is also caused by interference effects in thin film layer stacks although the effect is reduced if averaged over the FLA spectrum. Nevertheless, interference effects during FLA may result in reflectance variations in the order of 15% [45], and in case of thin film transistors (TFTs) they lead to an insufficient activation of junctions and a partially incomplete crystallization of the amorphous

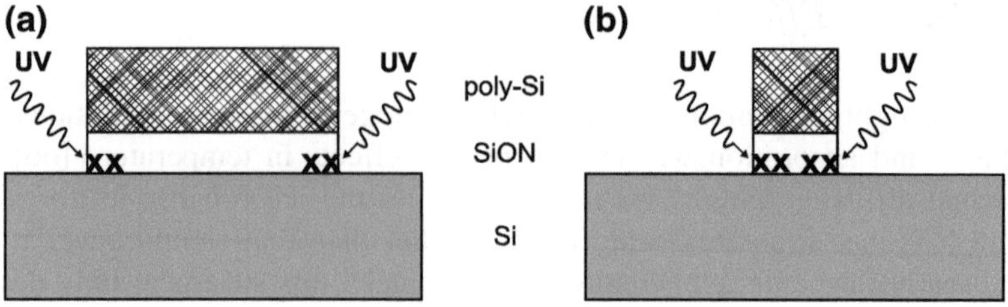

Fig. 3.34 The impact of lateral UV irradiation during FLA on the interface trap density of long (**a**) and short (**b**) channel MOSFETs. After [72]

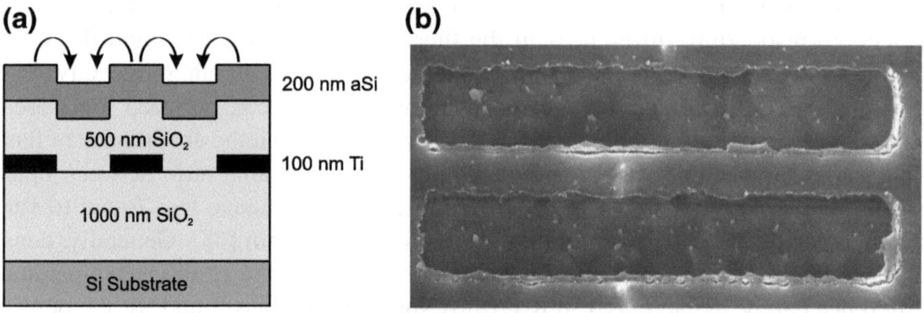

Fig. 3.35 Schematic layer structure (**a**) with embedded mirrors made of Ti. During FLA amorphous Si in areas with Ti mirrors melts, flows into the gap between two mirrors and forms large grains with dimensions in the order of 10 μm (**b**). The mirror size [dark areas in (**b**)] is $8 \times 48\ \mu m^2$. The SEM image was taken from [75]. Reprinted from [75], with the permission of AIP Publishing

Si [76]. In a study about the crystallization of amorphous Si on SiO_2 for single grain TFTs [75], interference effects were exploited to adjust the FLA temperature in such a way that aSi melts at locally defined areas. To do so, a layer stack as shown in Fig. 3.35a was deposited by using thermal oxidation (lower SiO_2 layer), e-beam evaporation (Ti and aSi) and ac sputtering (upper SiO_2 layer). Figure 3.36 displays the power density of the electromagnetic wave (which is proportional to the absolute value of the pointing vector) and the resulting absorption within the upper part of the layer system shown in Fig. 3.35a without (a) and with a Ti mirror (b). The simulation was performed by using the wave transfer matrix method (Sect. 3.2.1). Although the major part of the light is already absorbed in the upper aSi layer, the introduction of the Al mirror has two consequences: (i) part of the incoming light is reflected and leads to an additional absorption in the aSi layer, and (ii) the remaining light is absorbed by the Ti mirror itself. As a consequence, the local maximum temperature is higher in areas with a mirror than in areas without. A quantitative analysis reveals that the second effect is dominating: the Ti mirror increases the absorption in the aSi layer by 1.6%, whereas the Ti mirror itself absorbs another 5.2% of the light and additionally heats the aSi layer by thermal conduction.

For the right FLA conditions the aSi layer above the mirror melts, and there is a lateral mass transport (as schematically indicated in Fig. 3.35a) of molten Si towards the

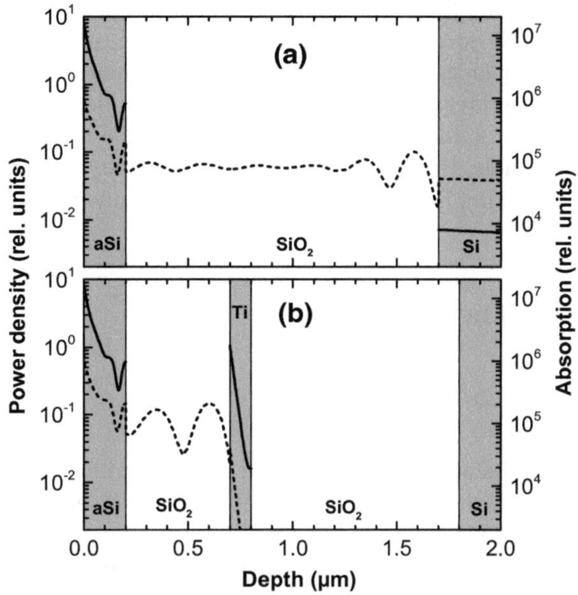

Fig. 3.36 Power density of the electromagnetic wave (dashed line, left scale) and the corresponding absorption (solid line, right scale) for a system (**a**) without and (**b**) with a 100 nm Ti layer during FLA processing. After [75]

space between two mirrors where it starts to solidify. As shown in the scanning electron microscope (SEM) image of Fig. 3.35b, there are two large, elongated Si grains in the gap between the two mirrors. Cross-section transmission electron microscopy (not shown) revealed that there is no grain boundary between the underlying, original aSi layer and the new Si layer which implies that the original aSi layer has also been fused [75]. A probable reason is that the FLA conditions are also sufficient to melt the aSi layers in areas without mirror, but that the melting temperature is reached later in time on the rising edge of the flash pulse. Furthermore, the release of latent heat and thermal conduction from the molten Si can give an additional contribution. It is assumed that solidification starts outside the gap, and that a solidification front runs from the outside to the middle of the gap forming a pronounced grain boundary somewhere in the middle of the gap. This grain boundary appears white in Fig. 3.35b.

Finally, there are couple of strategies to avoid pattern effects. If FLA is combined with preheating or RTA, the reduction of the FLA energy and an adequate increase of the preheating temperature will reduce the pattern effect. This was quantified by comparing the emissivity of FLA and spike RTA for different structures on the wafer [77]. If not applicable, the introduction of dummy structures on the chip can help to homogenize the reflectance of the wafer [67]. Another approach is the deposition of transparent and isolating capping layers which also make reflectance more uniform. Using this method it was possible to strongly reduce the variation in junction depth of ultra-shallow junctions which arise by corresponding temperature fluctuations [69]. The deposition of an opaque capping layer also leads to a more homogeneous absorption and a shielding of critical areas against a direct FLA exposure. In the aforementioned example of fully silicided gate MISFETs [73] the introduction of a

Si capping layer absorbing almost the entire flash light avoided the degradation of the NiSi Source/Drain areas. They are still heated by thermal conduction from the capping layer, but to a lesser extent than without a capping layer. Usually, an opaque capping layer also reduces the penetration depth of the light and absorption takes place in a thinner surface region. This issue is discussed in more detail in [45]. If applicable, FLA from the backside as already proposed in [78] and experimentally verified in [79] will completely eliminate any pattern effect.

3.4 Thermal Stress

3.4.1 Silicon

The influence of strain and stress to the material properties of Si is known since the 50s [80], and the problem of thermal stress during annealing became already apparent in the 60s [81] in the cooling phase of annealing. In this case the temperature gradients evolved due to the different cooling rates between the centre and the edge of the wafer. Anyway, a temperature gradient will lead to a different thermal expansion of the material and in turn to the development of strain, described by the strain tensor ε_{kl}, if the material cannot relax. According to the well-known Hooke's law this generates stress which can be described by

$$\sigma_{ij} = C_{ijkl}\varepsilon_{kl} \tag{3.31}$$

where σ_{ij} is the stress tensor and C_{ijkl} is Young's modulus which in fact is a fourth-order tensor. The different components of the stress tensor will trigger different material failure mechanisms [81]: The diagonal or normal stress components are associated with a volume change (they squeeze or enlarge the lattice), interact with point defects and are finally responsible for crack formation. The non-diagonal or shear stress components are not connected with a volume change, but trigger the movement of dislocations and plastic deformations. Hooke's law is valid for the elastic regime, i.e. small strains, where the material returns to its original shape if the strain is falling back to zero. However, if the stress exceeds the so-called yield stress, the material reacts with defect creation and plastic deformation. For brittle materials like Si at RT, the yield stress is rather high, and exceeding this limit will lead to breakage or crack formation rather than plastic deformations. However, the yield stress of Si drastically drops down from values >1 GPa for RT to values <1 MPa for temperatures close to the melting point [82]. In case of sheer stress the critical value for the onset of dislocation movements, called the Peierls stress, is the lowest for the highest interplanar spacing and the shortest interatomic distance, which is the group of (111) planes and the [110] direction in Si [80]. With increasing temperature, this limit is also significantly lowered. Fischer et al. [83] calculated the maximum temperature ramp rates for 300 mm wafers which are allowed to avoid slip formation. For annealing temperatures above 1000 °C these ramps are well below 1 K/s, a condition which is diametrically opposed to the physical situation in short time annealing. However, FLA has the advantage that these uncomfortable conditions prevail only for a few ms. Thus, defect creation is unavoidable during FLA, and the task of stress management is to limit defect creation and to finally prevent a wafer breakage.

The strain and stress distribution which arises during FLA was already investigated in the early 80s. Heinig et al. [84] investigated the time profile of the maximum thermal stress of the wafer during FLA and found that the critical time is not the profile maximum, but the time at which surface temperature is largest and the yield stress is lowest. In a more detailed analysis [44] the stress profile was calculated by

$$\sigma_{xx} = \sigma_{yy} = \frac{\alpha E}{1-\nu}\left[-T(z,t) + \frac{1}{2h}N_{\mathrm{T}} + \frac{3z}{2h^3}M_{\mathrm{T}}\right] \tag{3.32}$$

$$N_{\mathrm{T}} = \int_{-h}^{h} T(z,t)\mathrm{d}z \quad M_{\mathrm{T}} = \int_{-h}^{h} zT(z,t)\mathrm{d}z \quad d = 2h \tag{3.33}$$

where α is the thermal expansion coefficient, E is Young's modulus, ν is the Poisson ratio, d is the wafer thickness, and $T(z,t)$ is the temperature profile. σ_{zz} and the non-diagonal components of the stress tensor are assumed to be zero if the flash light is homogeneous and the wafer is freely moveable. The situation is illustrated in Fig. 3.37 for a 525 μm thick Si wafer subjected to a 20 ms FLA pulse with $E_{\mathrm{D}} = 105\ \mathrm{J\,cm^{-2}}$ and a preheating temperature of 520 K. Figure 3.37a displays the surface temperature (left scale, solid line) and the stress according to (3.32) and (3.33) at the surface (right scale, open symbols) as a function of time. As seen, the maximum of the stress occurs at the rising edge of the temperature profile. In addition, the yield stress (right scale, dashed line) strongly decreases with temperature, and there is a time interval of ca. 20 ms in which the stress significantly exceeds the yield stress. In this time defect creation and plastic deformation must be expected which is, in this specific case, mainly due to the very low value of the yield stress. In conclusion, this situation can hardly be avoided if high temperatures are needed. In addition, if the flash pulse becomes shorter, the stress values will increase and prolong the critical time interval. Typical slip patterns induced by thermal stress during annealing can be found in [81, 83].

Figure 3.37b shows the thermal stress as a function of depth where positive and negative values stand for tensile and compressive stress, respectively. Starting at the surface, the stress is compressive (light grey) but decreases with depth, becomes tensile (dark grey) in the middle of the wafer, and changes again to compressive stress (light grey) at the backside. This interesting behaviour is caused by the bowing of the wafer, and the bending radius is such that the elastic energy of deformation is minimized. This energy can be roughly estimated by [85]:

$$U = \frac{1}{2}B\delta^2 \quad \varepsilon_{ii} \approx \frac{1}{3}\delta \quad B = \frac{E}{3-6\nu} \tag{3.34}$$

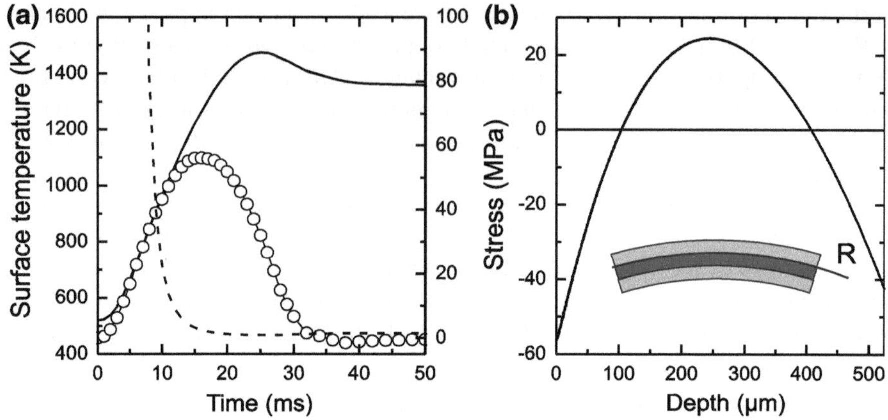

Fig. 3.37 **a** Surface temperature (solid line), thermal stress at the surface (open symbols), and yield stress (dashed line) as a function of time during a 20 ms flash pulse. **b** Stress profile through the wafer for $t = 16$ ms. A basic scheme of a bent wafer is shown as an inset. More details can be found in the text

where U is the energy density for the stored deformation energy, B is the bulk modulus, and δ is the volume expansion. The spherical curvature κ_T that evolves for a general temperature distribution can be expressed by [86].

$$\kappa_T = \frac{3}{2h^3} \int_{-h}^{h} \alpha(z) z T(z) \mathrm{d}z \tag{3.35}$$

The depth of ca. 100 μm may serve as a reference plane R in the following discussion (Fig. 3.37b). In this depth the strain ε_B caused by the wafer warpage equals the strain ε_T caused by thermal expansion, and the resulting stress is zero. Moving towards the surface, the lattice is widened because of the bowing of the wafer. However, this widening is smaller than that thermal expansion would cause, i.e. $|\varepsilon_T| > |\varepsilon_B|$, and the resulting stress is compressive. If we move in the opposite direction, the lattice is compressed due to wafer bowing, but once again the thermal shrinkage of the lattice would be larger, and the middle region of the wafer is under tensile stress. $|\varepsilon_T| > |\varepsilon_B|$ continues to be valid, but now with an opposite sign. Finally, the temperature gradient becomes smaller at the backside of the wafer. If it is small enough with $|\varepsilon_T| < |\varepsilon_B|$, lattice compression due to wafer warpage dominates leading to compressive stress at the backside. It should be noted that the extent of compressive stress on the cold backside sensitively depends on the difference of temperature gradients. If we assume only a small difference, or even a linear temperature profile from the front to the backside, the stress will monotonously change from compressive stress at the front side to tensile stress at the backside. Examples for this can be found

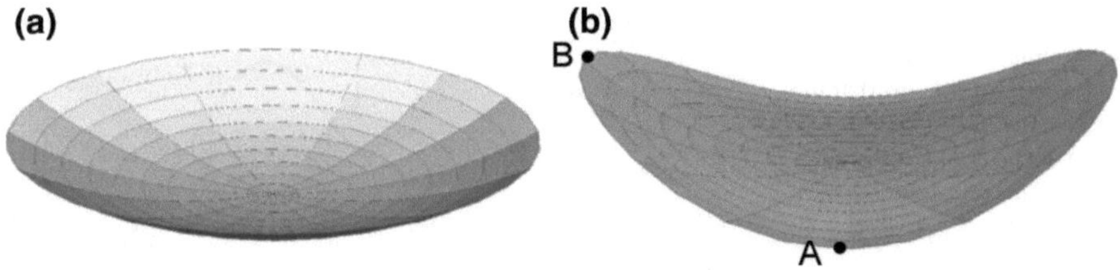

Fig. 3.38 Warpage behaviour of a wafer exposed to FLA from below. Spherical (**a**) and asymmetric, saddle-like mode (**b**) for low and high temperature gradients, respectively. Reprinted from [86], with the permission of AIP Publishing

in [87, 88]. Furthermore, if the wafer is unable to bow, monotonous stress profiles can also be obtained, but at significantly higher stress levels.

At the beginning, the wafer starts to bow in a spherical manner like a cap (Fig. 3.38a) due to the rotational symmetry of the wafer and the homogeneous illumination. However, with increasing bowing it stores an increasing amount of strain energy up to a turnover or bifurcation point where the deformation suddenly jumps in a second, energetically more favourable mode [86]. This mode is shaped like a saddle (Fig. 3.38b) and is now described by two curvatures κ_x and κ_y. The bifurcation point can be roughly estimated as the point where the spherical curvature is exceeding a critical threshold κ_T^* given by Smith et al. [86]:

$$\kappa_\mathrm{T}^* = \frac{8d}{r^2(1+\nu)^{3/2}} \tag{3.36}$$

where d and r are the thickness and the radius of the wafer. If we now monitor two points A and B at the edge of the wafer (see Fig. 3.38b for their location) during a 20 ms flash pulse with 520 K preheating temperature, the deflection of both points will rise together until the bifurcation point is reached (Fig. 3.39). At this point the warpage mode changes from spherical to saddle-like, and the deflection curves starts to diverge: whereas curve A stays roughly constant at a low level of deflection, curve B strongly increases and reaches a maximum deflection of approximately 10 times the wafer thickness. More details about the mathematical modelling and the resulting lateral stress distribution can be found in [41].

The present case is for 100 mm wafers, but deflections are correspondingly larger for 200 and 300 mm wafers. Equation (3.35) implies that the curvature decreases with increasing wafer thickness, but the wafer thickness increases only slightly when the wafer size extends to 200 or 300 mm. In addition, the temperature difference between front and backside will increase with increasing thickness. As seen in (3.36), the wafer size has a direct impact on the critical curvature for bifurcation: for large wafers this value is substantially lower than for small wafers. With respect to (3.32) and (3.33), the maximum thermal stress during FLA is independent of wafer size in a first approximation, but slight changes may occur due to the thickness dependence of curvature. Another point is that the edge of the wafer gets hotter because of

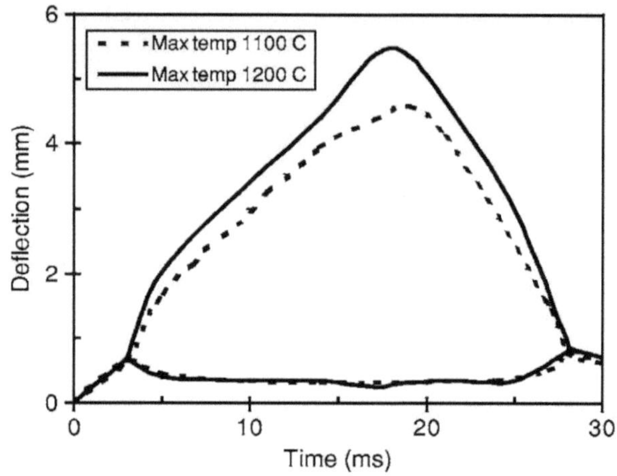

Fig. 3.39 The deflection of two points A and B (see Fig. 3.38b for location) at the wafer edge during a 20 ms pulse. Reprinted from [86], with the permission of AIP Publishing

the additional light absorption at the lateral edge. Thus, temperature gradients and occurring stress levels might be higher in this region. Anyway, large deflections are challenging for wafer mounting, and in most cases wafers are supported by a couple of quartz pins only. Wafer warpage in the range of ms can cause a jump of the wafer with the risk of destroying it in addition to the already existing risks due to the large stress.

There are a couple of investigations devoted to the questions which type of defects arise during FLA and under which conditions they become critical or even lead to wafer breakage. Foggiato and Yoo [89] investigated the onset of slip formation as a function of FLA energy density and preheating temperature for As-doped and BF_2-doped Si substrates. Thereby, the dependence on preheating temperature is complex: higher preheating temperatures lower both the maximum stress which can occur and the yield stress. So it was found that a preheating temperature of 650 °C can lead to easier wafer breakage than one of 450 °C [87]. There is also the expectation that pre-existing defects lead to higher stresses and an earlier wafer breakage. A recent study investigated the influence of a notch as an artificial defect on stress evolution and wafer breakage [90]. Both experimental observations and simulations reveal that the depth of the defect is the most important parameter: as long as the defect is in the upper half of the wafer it doesn't break. However, if the defect comes increasingly close to the backside, stress and breakage rate increases. Size and orientation of the defect played only a marginal role.

Another study [91] deals with the creation of bulk microdefects in Si during RTA and FLA which in fact is a result of oxygen precipitation. In the case of RTA the number of such defects is reduced for temperatures below 1200 °C, but rises for higher RTA temperatures. For FLA a slight reduction of the defect density was found for a 3 ms pulse, while a 20 ms pulse was able to reduce the defect density by more than two orders of magnitude. Moreover, the intrinsic point defect concentration after FLA, namely that of interstitials and vacancies, was completely different from that during RTA (more details in Sect. 4.1). A similar result was obtained when simulating the recrystallization of an amorphous Si layer during FLA after a Ge pre-amorphization implant followed by *B* implantation [92]. If only subjected to FLA the interstitial concentration in the region of end-of-range defects was higher compared to the case of spike RTA combined with FLA. By contrast, the number of larger defects, namely dislocation loops, was smaller in case of FLA. In addition, the diffusivity also depends on the stress during FLA: interstitials and vacancies are known to diffuse faster under tensile and compressive stress, respectively [80]. Finally, Kissinger et al. [47] investigated the diffusion of pre-existing dislocations in 900 μm thick Si during FLA and found that, depending on the temperature and stress profile, dislocations diffuse strongly from the front side into the wafer and only moderately from the backside, leaving a zone in a depth of 700 μm free of dislocations.

To minimize the adverse effect of thermal stress, temperature gradients should be as low as possible. However, at the same time the temperature dependence of the yield stress has to be considered. A preheating in the range of s or min is assumed to reduce the stress and the risk of wafer breakage, but this benefit can (but does not have to) be more than compensated by a low yield stress. Another possibility is the use of some kind of pre-flash, either by using a second lamp circuitry, a parallel connection of two LC sections (see Sect. 2.4.3) or a strong rise of the simmer current a few tens or 100 s of ms in advance of the flash. All these measures can act as a preheating in the ms range. If there is sufficient scope for different pulse lengths, longer pulses could be more advantageous with respect to adverse stress effects. Finally, strong asymmetric pulse shapes with a steep rising edge and a gently falling edge, as it is typical for the combination of a large capacitance and a small inductance, should be avoided.

3.4.2 Glassy Substrates and Thin Films

If using glassy substrates in FLA, the risk of plastic deformations is much more important than the risk of breakage. With increasing temperature the viscosity of the glass decreases by many orders of magnitude. If temperature exceeds the strain point, microscopic flows within the glass try to relieve strain which, however, occurs on a time scale of hours. If the temperature continues to rise and exceeds the annealing temperature, strain is relieved within minutes or even shorter, although the viscosity is still high enough to prevent deformations on a macroscopic scale. However, such deformations become possible if the temperature exceeds the softening point of the glass. Considering the FLA process, temperatures exceeding the strain point seem to be uncritical. This was verified in case of thin amorphous Si films on different glass carriers where no deformations were found after FLA, although the peak temperature was well above the strain point [93, 94].

However, the situation changes if the softening point is reached. The glass shows a viscoelastic behaviour which can be described by the Maxwell model [43, 95, 96]:

$$\frac{\mathrm{d}\varepsilon}{\mathrm{d}t} = \frac{\sigma}{\mu} + \frac{1}{E}\frac{\mathrm{d}\sigma}{\mathrm{d}t} \tag{3.37}$$

where μ is the viscosity of the glass and strongly temperature dependent:

$$\mu(T) = \mu_0 \exp\left(\frac{C_0}{T - T_0}\right) \tag{3.38}$$

μ_0, C_0, and T_0 are the strain point (Pa s), the annealing point (°C) and the softening point (°C), respectively, which, for the case of the glass Corning Eagle 2000, are given in [96]. The left side of (3.37) is a measure for the speed of deformation and strongly increases with temperature due to the temperature dependence of μ. If a glass substrate is homogeneously irradiated by the flash light, the substrate starts to bow in a similar way as a Si wafer (Fig. 3.40a). However, if the temperature exceeds the softening point, plastic deformations try to relieve the stress and thus the bowing of the substrate. As a consequence, the curvature of the glass substrate decreases and the bending radius increases (Fig. 3.40b). These deformations are permanent, and in the following phase of cooling the substrate first bends back followed by a bowing in the other direction (Fig. 3.40c). The situation is more complex if large glass carriers are considered. In this case the flash lamp irradiates only a part of the substrate which is moved underneath the lamp bank. A certain area on the glass surface will face several flash pulses, there are both lateral and vertical temperature gradients, and stress forces the wafer to bow upward whereas gravity acts in the opposite direction. As a result double peak structures as shown in Fig. 3.40d may evolve [43].

Using the right process conditions the plastic deformations can be limited to the edges of the glass substrate. In the specific case investigated in [96] the deformations at the edges of the glass substrates were in the range of 1–2 mm. In contrast to the case of Si wafers, short pulses will lead to less warpage than long ones, and there are mainly two reasons for that. At first, plastic deformations need time to evolve. Equation (3.37) gives a measure for the speed a deformation is developing, but the pulse length determines how long these conditions will last. Secondly, at short pulses and considering the low thermal conductivity of glass, only the surface region is heated and meets the conditions for plastic deformations. However, if the major part of the substrate remains stiff, the risk of plastic deformations at the surface is strongly reduced. The dependence of warpage on pulse length is comprehensively discussed in [43].

Thermal stress is also important to thin films as it involves the risk of delamination. Thus, stress management is an issue for various deposition techniques, but quantitative studies about thermal stress in thin film systems during FLA are rare.

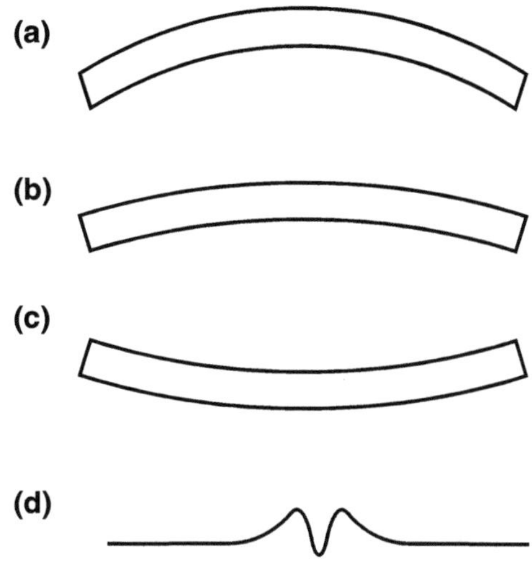

Fig. 3.40 Warpage of a glass substrate during the heating phase and below the softening point (**a**), above the softening point (**b**) and after cooling when permanent deformations force the substrate to bend up (**c**). For large substrates where the flash lamp works in a continuous mode more complex warpage patterns may occur (**d**). After [43]

However, there is a recent publication which discusses this topic by the examples of a thin copper foil and a three-layer system made of a silver and borosilicate glass layer on a PET foil [97].

3.5 Further Effects

Reproducibility is an important performance parameter for most industrial applications. This issue has already been addressed at appropriate positions in the text, for which reason this paragraph is only a short summary. For FLA, reproducibility mainly relates to the reproducibility of subsequent flashes and to drifts during long time operation. Subsequent flashes can differ due to the initial, random arc formation or due to the warming up of the flash lamp during the first flash pulses. If one lamp of a lamp field does not ignite, for whatever reason, this strongly affects reproducibility. One of the advantages of a simmer as discussed in Sect. 2.4.2 is an improved reproducibility because there is no initial arc formation.

To illustrate this, Fig. 3.41 displays the intensity distributions of two series of flash pulses with and without simmer in a box-whisker plot (data points taken from [46]). Each series comprises 1000 flash pulses with a break of two seconds between subsequent pulses. As seen, the distribution with simmer is much narrower than that without simmer: 1% compared to 8% if the whiskers (5–95%) are taken as a measure. In addition, the distribution with simmer is more symmetric. Further effects which affect the reproducibility are incomplete discharges of capacitors (only relevant for small input energies) and the long-term decrease of light intensity due to the deposition of material onto the inner wall of the lamp envelope or electrode aging.

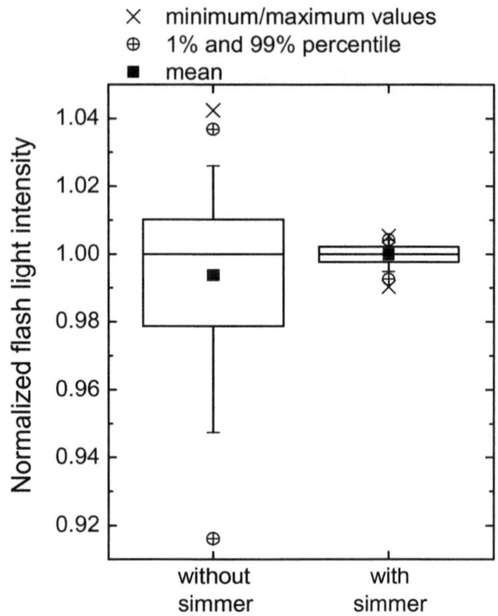

Fig. 3.41 Box-whisker plot of the light intensity of two sets of flash pulses with and without a simmer circuit. The box boundaries and the whiskers mark the 25, 50, and 75% as well as the 5 and 95% percentiles, respectively. Further symbols are explained in the legend. Without simmer, the minimum is about 0.83 (not shown)

Another critical issue is outgassing or evaporation of volatile compounds during the FLA process. Outgassing can lead to a couple of unwanted effects like bubble formation, crack formation, rough surfaces or layer delamination. This especially applies to thin films deposited by methods which use precursors or solvents like plasma-enhanced chemical vapour deposition (PECVD), atomic layer deposition or ink jet printing. In a previous study [98] the hydrogen content in SiON layers deposited by PECVD on a thermal SiO_2 layer on Si was measured by nuclear reaction analysis (NRA) before and after FLA. As shown in Fig. 3.42, the as-deposited hydrogen profile in SiON is characterized by a constant value of 18% before it drops down at the SiON-SiO_2 interface. Applying FLA at 1000 °C for 20 ms or furnace annealing at 1000 °C for 30 min the hydrogen concentration can be lowered to 3 and 0.2%, respectively.

In this specific example no disadvantageous effects were observed, but this may not be always true. In case of thin amorphous Si films which were intended to get crystallized by FLA, e-beam evaporation was superior compared to PECVD due to the different hydrogen concentration [75]. In order to get a good film quality, maximum tolerable hydrogen concentrations of 3 and 5% were estimated for FLA of amorphous Si deposited by catalytic chemical vapour deposition [99] and for LA of amorphous Si deposited by PECVD [100], respectively.

However, hydrogen can also have a beneficial role for photovoltaic applications. Bregolin et al. [101] investigated the evolution of the hydrogen content in crystalline Si after plasma immersion ion implantation using a PH_3/H_2 precursor gas followed by FLA under various conditions. During FLA the major part of hydrogen gasses out, but a part of the remaining hydrogen will be used to terminate dangling bonds,

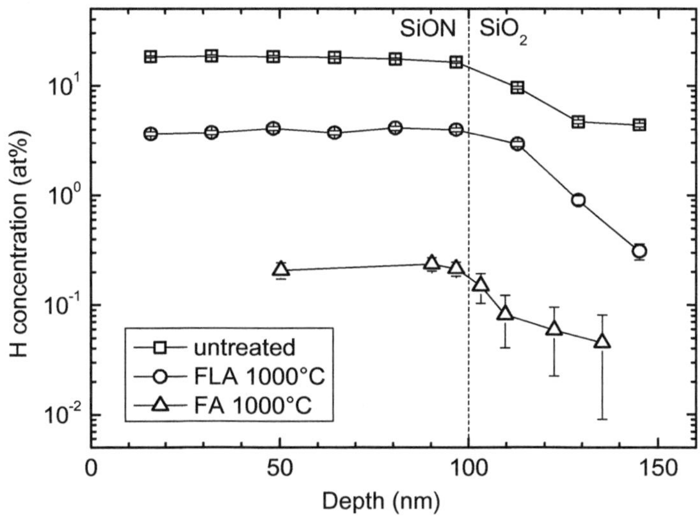

Fig. 3.42 Elemental depth profiles of hydrogen in SiON deposited by PECVD as derived by NRA. To get the depth scale, a SiO_2 density of 2.2 g cm^{-3} and an average energy loss of 1.55 keV/nm had been assumed. After [98]

especially in the surface region. Moreover, as shown in Fig. 3.43 and measured by ERDA, the remaining amount of hydrogen strongly depends on the FLA parameter. At first, the hydrogen concentration is strongly reduced if the pulse length increases from 3 to 20 ms. At second, higher input energies and thus higher peak temperatures will lead to stronger outgassing, although the difference is less than expected. Finally, there is a strong dependence on the annealing gas. In case of Ar the hydrogen loss is dramatic compared to N_2 which is explained by the formation of a very thin SiN layer at the surface which serves as a diffusion barrier. In case of forming gas (95% N_2, 5% H_2) the hydrogen loss is even less as the high partial pressure of hydrogen in the annealing gas additionally counters the outgassing of hydrogen.

There is a recent, comprehensive investigation about crack formation in copper zinc tin sulfide thin films during FLA [102]. In this study the pulse duration and break between subsequent pulses were fixed at 3.5 ms and 2.3 s, respectively, whereas the energy density and the number of flash pulses were varied. In case of single flash operation it was found that no crack formation was observed up to a certain threshold. With increasing energy density or flash number small cracks with a high areal density appear, followed by the development of large cracks and blisters with a low areal density up to a substantial delamination of the film. Typically, the increase of the thermal load will not lead to an increase of the crack density, but to an increase of the crack width. Whereas the formation of high density cracks are assumed to be due to the thermal mismatch between the film and the substrate, low density cracks evolve as a result of material decomposition in the interface region between film and substrate [102]. Increasing temperatures or an increasing number of flashes

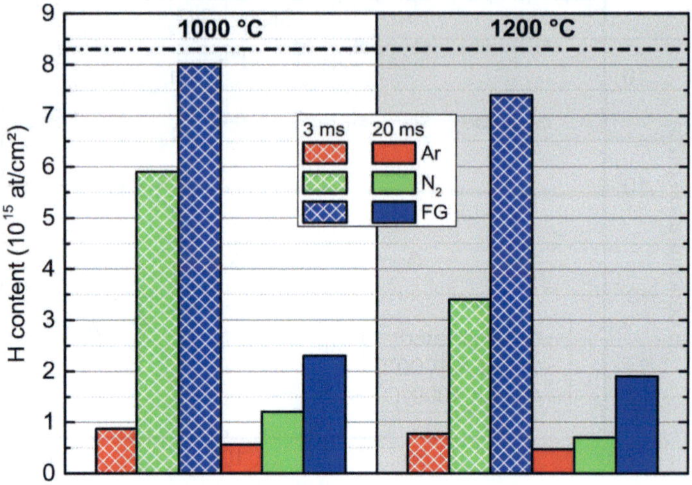

Fig. 3.43 Hydrogen concentration after FLA under different conditions (pulse time, annealing gas) as measured by ERDA. The as-implanted hydrogen concentration is given as a horizontal dashed line. Reprinted from [101], with the permission of AIP Publishing

increases the amount of gaseous species produced in this decomposition. This leads at first to bubble formation which grow and built up more and more stress in the surrounding. Finally, these species will be released in a local burst. To avoid the first crack formation mechanism, buffer layers with a thermal expansion coefficient between that of the film and that of the substrate can be used.

Finally a few words to potential hazards, although this paragraph is far from being a safety instruction. Probably the largest potential cause of accidents is the handling of high voltage, namely the danger that a capacitor has not or not completely discharged. However, high voltage handling is not uncommon in industry which is why the generally applicable safety measures should be sufficient. Being in the direct line of sight to a flash lamp when fired is another unpleasant threat. This scenario is probably uncritical for closed sheet-to-sheet systems, but requires higher attention for R2R tools where a light-proof cover is more difficult. Once again, the standard rules when working with high power lasers should meet the safety requirements. Besides the danger for persons, reflections inside the chamber can guide light to unwanted positions with adverse consequences, e.g. the early aging of rubber or plastic seals. Furthermore, each flash produces a certain amount of ozone which is probably not an immediate threat but harmful under long-term exposure. Thus, a suitable exhaust system is needed. Finally, certain types of samples represent a danger by releasing toxic, corrosive or contaminating species during the flash. Possible measures to counter this risk are, once again, an appropriate exhaust system or the use of transparent, closed protective containers made of quartz, although the latter solution is not very application-oriented. In case of multiple flash operation over long periods noise generation can become significant.

References

1. G. Kirchhoff, Ueber das Verhältniss zwischen dem Emissionsvermögen und dem Absorptionsvermögen der Körper für Wärme und Licht. Ann. Phys. **185**, 275 (1860)
2. T. Sato, Spectral emissivity of silicon. Jap. J. Appl. Phys. **6**, 339 (1967)
3. P.J. Timans, Emissivity of silicon at elevated temperatures. J. Appl. Phys. **74**(10), 6353 (1993)
4. B. Sopori, W. Chen, J. Madjdpour, N.M. Ravindra, Calculation of emissivity of Si wafers. J Elec. Mater. **28**(12), 1385 (1999)
5. D. Reichel, *Temperature Measurement During Millisecond Annealing* (MatWerk by Springer Fachmedien) ed. by F.O.R. Fischer (Springer, Wiesbaden, 2015)
6. M. Levinstein, S. Rumyantsev, M. Shur (eds.), *Handbook Series on Semiconductor Parameters*, vol. 1–2 (World Scientific, London, 1996)
7. D. Meschede (ed.), *Gerthsen Physik* (Springer, Berlin Heidelberg, 2015)
8. S. Adachi (ed.), *The Handbook on Optical Constants of Semiconductors* (World Scientific, Singapore, 2012)
9. R. Illgen, S. Flachowsky, T. Herrmann, T. Feudel, D. Thron, B. Bayha, W. Klix, M. Horstmann, R. Stenzel, A comparative study of non-melt laser spike annealing and flash lamp annealing in terms of transistor performance and pattern effects on SOI-CMOSFETs for the 32 nm Node and Below, in *10th International Conference on Ultimate Integration of Silicon (ULIS), Aachen* (2009), p. 157
10. D. Reichel, W. Skorupa, W. Lerch, J.C. Gelepy, Temperature measurement in rapid thermal processing with focus on the application to flash lamp annealing. Crit. Rev. Solid State Mater. Sci. **36**(2), 102 (2011)
11. D.M. Camm et al., Temperature measuremaent and heat-treating methods, USA Patent US 7616872 B2 (2009)
12. J.A. Curcio, C.C. Petty, The near infrared absorption spectrum of liquid water. J. Opt. Soc. Am. **41**, 302 (1951)
13. H.S. Grover, Modelling and control of a vortex arc lamp for RTP applications, PH.D. thesis, University of Toronto, Graduate Department of Electrical and Computer Engineering, 2014
14. H. Walk, T. Theiler, The wafer temperature measurement in dual OH-band quartz tube, in *2nd International Conference on Advanced Thermal Processing of Semiconductors* (1994), p. 194
15. T. Bauer, *Thermophotovoltaics: Basic Principles and Critical Aspects of System Design* (Springer, Berlin, Heidelberg, 2011)
16. E.B. Salomon, Energy levels and observed spectral lines of xenon, Xe I through Xe LIV. J. Phys. Chem. **33**(3), 765 (2004)
17. A.R. Striganov, N.S. Sventitski, *Tables of spectral lines of neutral and ionized atoms* (Springer Science+Business Media, New York, 1968)
18. J.L. Emmett, A.L. Schawlow, Enhanced ultraviolet output from double-pulsed flash lamps. Appl. Phys. Lett. **2**(11), 204 (1963)
19. M. Smith, R.A. McMahon, M. Voelskow, W. Skorupa, Modeling and regrowth mechanisms of flash lamp processing of SiC-on-silicon heterostructures. J. Appl. Phys. **96**(9), 4843–4851 (2004). https://doi.org/10.1063/1.1786650
20. J. Gelpey, S. McCoys, A. Kontos, L. Godet, C. Hatem, D. Camms, J. Chan, G. Papasouliotis, J. Scheuer, Ultra-shallow junction formation using flash annealing and advanced doping techniques, in *8th International Workshop on Junction Technology (IWJT '08)*, Shanghai, (2008), p. 82
21. H.S. Grover, F.P. Dawson, D.M. Camm, Y. Cressault, M. Lieberer, Application of a plasma arc lamp for thermal processing of semiconductor wafers. IEEE Trans. Ind. Appl. **51**(6), 4808 (2015)
22. Y. Yamada, J. Ishii, In situ silicon-wafer surface-temperature measurements utilizing polarized light. Int. J. Thermophys. **32**(11–12), 2304 (2011)
23. C.W. Schietinger, B.E. Adams, Non contact technique for measuring temperature of radiation-heated objects, USA Patent US 5318362 (1994)

24. D. Reichel, W. Skorupa, Precise millisecond annealing for advanced material processing. Phys. Status Solidi C **9**(10–11), 2045 (2012)
25. T. Kubo, T. Sukegawa, E. Takii, T. Yamamoto, S. Satoh, M. Kase, First quantitative observation of local temperature fluctuation in millisecond annealing, in *15th IEEE International Conference on Advanced Thermal Processing of Semiconductors*, Catania, (2007), p. 321
26. M. Rabus, A.T. Fiory, N.M. Ravindra, P. Frisella, A. Agarwal, T. Sorsch, J. Miner, E. Ferry, F. Klemens, R. Cirelli, W. Mansfield, Rapid thermal processing of silicon wafers with emissivity patterns. J. Elect. Mater. **35**(5), 877 (2006)
27. W. Skorupa, T. Schumann, L. Rebohle, Millisecond thermal processing using flash lamps for the advancement of thin layers and functional coatings. Surf. Coat. Technol. **314**, 169 (2017)
28. Y. Jo, S.-J. Oh, S.S. Lee, Y.-H. Seo, B.-H. Ryu, J. Moon, Y. Choi, S. Yeong, Extremely flexible, printable Ag conductive features on PET and paper substrates via continuous millisecond photonic sintering in a large area. J. Mater. Chem. C **2**(45), 9746 (2014)
29. D.P. Ceperley, A.R. Neureuther, A.H.X. Wang, M. Shen, Y. Wang, Wavelength polarization dependent absorption effects in millisecond annealing of metal gate structures, in *16th International Conference on Advanced Thermal Processing of Semiconductors*, Las Vegas, (2008), p. 211
30. T. Henke, M. Knaut, C. Hossbach, M. Geidel, L. Rebohle, M. Albert, W. Skorupa, J.W. Bartha, Flash-lamp-enhanced atomic layer deposition of thin films. ECS Trans. **64**(9), 167 (2014)
31. E.D. Palik (ed.), *Handbook of Optical Constants of Solids* (Academic Press, San Diego, 1998)
32. S. Babar, J.H. Weaver, Optical constants of Cu, Ag and Au revisited. Appl. Opt. **54**(3), 477 (2015)
33. K.M. McPeak, S.V. Jayanti, S.J.P. Kress, S. Meyer, S. Iotti, A. Rossinelli, D.J. Norris, Plasmonic films can easily be better: Rules and recipes. ACS Photon **2**, 326 (2015)
34. S. Weller, M. Junghähnel, Flash lamp annealing of ITO thin films on ultra-thin glass. Weinh. VIP-J. **27**(4), 29 (2015)
35. A.A. Assi, Conductive polymer dye sensitive solar cell (DSSC) for improving the efficiency, Ph.D. Thesis, University of Baghdad, 2014
36. B. Büchter, F. Seidel, R. Fritzsche, D. Lehmann, D. Bülz, R. Buschbeck, A. Jakob, S. Schulze, H. Freitag, H. Lang, M. Hietschold, D.R.T. Zahn, M. Mehring, Polycrystalline silicon foils by flash lamp annealing of spray-coated silicon nanoparticle dispersions. J. Mater. Sci. **50**, 6050 (2015)
37. B.K. Sun, X. Zhang, C.P. Grigoropoulos, Spectral optical functions of silicon in the range of 1.13–4.96 eV at elevated temperatures. Int. J. Heat Mass Tran. **40**(7), 1591 (1997)
38. M.A. Green, M. Keevers, Optical properties of intrinsic silicon at 300K. Progress Photovolt. **3**(3), 189 (1995)
39. M.J. Guillot, S.C. McCool: Effect of boundary condition approximation on convergence and accuracy of a finite volume discretization of the transient heat conduction equation. Int. J. Num. Methods Heat Fluid Flow **25**(4), 950–972 (2015). https://doi.org/10.1108/hff-02-2014-0033
40. S.J. Orfanidis, Electromagnetic waves and antennas (2002). www.ece.rutgers.edu/~orfanidi/ewa/. Revision date 2012. Accessed 18 July 2012
41. M.P. Smith, Modelling of the flash lamp processing of semiconductor materials. Dissertation, University of Cambridge, 2006
42. D.-H. Kim, B.-K. Kim, H.J. Kim, S. Park, Crystallization of amorphous silicon thin-film on glass substrate preheated at 650 °C using Xe arc flash of 400 μs. Thin Solid Films **520**(21), 6581–6588 (2012). https://doi.org/10.1016/j.tsf.2012.07.006
43. W.B. Jin, Y. Park, B.K. Kim, H.J. Kim, J.H. Hwang, H. Chung, J.H. Park, D.H. Kim, S. Park, Thermal warpage of a glass substrate during Xe-arc flash lamp crystallization of amorphous silicon thin film. Int. J. Thermal Sci. **83**, 25–32 (2014)
44. G.G. Bentini, L. Correra, Analysis of thermal stresses induced during xenon arc lamp flash annealing. J. Appl. Phys. **54**, 2057–2062 (1983)
45. D. Jennings et al., Dynamic surface anneal: activation without diffusion, in *12th IEEE International Conference on Advanced Thennal Processing of Semiconductors* (2004), p. 47

46. T. Gebel, L. Rebohle, R. Fendler, W. Hentsch, W. Skorupa, M. Voelskow, W. Anwand, R.A. Yankov, Millisecond annealing with flash lamps: tool and process challenges, in *14th International Conference on Advanced Thermal Processing of Semiconductors* (2006), pp. 47–55
47. G. Kissinger, D. Kot, M.A. Schubert, A. Sattler, Dislocation generation and propagation during flash lamp annealing. ECS J. Solid State Sci. Technol. **4**(7), P195–P199 (2015). https://doi.org/10.1149/2.0151507jss
48. H. Habuka, A. Hara, T. Karasawa, M. Yoshioka, Heat transport analysis for flash lamp annealing. Jpn. J. Appl. Phys. **46**(3A), 937–942 (2007). https://doi.org/10.1143/jjap.46.937
49. J.-W. Choi et al., Rapid activation of phosphorous-implanted polycrystalline si thin films on glass substrates using flash-lamp annealing. ECS J. Solid State Sci. Technol. **3**(11), P391–P395 (2014)
50. D. Bürger et al., Application of pulsed laser annealing to ferromagnetic GaMnAs. Phys. Rev. B **81**(11) (2010). https://doi.org/10.1103/physrevb.81.115202
51. A.G. Cullis, Transient annealing of semiconductors by laser, electron beam and radiant heating techniques. Rep. Prog. Phys. **48**, 1155–1233 (1985)
52. J.C. Conde, E. Martín, S. Chiussi, F. Gontad, C. Serra, P. González, Finite element simulation for ultraviolet excimer laser processing of patterned Si/SiGe/Si(100) heterostructures. Appl. Phys. Lett. **97**, 014102 (2010). https://doi.org/10.1063/1.3452341
53. K. Schroder, B. Wenz, D. Pope, I. Rawson, The photonic curing process for printed electronics with applications to printed RFID tags and thin film transistors, in *Digital Fabrication* (2012), p. 440
54. J.G. Edwards, Some factors affecting the pumping efficiency of optically pumped lasers. Appl. Optics **6**(5), 837 (1967)
55. Timans, Rapid thermal processing, in *Handbook of Semiconductor Manufacturing Technology*, 2nd edn., ed. by R. Doering, Y. Nishi (CRC Press, Boca Raton, FL, 2008), p. 11–1
56. L. Rebohle, M. Neubert, T. Schumann, W. Skorupa, Determination of the thermal cycle during flash lamp annealing without a direct temperature measurement. Int. J. Heat Mass Transfer **126**, 1–8 (2018). https://doi.org/10.1016/j.ijheatmasstransfer.2018.05.119
57. Y. Kim, S. Park, B.-K. Kim, H.J. Kim, J.-H. Hwang, Xe-arc flash annealing of indium tin oxide thin-films prepared on glass backplanes. Int. J. Heat Mass Transfer **91**, 543 (2015)
58. L. Rebohle, S. Prucnal, W. Skorupa, A review of thermal processing in the subsecond range: semiconductors and beyond. Semicond. Sci. Technol. **31**(10), 103001 (2016). https://doi.org/10.1088/0268-1242/31/10/103001
59. S.M. Jung, C.J. Park, H. Jeong, M.W. Shin, Effect of number of laser pulses on p+/n silicon ultra-shallow junction formation during non-melt ultra-violet laser thermal annealing. Mater. Sci. Semicond. Process. **60**, 34–39 (2017). https://doi.org/10.1016/j.mssp.2016.12.014
60. F. Terai, S. Matunaka, A. Tauchi, C. Ichimura, T. Nagatomo, T. Homma, Xenon flash lamp annealing of poly-si thin films. J. Electrochem. Soc. **153**(7), H147 (2006). https://doi.org/10.1149/1.2200291
61. A. Usami, Y. Tokuda, M. Katayama, S. Kaneshima, T. Wada, Studies of deep-level defects in flash lamp annealing of ion-implanted silicon. J. Phys. D Appl. Phys. **19**, 1079–1089 (1986)
62. W. Skorupa et al., Advanced thermal processing of semiconductor materials in the msec-range, in *13th IEEE International Conference on Advanced Thermal Processing of Semiconductors* (2005), p. 53. https://doi.org/10.1109/rtp.2005.1613684
63. O. Storbeck, D. Ganz, M. Stadtmuller, S. Frigge, W. Lerch, D. Graef, G. Obermeier, Influence of RTP flash anneal ramp rates on lithography overlay performance on 300 mm integrated wafers, in *13th International Conference Ion Implantation Technology Proceedings* (2000), p. 159
64. M. Braun, Elektrische und optische Eigenschaften von dotierten ZnO Dünnfilmschichten nach Temperung mit spektral modifizierten Blitzlampen, Master Thesis, Westsächsische Hochschule Zwickau, 2016 (in German)
65. K. Ohdaira, T. Fujiwara, Y. Endo, S. Nishizaki, H. Matsumura, Explosive crystallization of amorphous silicon films by flash lamp annealing. J. Appl. Phys. **106**(4), 044907 (2009). https://doi.org/10.1063/1.3195089

66. P. Timans, Z. Nenyei, R. Berger, Device scaling drives pattern effect solutions. Solid State Technol. **45**, 67 (2002)
67. P. Timans, J. Gelpey, S. McCoy, W. Lerch, S. Paul, Millisecond annealing, past, present and future. Mater. Res. Soc. Symp. Proc. **912**, 3 (2006)
68. K. Adachi et al., Issues and optimization of millisecond anneal process for 45 nm node and beyond, in *Symposium on VLSI Technology Digest of Technical Papers* (2005), p. 142
69. T. Ito et al., Minimization of Pattern Dependence by Optimized Flash Lamp Annealing, in *Extended Abstracts of the 5th International Workshop on Junction Technology* (2005), pp. S4–3
70. T. Miyashita et al., Study on millisecond annealing (MSA) induced layout dependence for flash lamp annealing (FLA) and laser spike annealing (LSA) in multiple MSA scheme with 45 nm high-performance technology, in *IEEE International Electron Devices Meeting* (2009), p. 27. https://doi.org/10.1109/iedm.2009.5424429
71. T. Kubo, T. Sukegawa, and M. Kase, Total temperature fluctuation of a patterned wafer in the millisecond annealing, in *16th IEEE International Conference on Advanced Thermal Processing of Seminconductors* (2008), p. 195
72. T. Sakoda et al., Characterization of millisecond-anneal-induced defects in SiON and SiON/Si interface by gate current fluctuation measurement, in *IEEE International Reliability Physics Symposium* (2010), p. 379. https://doi.org/10.1109/irps.2010.5488801
73. T. Matsuki et al., Area-selective post-deposition annealing process using flash lamp and Si photoenergy absorber for metal/high-k gate metal–insulator–semiconductor field-effect transistors with NiSi source/drain. Jpn. J. Appl. Phys. **45**(4B), 2939–2944 (2006). https://doi.org/10.1143/jjap.45.2939
74. T. Onizawa, S. Kato, T. Aoyama, Y. Nara, Y. Ohji, A proposal of new concept milli-second annealing: Flexibly-shaped-pulse flash lamp annealing (FSP-FLA) for fabrication of ultra-shallow junction with improvement of metal gate high-k CMOS performance, in *Symposium on VLSI Technology Digest of Technical Papers IEEE-2008* (2008), pp. 110–11
75. T. Henke, J.W. Bartha, L. Rebohle, U. Merkel, R. Hübner, M. Albert, W. Skorupa, Formation of regularly arranged large grain silicon islands by using embedded micro mirrors in the flash crystallization of amorphous silicon. J. Appl. Phys. **115**(3), 034301 (2014). https://doi.org/10.1063/1.4861398
76. A. Shima, R. Tsuchiya, T. Mine, K. Torii, Effects of thin film interference on junction activation during sub-millisecond annealing. Jpn. J. Appl. Phys. **46**(18), L427–L429 (2007). https://doi.org/10.1143/jjap.46.l427
77. T. Sanuki et al., High-performance 45 nm node CMOS transistors featuring flash lamp annealing (FLA), in *IEEE International Electron Devices Meeting* (2007), p. 281
78. R. Klabes, J. Matthäi, M. Voelskow, G.A. Kachurin, E.V. Nidaev, H. Bartsch, Flash lamp annealing of arsenic implanted silicon. Phys. Stat. Sol. (a) **66**, 261 (1981)
79. S. Prucnal et al., Ultra-doped n-type germanium thin films for sensing in the mid-infrared. Scientific Reports **6**, 27643 (2016). https://doi.org/10.1038/srep27643
80. P.R. Chidambaram, C. Bowen, S. Chakravarthi, C. Machala, R. Wise, Fundamentals of silicon material properties for successful exploitation of strain engineering in modern CMOS manufacturing. IEEE Trans. Electr. Dev. **53**(5), 944 (2006)
81. S.M. Hu, Stress-related problems in silicon technology. J. Appl. Phys. **70**(6), R53–R80 (1991). https://doi.org/10.1063/1.349282
82. J. Rabier, J.L. Demenet, Low temperature, high stress plastic deformation of semiconductors: the silicon case. Phys. Stat. Sol. (b) **222**, 63 (2000)
83. A. Fischer, H. Richter, W. Kürner, P. Kücher, Slip-free processing of 300 mm silicon batch wafers. J. Appl. Phys **87**(3), 1543–1549 (2000). https://doi.org/10.1063/1.372047
84. K.H. Heinig, K. Hohmuth, R. Klabes, M. Voelskow, H. Woittennek, Flash lamp annealing of ion implanted silicon. Radiat. Eff **63**(1–4), 115–123 (1982). https://doi.org/10.1080/00337578208222831
85. R. Gross, A. Marx, *Festkörperphysik* (Oldenbourg Wissenschaftsverlag, München, 2012) (in German)

86. M.P. Smith, K.A Seffen, R.A. McMahon, M. Voelskow, W. Skorupa, Analysis of wafer stresses during millisecond thermal processing. J. Appl. Phys. **100**, 063515 (2006). https://doi.org/10.1063/1.2337773
87. T. Ito et al., Low-resistance ultrashallow extension formed by optimized flash lamp annealing. IEEE Trans. Semicond. Manuf. **16**(3), 417 (2003)
88. K. Ohdaira, Y. Endo, T. Fujiwara, S. Nishizaki, H. Matsumura, Formation of highly uniform micrometer-order-thick polycrystalline silicon films by flash lamp annealing of amorphous silicon on glass substrates. Jpn. J. Appl. Phys. **46**(12), 7603–7606 (2007). https://doi.org/10.1143/jjap.46.7603
89. J. Foggiato, W.S. Yoo, Implementation of flash technology for ultra shallow junction formation: challenges in process integration. J. Vac. Sci. Technol. B **24**(1), 515 (2006). https://doi.org/10.1116/1.2151903
90. S. Govindaraju, C.-L. Shih, P. Ramanarayanan, Y.-H. Lin, K. Knutson, Advanced (Millisecond) annealing in silicon based semiconductor manufacturing. ECS Trans. **28**(1), 81–90 (2010)
91. G. Kissinger, D. Kot, W. von Ammon, Comparison of the impact of thermal treatments on the second and on the millisecond scales on the precipitation of interstitial oxygen. ECS J. Solid State Sci. Technol. **1**(6), P269–P275 (2012). https://doi.org/10.1149/2.008206jss
92. K.R.C. Mok, S.H. Yeong, B. Colombeau, F. Benistant, C.H. Poon, L. Chan, M.P. Srinivasan, Experimental and simulation study of the flash lamp annealing for boron ultra-shallow junction formation and its stability. Mat. Sci. Eng. B **154–155**, 14–19 (2008)
93. B. Pécz, L. Dobos, D. Panknin, W. Skorupa, C. Lioutas, N. Vouroutzis, Crystallization of amorphous-Si films by flash lamp annealing. Appl. Surf. Sci. **242**(1–2), 185–191 (2005). https://doi.org/10.1016/j.apsusc.2004.08.015
94. M. Smith, R. McMahon, M. Voelskow, D. Panknin, W. Skorupa, Modelling of flash-lamp-induced crystallization of amorphous silicon thin films on glass. J. Cryst. Growth **285**(1–2), 249–260 (2005). https://doi.org/10.1016/j.jcrysgro.2005.08.033
95. D.-H. Kim, W.-E. Hong, J.S. Ro, S.H. Lee, S. Park, Thermal deformation of glass backplanes during Joule-heating induced crystallization process. Vacuum **85**, 847–852 (2011)
96. J.-H. Hwang, H.J. Kim, B.-K. Kim, W.-B. Jin, Y. Kim, H. Chung, S. Park, Scanning multishot irradiations on a large-area glass substrate for Xe-arc flash lamp crystallization of amorphous silicon thin-film. Int. J. Thermal Sci. **91**, 1–11 (2015)
97. M.J. Guillot, Modeling thermal stress in thin films produced by photonic curing. Thin Solid Films **594**, 156–161 (2015). https://doi.org/10.1016/j.tsf.2015.10.012
98. L. Rebohle et al., Anomalous wear-out phenomena of europium-implanted light emitters based on a metal-oxide-semiconductor structure. J. Appl. Phys. **106**(12), 123103 (2009). https://doi.org/10.1063/1.3272781
99. K. Ohdaira, S. Nishizaki, Y. Endo, T. Fujiwara, N. Usami, K. Nakajima, H. Matsumura, High-quality polycrystalline silicon films with minority carrier lifetimes over 5 μs formed by flash lamp annealing of precursor amorphous silicon films prepared by catalytic chemical vapor deposition. Jpn. J. Appl. Phys. **46**(11), 7198–7203 (2007). https://doi.org/10.1143/jjap.46.7198
100. D. Toet, P.M. Smith, T.W. Sigmon, T. Takehara, C.C. Tsai, W.R. Harshbarger, M.O. Thompson, Laser crystallization and structural characterization of hydrogenated amorphous silicon thin films. J. Appl. Phys. **85**(11), 7914–7918 (1999). https://doi.org/10.1063/1.370607
101. F.L. Bregolin et al., Hydrogen engineering via plasma immersion ion implantation and flash lamp annealing in silicon-based solar cell substrates. J. Appl. Phys. **115**(6), 064505 (2014). https://doi.org/10.1063/1.4865737
102. B.A. Williams, M.A. Smeaton, C.S. Holgate, N.D. Trejo, L.F. Francis, E.S. Aydil, Intense pulsed light annealing of copper zinc tin sulfide nanocrystal coatings. J. Vac. Sci. Technol. A **34**(5), 051204 (2016). https://doi.org/10.1116/1.4961661

Chapter 4
Semiconductor Applications

Abstract This chapter gives an overview where flash lamp annealing is used in semiconductor applications. After a short introduction to defect engineering several use cases in the field of doping are discussed including ultra-shallow junctions and hyperdoping in silicon, doping and superconductivity in germanium, silicon carbide, III–V semiconductors, and diluted magnetic semiconductors. The following sections deals with the crystallization of semiconductors with the focus on thin amorphous silicon films and silicon compound semiconductors. The chapter closes with a section about flash lamp annealing of semiconductor nanostructures, namely group IV nanoclusters, III–V nanocrystals in Si and semiconductor nanowires.

4.1 Defect Engineering

Defect engineering is a very important issue in microelectronics as defects strongly influence the characteristics of potential devices. Defects can introduce unwanted energy levels in the semiconductor bandgap and lead to an increased leakage current and thus an increased power consumption. Defects are especially important for the doping of semiconductors as they have a significant effect on diffusion and activation of dopants. This section should give a short overview about the most relevant defects, their development during annealing, and their interaction with dopant atoms. The overview is focused on Si as the best known and dominating material in microelectronics, but many issues can be applied to other semiconductors as well. There are several reviews in literature [1–4], whereby the next sections follow the explanations given in [5] and references therein.

4.1.1 Defect Evolution During Annealing

Up to now, ion implantation is the most common doping technique as it gives an excellent control over the amount and distribution of dopants, especially in the context of ultra-shallow junction (USJ, see Sect. 4.2.1). If an ion penetrates a target material

L. Rebohle et al., *Flash Lamp Annealing*, Springer Series in Materials Science 288,
https://doi.org/10.1007/978-3-030-23299-3_4

it loses energy by electronic and nuclear processes causing a corresponding material damage. For typical implantation energies below 200 keV the nuclear stopping power and thus the damage is larger than the electronic one, except for light ions. For B as a very light dopant, the nuclear stopping power for 200 keV in crystalline Si is ca. 5% of that of the electronic one, but increases with decreasing implantation energy [6]. The nuclear damage is caused by nuclear collisions leading to the displacement of target atoms from their lattice sites and thus to the formation of Frenkel pairs, i.e. an interstitial and a vacancy. If the transmitted energy during the collision is high enough the displaced target atom can cause further displacements. Thus, the implanted ion can cause a collision cascade and the creation of several Frenkel pairs. Most of these Frenkel pairs quickly recombine already during the implantation or at the onset of annealing at the latest, leaving approximately one interstitial per cascade behind which cannot be annealed out immediately [7]. Because of the forward momentum of the ions, regions close to the surface are characterized by an excess of vacancies, whereas deeper regions are dominated by interstitials [8].

In many cases the implantation fluence is high enough to destroy the crystal lattice completely and to create an amorphous layer. Immediately behind this layer, in the tail of the implantation profile, there is a crystalline, but heavily damaged zone. The defects of this zone are called end-of-range (EOR) defects, and they are mainly composed of interstitials. The amorphous layer recrystallizes during annealing, and in the ideal case the regrown layer is nearly free of defects, except dopant atoms at a substitutional or interstitial position. However, the EOR defects undergo a defect development and are annealed out at high thermal budgets only. In Si, this development is as follows [5, 9]: Si interstitials form small interstitial clusters which evolve into large interstitial clusters, followed by {113} defects and finally dislocation loops. This development already starts during implantation, and is favoured by higher implantation fluences and higher thermal budgets.

Si interstitials agglomerate very quickly into small clusters with few atoms, whereby the formation energy oscillates between "magic" and "non-magic" numbers [10]. Clusters with a high formation energy delay the development, but if once passed, the clusters grow via Ostwald ripening up to a size of 100 atoms and more. At this size, {113} defects may start to evolve. {113} defects are made of long chains of Si interstitials along the $\langle 110 \rangle$ direction which are arranged in the {113} plane [5]. {113} defects grow by increasing their chain lengths [11] and finally evolve into dislocation loops. Based on this some kind of defect phase diagram in Si can be constructed [5, 9, 12] which sketches the behaviour of Si interstitial defects as a function of implantation fluence and thermal budget. As seen in Fig. 4.1, for very small fluences only small interstitial clusters are formed which can be easily annealed out. With increasing fluence larger clusters and {113} defects arise already during ion implantation, and for high fluences {113} defects and dislocation loops are dominating. Generally, the larger defects are thermally more stable and need higher thermal budgets to get eliminated by annealing. If this model is applied to flash lamp annealing (FLA) and rapid thermal annealing (RTA) as the main annealing techniques for doping, RTA is normally eliminating EOR defects, whereas the thermal budget of FLA is usually not sufficient for this. The defect development was confirmed for the

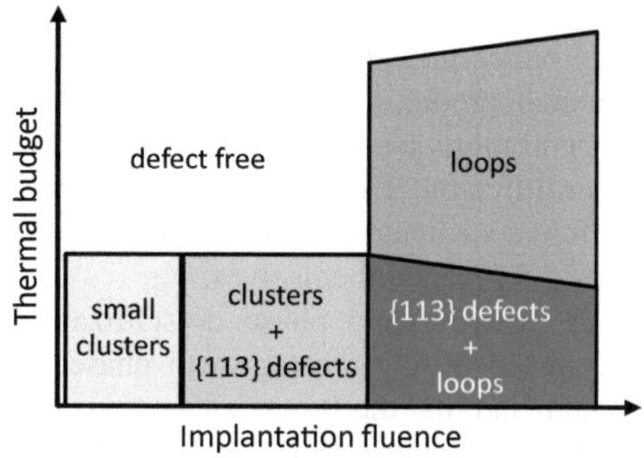

Fig. 4.1 Schematic model of the evolution of implantation defects based on Si interstitials. After [5]

case of FLA in [13]. The authors applied several flashes and showed that with increasing number of flashes the concentration of EOR defects decreases and that their type changes from small clusters and {113} defects to dislocation loops. This development of EOR defects was also found in [14]. However, both annealing techniques have to pay a price: remaining EOR defects increase the leakage current in case of FLA, and RTA may suffer from lower activation levels and junction broadening.

All the defect transformations mentioned above are mainly governed by the emission and capture of Si interstitials and thus on their diffusion properties. At least 4 different interstitial configurations have been identified in literature, and there are several potential diffusion mechanisms [15]. As shown, all configurations simultaneously exist with different concentrations, and diffusion occurs through permanent transformations between the different interstitial types. However, a significant part of the emitted Si interstitials is not used for the growth of larger defects but diffuse towards the surface where the defects annihilate. As a consequence the defects may grow in size during annealing, but the defect density decreases in a disproportionately high manner. This leads to a transient supersaturation of Si interstitials, i.e. their concentration is far above that of thermal equilibrium. The effect is a transient one because it is limited by the reservoir of the EOR defects and occurs only for a short time. According to the model of Rafferty [16], small and unstable defects cause a high degree of supersaturation for a short time, whereas supersaturation for larger defects is smaller but lasts for a longer time.

4.1.2 Boron Diffusion in Si

Diffusion phenomena in semiconductors are very diverse, for which reason this small section is limited to B diffusion in Si as a very relevant and probably the best investigated case. If one considers B diffusion or redistribution over the whole annealing process, it can be roughly divided into B diffusion in a-Si, B redistribution during the regrowth and B diffusion in c-Si. B diffusion in a-Si was found to be transient and concentration dependent, and thus not obeying Fick's law of diffusion.

According to the model proposed in [17] diffusion is governed by the interplay of threefold coordinated Si atoms (dangling bonds) and fivefold coordinated Si atoms. The main prediction, in accordance with experimental observations, is that higher B concentrations lead to a higher concentration of dangling bonds which promotes a faster diffusion. The effect is transient as with the onset of annealing threefold and fivefold coordinated Si atoms start to annihilate, and threefold coordinated Si atoms start to diffuse by themselves.

In case of solid phase crystallization (SPC) the B profile remains roughly unchanged during the regrowth phase, provided that the regrowth phase is short enough and that the B concentration is not above the solubility limit. In case of liquid phase crystallization (LPC) the diffusion constant of B in the liquid phase with a value of 2.4×10^{-4} cm^2 s^{-1} [18] is so large that B will have an approximately constant distribution over the thickness of the liquid layer. If the concentration of B is above the solubility limit, it is a matter of annealing time if the excess B can be incorporated or not. If the time scale of annealing is short enough as in case of FLA and laser annealing (LA), B can be incorporated in concentrations above the solubility limit. If not, excess B is pushed in front of the expanding crystalline-amorphous interface during recrystallization which is usually called the snow-plug effect. However, dopant concentrations above the solubility limit are metastable in a thermodynamic sense, which is why they are sensible to additional annealing steps.

In case of c-Si there is a close interaction between B and Si interstitials making the diffusion dynamics of B quite complex. At first, it has been found that B diffusion in Si is largely mediated by interstitials, known as transient enhanced diffusion [19]. There was a longer discussion about the dominating interstitial-mediated mechanism, but it seems that diffusion in intrinsic and p-type Si is due to mobile, neutral B-interstitial pairs B-I^0, whereas under high n-type doping B diffuses via negatively charged B-I^- pairs [4]. Thus, the extent of transient enhanced diffusion strongly depends on the degree of Si interstitial supersaturation (see previous section), which in turn depends on ion implantation and annealing parameters. As during annealing interstitials diffuse from the EOR defects towards the surface, transient enhanced diffusion of B affects the B concentration in this region and can lead to a significant broadening of the B profile. However, there is a contrary process which temporarily immobilizes and deactivates B: the formation of boron interstitial clusters. Their formation becomes dominant for B concentrations above a threshold value of about 10% of the solubility limit [20, 21]. This may lead to a situation where the peak concentration of a B profile remains unchanged during annealing (but with poor activation), whereas the falling edge of the profile undergoes transient enhanced diffusion. FLA and LA can minimize these effects by their short time scale and due to the fact that EOR defects are not dissolved and thus emit only low amounts of interstitials. But once again, the result is vulnerable to additional post-anneals which dissolves EOR defects and release a corresponding amount of interstitials.

4.1.3 Studies Related to Flash Lamp Annealing

There is only a small number of papers which focus on the creation, the development or the disappearance of point defects in Si during FLA. This topic is usually treated on the side with recrystallization mechanisms or dopant activation, or if point defects are responsible for undesired device trends like the increase of leakage currents. However, there are few reports from the 80s which already investigated the behaviour of point defects by selected electrical characterization techniques, namely *C-V* measurements [22] and deep-level transient spectroscopy (DLTS) [23]. Lysenko et al. [22] investigated the impact of FLA with a pulse length of 250 μs on metal-oxide-semiconductor structures made of Al, SiO_2 and Si and partly implanted with boron. The analysis of the shift of the flat band voltage revealed that the amount of fixed oxide charges generally reduces during FLA, albeit to a larger extent for the implanted case. DLTS measurements were used to trace the signal of the single and twice negatively charged divacancy as well as that of the vacancy-phosphorous pair [23]. Whereas the divacancy is typical for the structures annealed by FLA, the vacancy-phosphorous pair seemed to be favoured by furnace annealing (FA). Another DLTS study traced the energetic level and the depth profile of hole traps in P implanted Si which evolve during FLA and disappear after multiple FLA or other post-FLA annealing steps [24]. Finally, a more recent study deals with the formation of FLA-induced defects in SiON or at the SiON/Si interface and with the resulting impact on potential static random-access memory (SRAM) devices [25].

More recently, a detailed study about the behaviour of self-interstitials and vacancies in an unimplanted Si wafer during FLA and their impact on the precipitation of interstitial oxygen was published [26]. According to this, RTA leads to a dominance of vacancies over the whole depths due to their higher solubility compared to self-interstitials and the long annealing time being sufficient to balance their concentrations. Vacancies promote activation, but also the precipitation of oxygen which is an undesired process leading to higher leakage currents in power devices. However, the diffusion constant of vacancies is orders of magnitudes larger than that of self-interstitials, which is why vacancies can easily cross the wafer within ms during FLA whereas self-interstitials cannot. As a consequence, the upper part of the wafer features an excess of self-interstitials which effectively suppresses oxygen precipitation [26]. The situation becomes more complex if FLA is combined with ion implantation. Due to the forward momentum of the ions target atoms are preferably kicked towards the bulk creating a vacancy-rich region close to the surface as supposed in [8]. In contrast, the EOR region is rich on interstitials as mentioned in the previous sections. Interstitials and vacancies also play an important role in the crystallization process. In recent study about the crystallization of thin amorphous Si films it was found that vacancies and interstitials plays the dominant role in case of SPC and LPC during FLA [27]. In the solid phase crystallization mainly proceed via the site exchange of atoms with vacancies, whereas the kick-out mechanism of interstitials is dominant in case of LPC. In contrast to this, vacancies are dominant for both SPC and LPC in case of thin a-Ge films.

4.2 Doping

4.2.1 Ultra-Shallow Junctions in Silicon

For decades, the development of microelectronics is governed by the continuous reduction of device sizes and characteristic lengths in all dimensions. This also applies to dopant profiles and p-n junctions, and especially for source/drain contacts of metal-oxide-semiconductor field-effect transistor (MOSFETs). These so-called USJ require very shallow source/drain extensions at very high activation levels [28]. To do so, diffusion must be suppressed as much as possible, and the first way to do this is the reduction of annealing time at unwaveringly high temperature. In fact, the fabrication of USJ was one main driver for the development of short-time annealing which started in the 80s with RTA for USJ and changed after the turn of the millennium to FLA or a combination of both. More recent investigations which explore the possibilities of alternative annealing concepts also comprise LA [29, 30] and microwave annealing [31, 32], each one with its own specific advantages and disadvantages.

4.2.1.1 Sheet Resistance Versus Junction Depth

The aim for USJ is to achieve a sheet resistance R_S as low as possible and a junction depth x_J as small as possible, which is why the R_S–x_J plot has advanced to a figure of merit for USJ. Figure 4.2 displays such a plot with selected literature values comparing USJ annealed by RTA or spike RTA with those annealed by FLA or a combination of RTA and FLA. In all cases the dopant was introduced by ion implantation with one exception: in [33] an As-doped sol-gel coating was used as a dopant source.

Best values should be located in the lower left corner of the plot, but low R_S and small x_J are opposed demands. In fact, a constant specific resistance R_S is inversely proportional to x_J as depicted by the solid black line in Fig. 4.2. In order to compare different R_S-x_J combinations, the product $R_S \cdot x_J$ is sometimes used as a performance benchmark [28]. In addition, the pn-junction should be as abrupt as possible, which is quantified as abruptness, given in nm/dec. This value describes the distance in which the doping level (and thus majority carrier concentration) decreases by a factor of 10. Without doubt, this value should also be as low as possible. Similar to R_S, a plot abruptness versus junction depth can serve as a figure of merit, although it appears less frequently in literature.

In the very most cases the dopants are introduced by ion implantation, and only very recently monolayer doping (MLD) has emerged as a new doping technique (Sect. 4.2.1.3). Ion implantation determines, beside the type of dopant, the number and the profile of the dopants and thus the lowest R_S and x_J values which can be achieved. In fact, the variation in Fig. 4.2 within one annealing method is partly caused by different implantation and pre-amorphization conditions. In a very rough

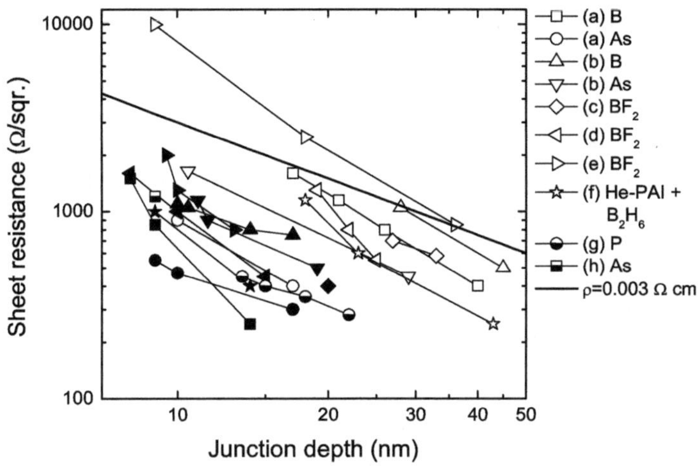

Fig. 4.2 R_S-x_J plot with selected literature values comparing USJ annealed by RTA or spike RTA (open symbols) with those annealed by FLA or a combination of RTA and FLA (solid and semi-solid symbols). The solid line is the theoretical dependence of a material with a constant specific resistivity of 0.003 Ω cm. Literature data for both FLA and RTA are from (a) [34], (b) [35], (c) [36], (d) [37], (e) [38], (f) [39], (g) [40], (h) [33]

(!) approximation someone can assume that for otherwise constant process conditions different temperatures shift the position in the R_S-x_J plot along the hyperbola of constant specific resistance (low R_S, but high x_J and vice versa), whereas suitable annealing times and implantation conditions can shift this hyperbola more close to the lower left corner. Please consider that this approximation visualizes only a tendency as many complex processes influence the final values of R_S and x_J: diffusion, activation, segregation, recrystallization, defect concentration and the reaction dynamics between defects and dopants.

If dopants are implanted in a crystalline material, a part of them are scattered into preferred crystal directions, called channelling. If once moving along such a crystal axis, the interaction cross section is reduced leading to an extended range of these ions. As a consequence, the profile broadens and the abruptness becomes worse. To avoid this, a pre-amorphization implant, mostly with Ge, is performed which eliminates channelling. Further alternatives to traditional ion beam implantation include molecular implantation, co-implantation of other elements and plasma-immersion ion implantation. The use of molecular species for implantation has several advantages, and various B_xH_y [41–43] or BF_x species [28, 30, 32, 34, 36, 44] were used to implant boron with a reduced projected range of the ions. Molecular species allow a higher degree of process control for implantation energies significantly lower than 1 keV and lead to amorphization already in the initial state of implantation [43]. Mizuno et al. [45] compared a Ge pre-amorphization implant followed by BF_2 implantation with plasma doping of B_2H_6 with He for pre-amorphization and found that the latter method achieved a better abruptness (1.5 nm/dec vs. 2 nm/dec) and a higher active B concentration. Details about this plasma doping can be found in [42]. In case of BF_x molecules F was found to reduce the diffusion of B in subsequent annealing steps with an optimum F:B ratio of 0.6 [28]. In addition, the combination of B and F can lead to higher solid solubility concentrations [46]. However, under certain conditions the subsequent implantation of B and F can be more advantageous as the implantation of BF_2 [46]. Another example is the co-implantation of In and C leading to higher activation and a reduced diffusion and segregation of In [47]. There are also ongoing speculations about the minimum junction depth one can obtain by ion implantation in general. Polley et al. claim 9 nm [48], Fig. 4.2 contains values around 8 nm, and there are few works with values slightly below.

Annealing has now the task to maximize activation without influencing abruptness and junction depth. Thus, the desired processes during annealing are activation, recrystallization and defect annealing, whereas deactivation, diffusion and segregation are processes to be avoided. There is a large number of publications which demonstrate that the implanted dopant profile broadens with RTA but not with FLA. This is not very surprising when one considers that the diffusion length x_D depends, in the easiest case, exponentially on temperature and square-root-like on time according to:

$$x_{\mathrm{D}} \approx 2\sqrt{Dt} \quad D = D_0 \exp\left(\frac{E_{\mathrm{A}}}{k_{\mathrm{B}}T}\right) \tag{4.1}$$

Figure 4.3 shows the diffusion length of B, P and As in Si as a function of annealing temperature and annealing time for constant D_0. Although the real diffusion length can significantly differ due to the dependence of diffusion on defect concentration, carrier concentration and other parameters, the diffusion length is much shorter for FLA compared to RTA. In polycrystalline Si the problem of too large diffusion lengths becomes even more severe as diffusion is usually enhanced by grain boundaries [46]. To give an example, a USJ depth of 10 nm will not tolerate diffusion lengths above 3 nm [49] which is difficult to achieve with RTA alone if temperatures in excess of 1000 °C are needed. Figure 4.3 also reveals that considerable diffusion and profile broadening can even occur during FLA when temperatures are high enough [50] or several flashes are applied [51]. So at a peak temperature of 1350 °C even a 3 ms flash with a preheating of 650 °C will cause a junction depth broadening by 15 nm for Si implanted with BF_3 [52]. In a similar way a reduction of the B concentration under the source/drain junction interface after each FLA step in the process flow was observed [53].

However, to focus only on diffusion lengths does not consider the complex nature of interactions between dopants, defects and the host matrix. There is especially a close interaction between the defect development as discussed in Sect. 4.1.2 and the final dopant profile. So it was found in multi-flash experiments performed with B implanted Si that in case of pre-amorphized materials the first flash leads to transient enhanced diffusion, whereas during the following flashes or in the crystalline case only ordinary diffusion takes place [51]. Typically, the profile can be divided into a surface region and a bulk region showing mostly less and more diffusion, respectively.

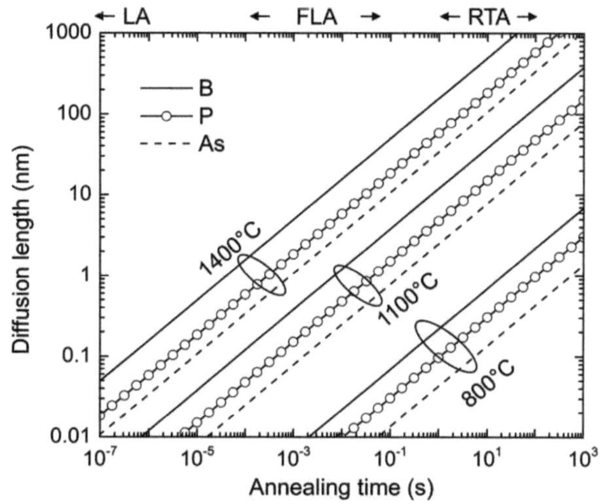

Fig. 4.3 Diffusion lengths of selected dopants (B, P and As) in crystalline Si as a function of annealing temperature and annealing time. The diffusion coefficients were extracted from [54]. Please note that the real diffusion length can additionally depend on many other parameters like defect concentration, carrier concentration, stress distribution, interfaces etc.

B from the surface region tends to diffuse towards the surface and agglomerates there which, under certain conditions, can lead to a significant loss of B [43]. In contrast, B from the bulk region diffuses further into the bulk which often leads to the development of a kink in the profile separating the two regions. It was also argued that the kink is often in a depth where remaining EOR defects are located which tend to trap more B than elsewhere [13]. Due to ion implantation, there is an excess of vacancies in the surface region which is attractive for boron, and consequently, there is considerable boron diffusion towards the surface in this region [8]. Vacancies promote the incorporation of boron atoms into a substitutional site, and if once there, it is much more immobile than interstitial boron. A comparison of different FLA conditions revealed that diffusion is very sensitive to the peak temperature during FLA, but rather insensitive to the pulse length [55]. Moreover, it is completely independent of the implantation temperature. Finally, the annealing ambient can strongly influence the diffusion behaviour, e.g. by introducing oxygen enhanced diffusion or by the formation of thin protective layers reducing a potential loss via the surface [51]. In general, the discussed diffusion behaviour also applies to other dopants like P or As, but probably with slightly different parameters. Thus, Fig. 4.4 displays As profiles after different thermal treatments to illustrate the different diffusion processes.

4.2.1.2 Activation

Activation is probably even more important than the complete suppression of diffusion. In many cases the fraction of activated dopants depends linearly on time and exponentially on the activation energy before it saturates [50]. Thus, annealing is often a compromise between high activation (to get a low R_S) and low diffusion (to get a small x_J). To address both issues, the common strategy of using higher

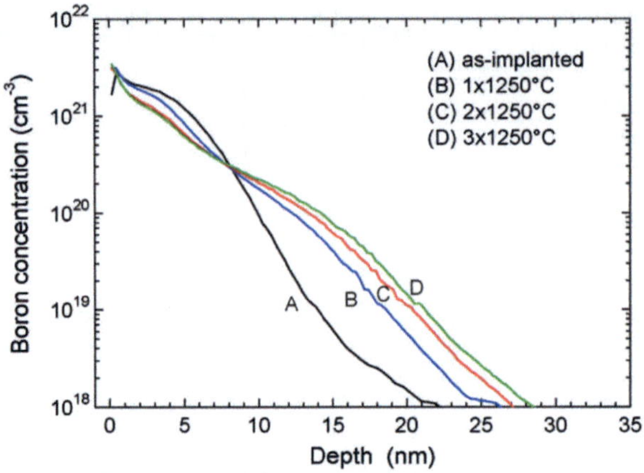

Fig. 4.4 Secondary ion mass spectroscopy profile of as-implanted 1 keV As in crystalline Si (A) as well as after single (B), double (C), and triple (D) flash with a pulse length of 1.6 ms. The annealing was carried out in nitrogen with 100 ppm O_2 and a preheating temperature of 750 °C. The different diffusion behaviour in the surface and bulk region is well visible. Reprinted from [51], Copyright (2008), with permission from Elsevier

temperatures at shorter times has evolved. However, matter is more complex than this. Considering the typical dopant profiles as shown in Fig. 4.4, FLA was found to be superior for activation in the surface region, whereas spike RTA yields better activation values in the bulk region [56]. In case of silicon-on-insulator (SOI) wafers it was also shown that, providing the right anneal parameters, FLA can reach similar activation levels as RTA [8].

However, FLA is also suspected to lead—at least in parts—only to a metastable activation. This statement is based on the observation that the first flash achieves a high activation, but that subsequent flashes or other post-processing annealing steps may lead to serious deactivation [46, 57, 58]. In case of an amorphous layer it is assumed that the dopant is already incorporated into a lattice site during recrystallization, probably in the first part of the pulse, and that the following deactivation is due to the interaction of dopants with defects, e.g. during the dissolution of EOR defects [59] (Sect. 4.1.2). In fact, FLA will not anneal out all defects completely and leaves some EOR defects behind which usually leads to the formation of defect complexes like vacancy-As complexes and Si-As clusters in case of As [51, 60] or B-Si interstitial complexes in case of B [61]. The vulnerability to deactivation was tested by an additional post-annealing RTA step at 800 °C for 60 s [13]. As shown in Fig. 4.5 and traced by Hall and sheet resistance measurements deactivation decreases with increasing number of flashes. There is no deactivation if FLA is replaced by spike RTA plus one flash, but in this case the maximum activation level is also much lower than in case of FLA. The authors explain the deactivation by the formation of inactive B-interstitial clusters which evolve when free interstitials from the EOR region diffuse towards the surface during post-annealing.

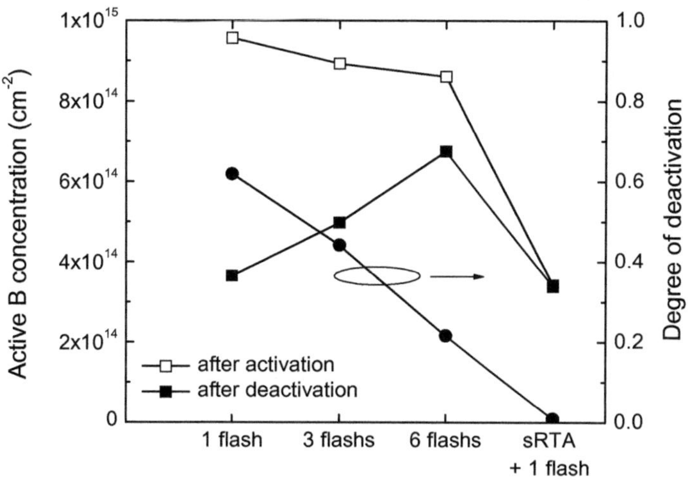

Fig. 4.5 Active B concentration after activation (open squares, left scale) and after deactivation (solid squares, left scale) as well as the resulting degree of deactivation (solid circles, right scale) after tabulated data published in [13]. Please note that concentration is given as an areal density

If the dopant concentration is even above the solubility limit, thermal exposure can easily lead to deactivation by segregation [37]. There is a detailed study about boron diffusion and activation during FLA and RTA in which a model is proposed where clustering or activation depends on the relation between diffusion lengths and the average cluster distance [61]. However, in case of nanostructures like nanowires (NWs) not even RTA can remove all EOR defects caused by implantation [62].

Quite soon it became apparent that the best results will be obtained by the right combination of FLA and RTA, whereas the latter serves as preheating [63]. Thereby, the temperature during RTA must be low enough to be insignificant for diffusion and segregation, but high enough to anneal existing defects and prevent the creation of new ones by FLA. In fact, there is an optimum FLA process window for that [46, 58] as sketched in Fig. 4.6: if the combined thermal budget of RTA and FLA is too low, only a small fraction of dopants will be activated; is it too high, there is the risk of crack or slip formation, most probably caused by thermal stress (Sect. 3.4). If the preheating temperature is too high, recrystallization with a low degree of activation already occurs before the flash, which leads to a reduced R_S and an increased x_J [64]. Finally, a similar optimum parameter range as shown in Fig. 4.6 was observed if FLA is used for Ni silicidation in sub-35 nm MOSFETs [65].

However, the usefulness of FLA is finally determined by the specific impact on device parameters. There is a general concern that FLA increases the leakage current to a probably critical extent. In fact, investigations performed on 45 nm p-MOSFETs with strained SiGe source/drain junction revealed that the leakage current increases by 2–3 orders of magnitude by using FLA [66]. Based on temperature-dependent measurements the authors concluded that the higher effective activation in case of

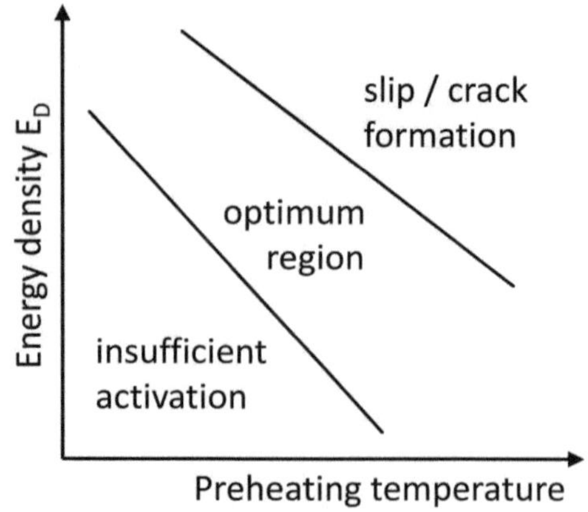

Fig. 4.6 Basic scheme for the optimum FLA process window for the activation of dopants in Si

FLA leads to a reduction of the depletion layer width which makes band-to-band tunnelling more likely, especially at higher reverse biases. In contrast to this, in case of SOI transistors a leakage current increase by only 30% was observed while the sheet resistance decreased by 21% [63]. Illgen et al. found even no difference between FLA and RTA [67]. To summarize these controversial findings, FLA seems to have the risk of causing an increased leakage current, but it can obviously be reduced by using suitable annealing and design parameters.

On the other hand, there are a couple of studies reporting the beneficial role of FLA with regard to important device parameters. So it was found that pMOSFETs annealed with FLA exhibit a better short channel effect behaviour and a better I_{ON}/I_{OFF} ratio than those annealed with spike RTA [38, 39]. In this case FLA is assumed to effectively suppress the increase in the surface As concentration without increasing the leakage current, which resulted in a doubling of the channel mobility. Furthermore, FLA was found to double the drive current compared to spike RTA with no substantial change of the subthreshold slope or the high field mobility [68, 69]. To be precisely, the electron mobility in nFETs decreased in the order of 10–15%, whereas the hole mobility in pFETs showed only insignificant changes for electric fields above 1 MV cm^{-2} [69]. Drive current improvements up to 11% have also been reported if FLA is integrated into the process chain, mainly in combination with spike RTA [70]. It was also reported that the fail bit count of embedded dynamic random-access memory cells are lower if processed by FLA than by RTA [71]. Finally, it should be noted that there is a theoretical study which simulated the dopant distribution in very small volumes with dimensions of a few nm only and in which the number of dopant atoms is very low [72]. They came to the conclusion that the variation of the number of dopant atoms in case of FLA is larger (and thus more critical) than in case of LA.

4.2.1.3 Monolayer Doping

This technique, sometimes called δ-doping, has evolved in the last years as a modern revenant of the old school dopant indiffusion of the 50s. Typically, MLD comprises 3 steps: providing the dopant at the Si surface, incorporation into Si, and diffusion into the bulk. In the first step the dopant or a dopant-containing species is attached or deposited at the Si surface, and the reaction is limited in such a way that only a monolayer or part of a monolayer is provided as a diffusion reservoir. The dopant can be provided by an organic precursor [73, 74] or a reactive gas like PH_3 [48, 75]. The following two steps of MLD require annealing to decompose the precursor and to drive the dopants into Si. To avoid a substantial loss during annealing by evaporation or sublimation, the monolayer is often encapsulated by a suitable capping layer, mostly SiO_2. Another method is the combination of spin-coating which usually results in layers much thicker than a monolayer with FLA [76, 77]. In this case the "monolayer approach" is achieved by the ms annealing time which limits the indiffusion of the dopant. Further approaches include the deposition of thin, dopant-containing layers by atomic layer deposition [78] or sol-gel coating [33].

MLD has the advantage of avoiding the typical ion implantation damage. In case of a self-limiting surface reaction it also allows a conformal deposition which is especially interesting for the doping of nanostructures [74]. In addition, it can be often combined with techniques of nanoimprint lithography to obtained patterned doping [79]. The use of precursors unavoidably leads to the indiffusion of carbon or hydrogen, but so far no negative effects are known by this [74].

However, there is probably a serious drawback with respect to USJ because of the dopant profile is formed by diffusion, regardless of the specific annealing time or temperature. O'Connell et al. [80] investigated As MLD with RTA for 5 s and found well-shaped junction profiles but with junction depths between 50 and 100 nm depending on annealing temperature. Moreover, the abruptness achieved values in the order of 25 nm/dec which are one order of magnitude worse than those known from ion implantation. A recent study [76] using FLA and different levels of preheating makes the dilemma even clearer: in case of no preheating FLA leads to narrow junctions, but with quite low dopant concentrations. In contrast, a preheating temperature of 300 °C results in high dopant concentrations exceeding the level of 10^{20} cm^{-3} by far, but now with junction depths in the order of 100 nm. More seriously, there was not a single case in which the abruptness had a value better than 20 nm/dec.

In contrast, ultra-narrow B and P profiles with $x_J < 5$ nm were reported using MLD in combination with RTA [73]. A more recent investigation also combining MLD of P with RTA [81] reveals that there is a very narrow transition region of 2–3 nm containing C, O and P, the latter one in concentrations of 10^{20} cm^{-3} as measured by atom probe tomography. Comparing the annealing temperatures of 825 and 1025 °C this peak does not change very much, whereas at 1025 °C strong diffusion of P into Si takes place. As a result one can assume that this transition region builds up quite early and depends on the surface chemistry. It should be noted that secondary ion mass spectroscopy overestimated this surface peak by one order of magnitude. Another promising approach involves the deposition of a self-limiting monolayer of P from a PH_3 ambient followed by Si overgrowth steps [48]. The result is a high density, highly abrupt P profile with x_D values between 2 and 6 nm. However, to get a clean Si surface with a Si(100) 2 × 1 surface reconstruction, samples had been heated in ultra-high vacuum to 1200 °C in advance of the P deposition.

In summary, MLD, especially in combination with short time annealing, is useful for applications were ion implantation is critical and where conformal doping is desirable, e.g. for the doping of NW transistors. If the junction depth is not limited by geometry or other processes, USJ are more difficult to fabricate. A potential exception could be the complete suppression of diffusion into the bulk and the restriction to the surface transition zone, if this is applicable.

4.2.2 Hyperdoping in Silicon

4.2.2.1 Dependence on Doping Technology and Annealing

The internal properties of semiconductors can be modified by doping and strain engineering i.e. the introduction of foreign atoms and the modification of lattice parameters, respectively. Depending on the type of the doping element and on the concentration, one can distinguish between conventional doping, hyperdoping and alloying. Conventional doping is considered when shallow dopants are incorporated into semiconductors with a concentration lower than the equilibrium solid solubility. Such dopants define the type of conductivity, and in heavily doped semiconductors they can slightly change the optical bandgap due to the bandgap renormalization or the Burstein-Moss effect like Zn or Si in III–V compound semiconductors or group III and V elements in group IV semiconductors. Hyperdoping describes semiconductors containing impurities with a concentration much higher than the solid solubility. The term of hyperdoping can be used both for shallow and deep dopants. In the case of shallow dopants like Ga or P in Ge the hyperdoping is used to obtain a carrier concentration at least one order of magnitude higher than the solid solubility limit [82, 83]. Exotic dopants (deep dopants) like group VI elements or transition metals are considered for hyperdoping of Si or Ge when the concentration is few orders of magnitude higher than the solubility limit [84, 85]. To realize hyperdoping of semiconductors, strong non-equilibrium processing is needed. Therefore, the hyperdoped semiconductors are made by epitaxy techniques or ion implantation followed by sub-second thermal annealing. Figure 4.7 shows the typical range for conventional doping, for solid solubility limits and for hyperdoping in semiconductors.

Hyperdoping can be performed either in situ during the growth of semiconductors e.g. using molecular beam epitaxy (MBE) or vapour-liquid-solid (VLS) growth of nanostructures [82] or ex situ by using ion implantation and thermal annealing. The ion implantation is a strong non-equilibrium process which allows to introduce any kind of elements into solids with concentrations exceeding the solubility limit by far. Unfortunately, ion implantation is destructive for crystals, and the crystalline layer implanted with high ion fluences becomes amorphous which requires

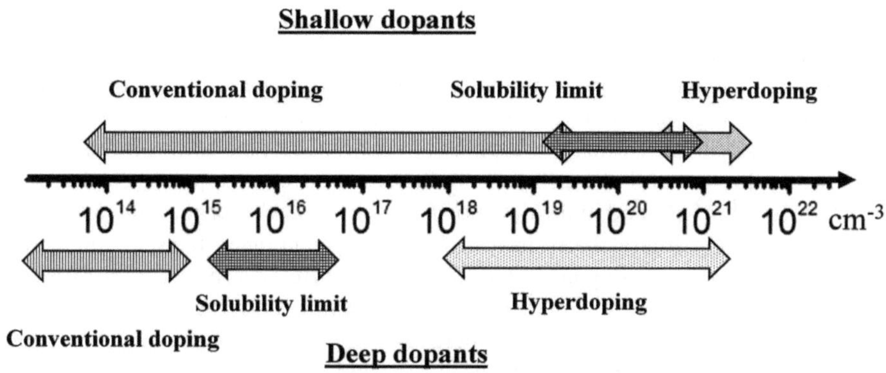

Fig. 4.7 Typical concentration range of conventional doping, solid solubility limits and hyperdoping of shallow and deep dopants in semiconductors

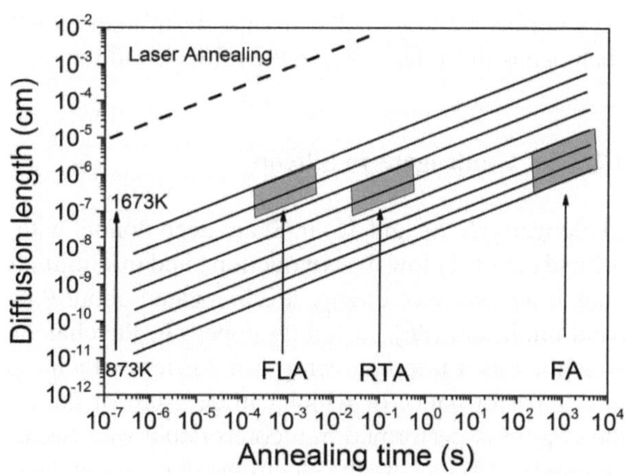

Fig. 4.8 Diffusion length as a function of annealing time for Se in Si in the solid phase. The dashed line shows the diffusion length of Se in Si in the liquid phase. After [84]

post-implantation annealing. To avoid dopant segregation and diffusion, the post-implantation annealing must also strongly run in non-equilibrium. The recrystallization of the implanted layer can be performed either via liquid phase epitaxy (LPE) or via ms-range solid phase epitaxy (SPE). LPE usually takes place during nanosecond range pulsed laser melting (PLM) of the implanted layer. Using millisecond FLA the implanted layers recrystallize mostly via SPE. Which annealing technique and recrystallization process are preferable for hyperdoping of semiconductors is determined by the diffusion coefficient of dopants and the solid solubility of the dopants of interest. As example, Fig. 4.8 shows the diffusion length as a function of the annealing time for Se in Si at different temperatures. Se implanted Si wafers were annealed using conventional RTA and FA or non-equilibrium FLA in the millisecond time scale. The diffusion lengths of Se in Si during LPE, as it is typical for LA, is shown as well (dashed line).

In general, the diffusion coefficient in molten semiconductors is few orders of magnitude higher than the diffusion coefficient in the solid phase. The precise control of the diffusion length of dopants in semiconductors is crucial for the uniform doping of the junction region, to control the junction depths and to avoid secondary phase formation, e.g. clusters, when the doping is in the range of the solid solubility limit or higher. In the case of Si the diffusion coefficient of different dopants is in the range of 10^{-11}–10^{-18} $m^2\ s^{-1}$ and in the range of 10^{-5}–10^{-7} $m^2\ s^{-1}$ for the solid and liquid phase, respectively. Interestingly, the diffusion coefficient in molten semiconductors is very similar for most dopants. Due to the huge difference between diffusion in the solid and liquid phase the annealing time and temperature is critical for the final microstructural and optoelectronic properties of doped semiconductors. According to Fig. 4.8 the diffusion lengths of Se in Si is roughly the same when the material is annealed at 873 K for 30 min in a conventional furnace or at 1273 K for 3 ms using FLA. But after PLM for 30 ns the diffusion length of Se in Si is about two orders of magnitude bigger than that during FLA at 1237 K. This means that the LPE in the nanosecond range will cause much stronger redistribution/diffusion of implanted chalcogens in Si than SPE in the millisecond range.

4.2.2.2 Chalcogens in Silicon

Chalcogens (S, Se and Te) in Si are deep donors with energy levels located between 0.2 and 0.3 eV below the conduction band minimum for a singly ionized dopant. The double donors have energy levels located about 0.4–0.6 eV below the conduction band minimum [86]. In lightly doped Si, the chalcogens form single energy levels which act as a trapping centre for carriers which mainly cause the degradation of device performance or which can be used for the net carrier compensation. But if the doping is performed in a controllable way Se in Si^{28} can be used for quantum computing [87] where the electronic devices can be controlled via the manipulation of donor spins. The use of chalcogen doped Si for quantum technology requires a doping level in the range of 10^{14}–10^{16} cm^{-3}. The increase of the Se concentration to about 10^{17}–10^{18} cm^{-3} introduces an impurity band in the bandgap making Si n-type doped. However, the Si:Se system is still like intrinsic Si because carriers are localised within the impurity band. Moreover, the Mott limit for the deep donors in Si is higher than 10^{19} cm^{-3}. The insulator-to-metal (IMT) transition for the chalcogens in Si appears at a dopant concentration in the range of 10^{20} cm^{-3} or higher. The evolution of the impurity band in chalcogen doped Si is shown schematically in Fig. 4.9. The continuous increase of the chalcogen concentration in Si merges the impurity band with the conduction band minimum. This leads to significant reduction of the optical bandgap of Si from 1.1 eV to about 0.2 eV and the IMT takes place [88].

The IMT in chalcogen doped Si using ion implantation and ms-range SPE was investigated in details by Liu et al. [89]. Se implanted Si wafers were recrystallized using FLA for 1.3 ms, and the concentrations of Se were 2.5, 6.3 and 13×10^{20} cm^{-3} (the corresponding ion fluences are given in the legend of Fig. 4.10). Microstructural

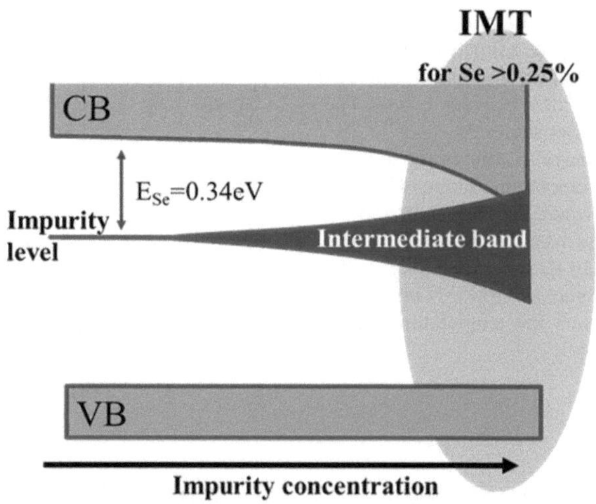

Fig. 4.9 The evolution of the impurity band in Se doped Si with increasing Se concentration. The IMT is usually achieved for Se concentrations higher than 0.25%

investigations, namely micro-Raman and Rutherford backscattering spectrometry (RBS), show that after FLA the implanted layer regrows epitaxially and Se is incorporated into the lattice site of Si. For the lowest and medium Se concentration almost 100% of Se substitutes Si in the crystal. For the highest concentration at least 80% of Se replaces Si. Simultaneously, the diffusion of Se during FLA is fully suppressed. Figure 4.10 shows the change of the sheet resistance (R_S) as a function of temperature in an as-implanted (1.3×10^{21} cm^{-3}, not annealed) sample and after FLA for three different Se concentrations. R_S in the as-implanted sample increases above the sensitivity range of the measurement system already at 150 K. The annealed samples can be divided to two groups: (i) samples with doping levels lower than 3×10^{20} cm^{-3} where the temperature-dependent R_S exhibits the typical behaviour for lightly or medium doped semiconductors and (ii) samples with Se concentrations higher than 3×10^{20} cm^{-3} when R_S is almost temperature-independent. According to the room temperature (RT) Hall effect measurements the three Se concentrations 2.5×10^{20}, 6.3×10^{20}, 1.3×10^{21} cm^{-3} correspond to an effective electron concentration of about 5×10^{18}, 7×10^{19} and 9×10^{19} cm^{-3}, respectively. Thus, the IMT in Se doped Si using ion implantation and SPE during ms-range FLA appears for Se concentration in the range of 6×10^{20} cm^{-3} or higher and for active carrier concentrations above $7 \pm 1 \times 10^{19}$ cm^{-3} [89].

Zhou et al. have investigated the influence of different annealing techniques on the IMT and the Se distribution in ion implanted Si. The implanted samples were recrystallized either via LPE during ns-range PLM or via SPE during ms-range FLA [84]. Figure 4.11 shows the distribution of Se in Si obtained with the SRIM code [6] and experimental RBS spectra obtained from Si:Se annealed either by FLA for 1.3 ms or by PLM for 30 ns. The power of the laser pulse was sufficiently high to melt an about 120 nm thick Si layer. The solidification process of the molten layer during PLM is in the μs range. During FLA the maximum temperature is lower than

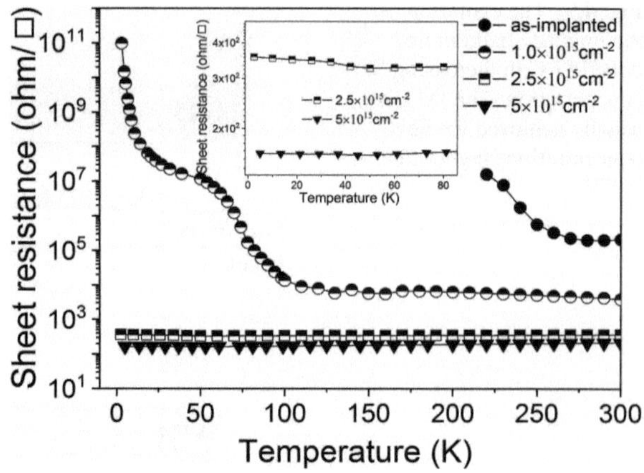

Fig. 4.10 Sheet resistance (R_S) as a function of temperature for Se implanted Si after FLA for 1.3 ms. The inset shows the magnification of the low temperature range of the R_S where the IMT is directly proven by a stable R_S below 10 K. In semiconductors R_S increases towards infinity at very low temperature. After [89]

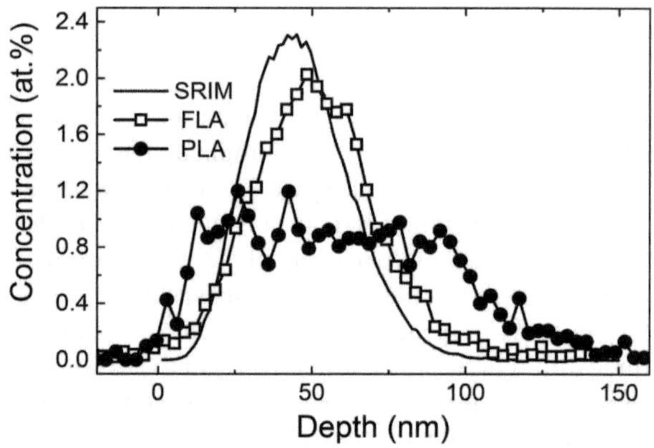

Fig. 4.11 Simulated as-implanted depth distribution and experimental depth distribution of Se in Si obtained after FLA and PLM as derived from RBS. After [84]

the melting temperature of amorphous Si (<1200 °C) causing SPE of the implanted layer. Ion implantation performed with a single energy always leads to a Gaussian-like distribution of the implanted elements in the solids. This is confirmed both by simulation and by experiments. Surprisingly, after FLA the Se distribution in Si looks similar to that in the as-implanted stage. This means that the diffusion of Se during ms-range SPE is fully suppressed. During FLA for 1.3 ms the amorphous/crystalline interface moves towards the surface with a speed of about 10 m/s which is much faster than the diffusion of Se in Si in the solid phase. This implies that during the epitaxial regrowth of the implanted layer Se is trapped into the lattice of Si and frozen there. Moreover, the snow-plough effect typically observed in heavily implanted semiconductors after FLA is not present.

After PLM the depth distribution of Se in Si looks completely different. Instead of a typical Gaussian distribution the Se profile has a box-like shape and the peak Se concentration drops down by half. Simultaneously, the thickness of the doped layer increases from about 60 to 120 nm, and the Se doped layer overlaps with the molten Si layer. During PLM Se diffuses towards both the surface and the solid/molten

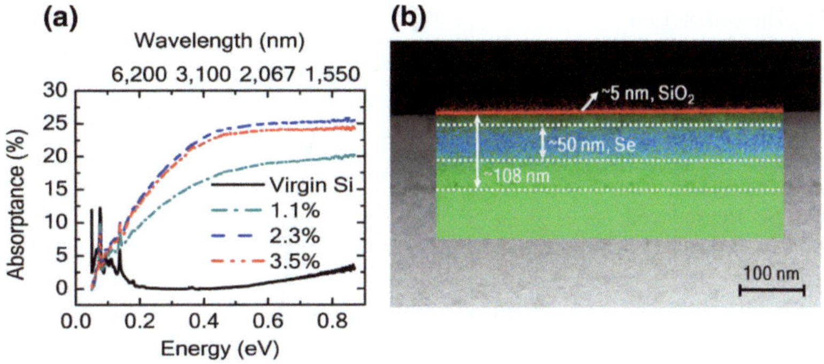

Fig. 4.12 Absorptance of hyperdoped Si for different Se concentrations (**a**) and cross-sectional HAADF-STEM image of Se implanted and annealed Si together with the Se (blue), Si (green) and O (red) distribution obtained by EDX mapping (**b**). Reproduced from [90] (CC BY-NC 4.0)

interface. Although the pulse length of the laser is only 30 ns the annealed Si layer is liquid for few microseconds. Taking into account the diffusion coefficient of Se in the liquid phase, the diffusion length of Se during PLM in Si is in the range of 1 μm which is much more than the molten layer. This explains the homogeneous distribution of Se over the molten Si layer.

In summary, the SPE of Se implanted Si (ion implantation followed by FLA) allows the full control of the thickness of the hyperdoped Si layer and the chalcogen concentration by adjusting the ion implantation parameters. Using LPE (ion implantation followed by PLM) the thickness of the molten layer strongly influences the thickness of the hyperdoped Si. Therefore, from the application point of view the SPE is more desired. Berencen et al. have fabricated a mid-infrared Si photodetector using Se implantation into Si followed by FLA for 1.3 ms [90]. Figure 4.12a shows the absorptance as a function of energy/wavelength for Si:Se layers containing 1.1, 2.3 or 3.5% of Se. The absorptance of virgin Si is shown as well. Due to the Se doping the effective optical bandgap of Si:Se is reduced to about 0.3 eV. Figure 4.12b shows a cross-sectional high-angle annular dark-field scanning transmission electron microscopy (HAADF-STEM) image combined with energy-dispersive X-ray spectrometry (EDX) mapping of Se implanted Si after FLA (green: Si, red: O, blue: Se). After FLA Se is fully incorporated into the Si matrix, no Se clusters were detected.

The fabricated Si:Se layers with unique optical and crystalline properties are perfectly suited for complementary metal–oxide–semiconductor (CMOS) compatible, short-wavelength infrared (IR) detectors. Figure 4.13 shows the photocurrent measured in a Si:Se detector at a reverse bias of 1 V and under the illumination of a 5 W tungsten halogen lamp (Ocean Optics HL-2000 series) that is filtered by a monochromator. Similar optical and electrical properties of hyperdoped Si are expected if implanted with S and Te. Wang et al. have investigated the influence of Te doping on the mid-infrared absorption of hyperdoped Si [91]. In contrast to Se, in Te hyperdoped Si fabricated by ion implantation and FLA the plasma frequency

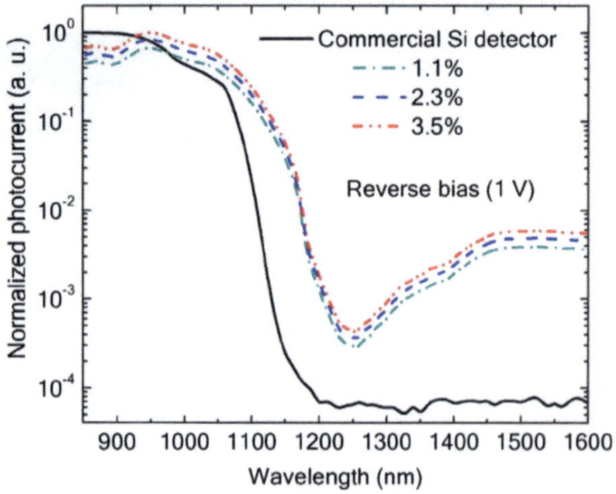

Fig. 4.13 The comparison of the photocurrent collected under the illumination of a 5 W tungsten halogen lamp from a commercial Si detector and a short-wavelength IR detector made of Si:Se. Reproduced from [90] (CC BY-NC 4.0)

can be tuned by the carrier concentration which is directly proportional to the Te concentration.

In summary, the hyperdoping of Si with chalcogens opens different application fields for Si as (i) the effective bandgap of Si is reduced to 0.2 eV which is attractive for long wavelength photonics, (ii) the IMT is obtained for a chalcogen concentration in the range of 3×10^{20} cm^{-3}, (iii) the electron effective mass is significantly reduced due to the enhancement of electron mobility.

4.2.2.3 Transition Metal Hyperdoped Silicon

In contrast to chalcogens the transition metal impurities in Si can be both deep donors (Ti, V, Au, etc.) and deep acceptors (Pt). But similar to chalcogens the transition metals in Si are fast diffusers and the solubility limit is in the range of 10^{15}–10^{16} cm^{-3}. The hyperdoping of Si with transition metals can be performed utilizing non-equilibrium processing like ion implantation followed by either LPE after PLM or ms-range SPE. During PLM the recrystallization of transition metal implanted Si is accompanied by the cellular breakdown [92] where the transition metals segregate in the form of filaments (see Fig. 4.14a). The green filaments visible in the inset of Fig. 4.14a are Ti rich (more than 10%) and amorphous. The concentration of Ti between filaments is in the range of the solid solubility limit of Ti in Si. Using ms-range SPE the cellular breakdown can be fully suppressed (see Fig. 4.14b). The excess of Ti diffuses towards the surface and is agglomerated there within small clusters.

Optically, Ti doped Si has similar properties like chalcogen hyperdoped Si where sub-bandgap energy levels are responsible for the light absorption below the bandgap of Si. Therefore, it was proposed that Ti doped Si can be attractive for intermediate band solar cells. However, the electrical properties of the Si:Ti

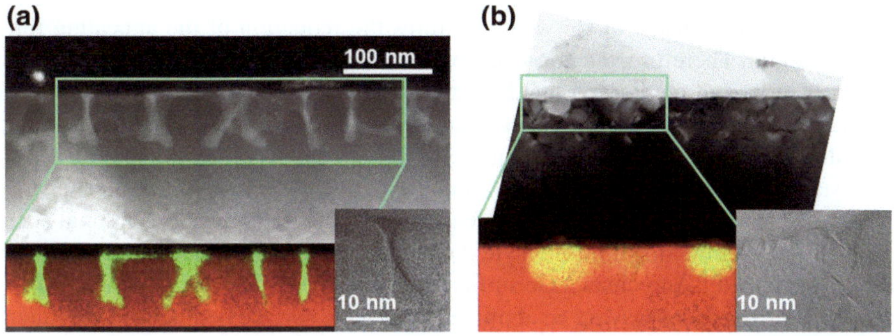

Fig. 4.14 Cross-section bright-field and high-resolution transmission electron microscopy (TEM) micrographs (right insets) of Ti implanted Si after PLM (**a**) and FLA (**b**). The EDX mappings of the Ti and Si distribution originate from rectangular regions marked in the main images. After [92]

system differs significantly from that of chalcogen doped Si. Independent of the Ti concentration and the used annealing technique the IMT was not observed [93]. In the case of samples recrystallized via LPE the electrical transport is caused by hopping conduction between Ti rich cellular walls. After FLA the Ti doped Si is n-type but most carriers are localized within the impurity band.

4.2.3 Doping and Superconductivity in Germanium

The next generation of high-performance nanoelectronic devices needs the monolithic integration of foreign semiconductors characterized by a carrier mobility much higher than that of Si. Furthermore, the incorporation of different functional optoelectronic elements on a single chip enables a significant performance progress. Due to its chemical similarity to Si and the high hole mobility, Ge is an obvious candidate. The efficient and cost-attractive integration of Ge into CMOS technology for the future high-speed optoelectronics can be done by using only a thin Ge layer placed on cheap substrates. In order to improve the device performance and to fully exploit the unique properties of Ge, Ge-on-insulator and Ge-on-Si structures are mainly used.

Another important issue for the creation of high-performance Ge-based nanoelectronic devices is the fabrication of highly-doped p- and n-type ultra-thin Ge layers and Ge NWs. In general, it has been proven that n-type doping of Ge is a key bottleneck in the realization of advanced MOSFETs. Having in mind the Roadmap 2015 [94] for the next generation of nanoelectronics, the device architecture should focus on non-planar field effect transistors (FETs) using a NW- or fin-based design.

To this day the most developed doping technique for semiconductors is ion implantation followed by thermal annealing. The post-implantation annealing has two functions: (i) electrical activation of dopants and (ii) to restore the initial properties of the implanted crystal structure. In contrast to Si group V elements (n-type dopants in Ge) are fast diffusers in Ge which requires the reduction of the annealing time to a minimum. Therefore, the non-equilibrium post-implantation thermal processing becomes more and more important for nanoelectronics. One of the first applications of FLA for Ge processing was presented by Cohen et al. [95]. They have shown that ultra-short flash pulses with a duration of 50 μs and a preheating of 300 °C can recrystallize Ge implanted with Te to a fluence of 2×10^{15}cm^{-2} and with an energy of 50 keV. Under these conditions, at least 50% of the Te is on substitutional sites, and the RBS channelling minimum yield was comparable to virgin Ge (4.5%). Hovis et al. have used FLA to clean a Ge surface [96] and showed that FLA is able to efficiently remove oxides leaving an atomically clean Ge surface behind for subsequent processing steps. Recently, ion implantation of P and As in combination with FLA has been used to fabricate USJ in Ge with a junction depth of 10–20 nm [97–99]. In addition, the co-implantation of N and the use of an Al_2O_3 capping layer were found to suppress P diffusion in Ge during FLA and to lead to lesser damage of the Ge surface [100].

4.2.3.1 n-Type Germanium: Electrical Properties

Similar to Si, the main doping techniques relay on diffusion or ion implantation. Since in Ge the diffusion coefficient for group V elements is few orders of magnitude higher than for group III elements (10^{-9} vs. 10^{-13} cm^2/s) the application of diffusion for shallow doping of Ge is limited [101]. The main reason is that the precise control of the carrier concentration and the junction area using diffusion is very difficult. Nevertheless, MLD of Ge has been recently considered as an alternative to ion implantation for future nanoelectronics [102]. Similar to the case of Si (Sect. 4.2.1), the advantage of diffusion doping is that the doped layer is more or less free of defects while ion implantation always introduces some damage to the crystal. Scappucci et al. have developed the inter-diffusion process for ultra-doped n-type Ge using δ-doped layers as a source of P [103]. Figure 4.15 shows the concept for the fabrication of n-type Ge using in situ δ-doping during the growth of the Ge layer by MBE. It was shown that the maximum carrier concentration depends on the thickness of the grown layer. The maximum electron concentration decreases from 2×10^{20} cm^{-3} in a monolayer to 6–8 $\times$ 10^{19} cm^{-3} for thicker layers. The deactivation of P with increasing thickness of the doped layer is mainly due to the formation of P-P dimers, i.e. the incorporation of P atoms into the neighbouring position in the Ge lattice.

However, also other doping techniques achieved similar or higher carrier concentrations. Kim et al. have investigated the influence of multi-implantation steps on the maximum carrier concentration in Ge [104]. Instead of single high fluence ion implantation which leads to the amorphization of the matrix they have implanted P into Ge using multi-step implantation followed by RTA after each implantation step. The maximum electron concentration obtained in this way is in the range of 1×10^{20} cm^{-3}. Figure 4.16 shows an overview of the maximum electron concentration in Ge obtained by different doping techniques. The highest ever published electron concentration in Ge was achieved after ion implantation followed by non-equilibrium

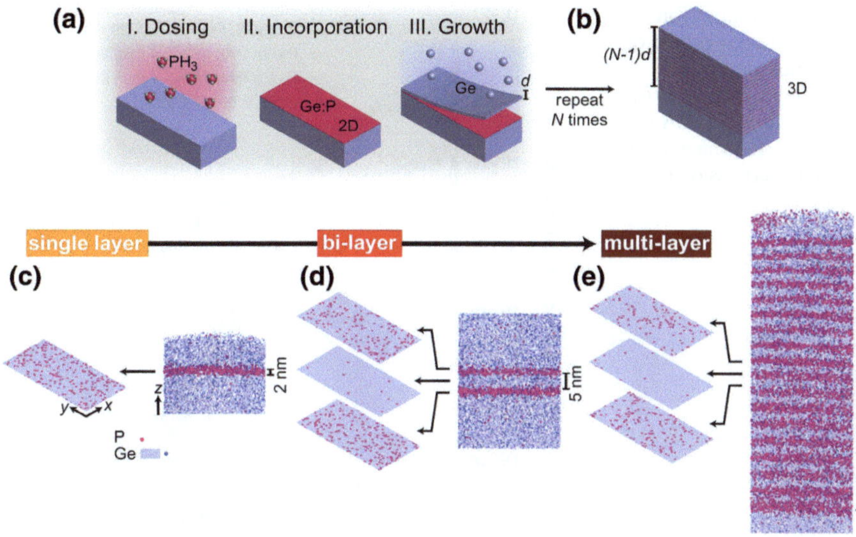

Fig. 4.15 Three-dimensional assembly of atomically sharp doping profiles using a bottom-up approach. Reproduced from [103] (CC BY-NC 4.0)

Fig. 4.16 Maximum carrier concentration in Ge obtained by different doping techniques [104–109]

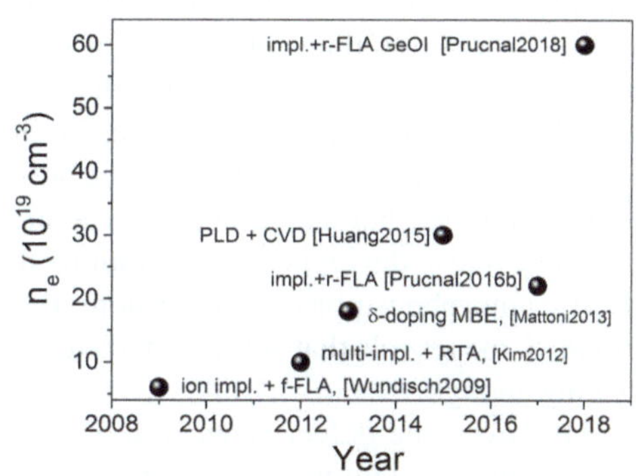

thermal processing: 2×10^{20} cm^{-3} using millisecond range FLA [105] and 3×10^{20} cm^{-3} using nanosecond PLM [106].

Both annealing techniques are working far away from equilibrium. In the present case the main difference between FLA and PLM is that during PLM the implanted layer recrystallizes via LPE while during FLA the implanted layer recovers its initial properties via SPE. But, in both cases, the crystalline/amorphous (for FLA) or the crystalline/liquid (for PLM) interface moves towards the surface with a speed of about 10 m/s [110]. To this day PLM is mainly used in laboratories because the annealed area is limited, usually to 10×10 mm^2 and below. Therefore, it is not relevant for industry. In contrast to PLM, FLA systems are commonly used by large semiconductor companies. The main application of non-equilibrium annealing is to activate dopants without the change of their depth distribution [111]. Satta et al.

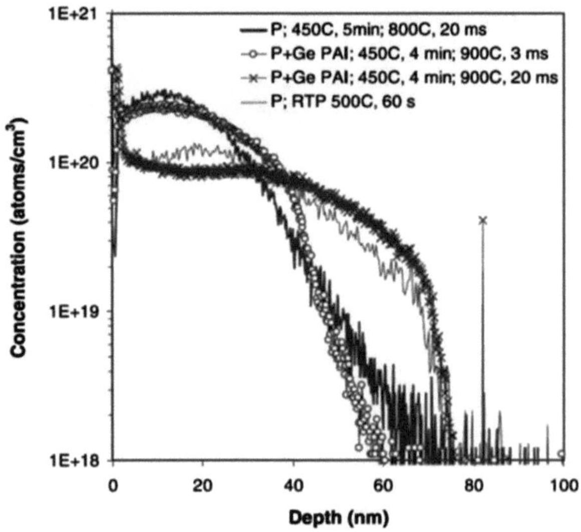

Fig. 4.17 Secondary ion mass spectroscopy profile of P in Ge after ion implantation with an energy of 15 keV and after FLA for 3 and 20 ms. Reprinted from [112], with permission from Elsevier

have used FLA to fabricate shallow junctions in P implanted Ge [112]. Prior to FLA samples were preheated to 400–450 °C and the annealing time was up to 20 ms. Figure 4.17 shows the depth profile of P in Ge with and without pre-amorphization before and after FLA. It was shown that, independent of the annealing time, FLA up to 800 °C does not activate the diffusion of P. However, significant P diffusion was observed after annealing at 900 °C. The pre-amorphization with Ge does not affect the distribution of P but increases the concentration of electrically active P. This is most probably due to the higher incorporation efficiency of P during recrystallization from a fully amorphous phase. Furthermore, the comparison between FLA performed at 3 and 20 ms reveals that significant diffusion in Ge can also occur if the pulse length is long enough.

Moreover, the defect concentration plays an important role for the final effective carrier concentration due to the formation of defect-dopant complexes. After annealing at high temperature, dopant-vacancy complexes are formed. n-type dopants bonded with a vacancy create an acceptor level in Ge which significantly reduces the effective electron concentration. Details about the interactions between defects and n-type dopants like P, As and Sb during equilibrium annealing can be found in [113]. The point-defect engineering in ion implanted Ge via non-equilibrium thermal processing was investigated by Simeon et al. and later by Wündisch et al. [101, 107, 114]. They have shown that during ms-range FLA the concentration of Ge-vacancies can be significantly reduced. Although the concentration of P incorporated into the lattice site of Ge was close to the solid solubility limit of P in Ge (2×10^{20} cm^{-3}) the effective carrier concentration did not reach the level of 10^{20} cm^{-3}. Moreover, it was found that for an implanted layer thicker than 100 nm the FLA process leads to the formation of a nanocrystalline layer close to the surface and only to a partial epitaxial regrowth of the doped layer. The thickness of the nanocrystalline layer is mainly defined by the annealing time and the thickness of the implanted layer but is independent of the implanted element.

The full integration of Ge with CMOS technology using ion implantation and FLA was presented by Ootsuka et al. [69]. To fabricate a 45 nm metal/high-k FET the p-type and n-type part of the junction was realized by low energy implantation of boron and arsenic, respectively, followed by FLA for 1 ms. The junction depth for the n- and p-type channel was 26 and 13 nm, respectively. The high device performance was mainly achieved due to the fact that during FLA the density of vacancies was reduced, the depletion problem and mobility degradation was suppressed, and a low threshold voltage was obtained. Usuda et al. have successfully utilized FLA to fabricate tri-gate junction less transistors using both p- and n-type poly-Ge [115]. The poly-Ge was made by explosive solid phase recrystallization of amorphous Ge deposited on a SiO_2/Si substrate. Because of the as-fabricated Ge layer was p-type the pMOSFETs were fabricated without external doping. The n-MOSFETs were realized by P implantation and double-pulse FLA; the first flash was used to recrystallize Ge and the second flash was applied to activate the implanted P. Despite that the Ge layer was polycrystalline the carrier mobility both for holes and electrons was much higher than that in Si for the same doping level.

Significant enhancement of the carrier mobility in poly-Ge on insulator was also obtained by alloying of Ge with Sn [116]. It was shown that even small amounts of Sn can significantly enhance the hole and carrier mobility. Like pure poly-Ge the non-intentionally doped poly-GeSn formed via millisecond range epitaxial solid phase regrowth is p-type with a carrier concentration in the range of 5×10^{18} cm^{-3} and a hole mobility higher than 200 cm/Vs. After P implantation the poly-GeSn is n-type with an electron concentration as high as 2.5×10^{19} cm^{-3} and an electron mobility above 150 cm/Vs which is comparable with Ge epitaxially grown on virtual substrates [116].

4.2.3.2 Rear Side FLA

Independent of the doping technique the maximum achievable electron concentration in Ge is in the range of 3×10^{19} cm^{-3} if using equilibrium conditions only [117]. The metastable doping close or slightly higher than the solid solubility limit for P in Ge can be obtained via non-equilibrium processing. Using the conventional approach, the highest electron concentration in P implanted Ge was 6×10^{19} cm^{-3} for a P concentration of 3×10^{20} cm^{-3} which means that only 20% of implanted P is electrically active [107]. In order to avoid the formation of a nanocrystalline Ge layer at the surface of the implanted wafer Prucnal et al. have developed the rear-side FLA (r-FLA) process [105]. Crystalline Ge can be easily amorphized during high fluence ion implantation. The RT threshold ion fluence for amorphization of Ge for n-type dopants is in the range of 5×10^{14} cm^{-2} for P and decreases with increasing atomic mass of the implanted element. The recrystallization becomes more and more difficult with increasing ion fluence. Figure 4.18 shows the differences for the recrystallization process in ion implanted Ge using conventional front-side FLA

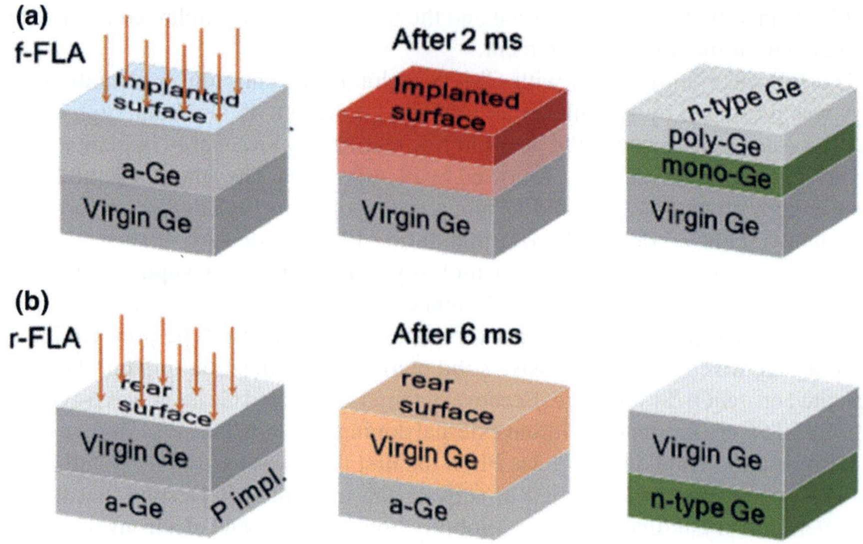

Fig. 4.18 Schematic representation of the recrystallization mechanism of ion implanted Ge during conventional FLA (**a**) and r-FLA (**b**). The times of 2 and 6 ms denote the time where surface temperature is at maximum (**a**) or has reached the equilibrium temperature (**b**). Reproduced from [105] (CC BY-NC 4.0)

(a) and r-FLA (b). The main optical emission of Xe lamps used in the FLA system is in the ultraviolet (UV) and visible spectral range (see Fig. 2.23). Taking into account the optical properties of Ge the main part of the energy deposited to the sample surface during FLA is absorbed within the first 50 nm. This leads to the temperature gradient between the sample's surface and the crystalline/amorphous interface. Moreover, the minimum energy needed for the crystalline seed nucleation within an amorphous layer is slightly lower than the energy needed to start the epitaxial regrowth of the implanted layer [118].

Crystallization requires the existence of crystalline seeds from which crystallization can start. In case of the amorphous/crystalline interface the crystalline part of Ge can act as epitaxial pattern, and crystallization can start immediately with a growth rate strongly depending on temperature and probably assisted by dopants. However, if there are no seeds at the surface, a certain time, called the incubation time, will pass before first seeds occur and a crystallization front can start from the surface. The values for the incubation time and the crystallization velocity differ in literature, depending on whether they were obtained by long time annealing [119, 120] or LA [121]. As a consequence, a completely monocrystalline layer can be grown if, and only if, the layer thickness is small enough to complete crystallization within the incubation time. In case of front-side FLA the temperature at the surface is somewhat higher than at the interface, the incubation time is shorter, and a completely monocrystalline layer can only be achieved for very thin layers. For thicker layers, like 100 nm and more, a second crystallization front can start from the surface, and

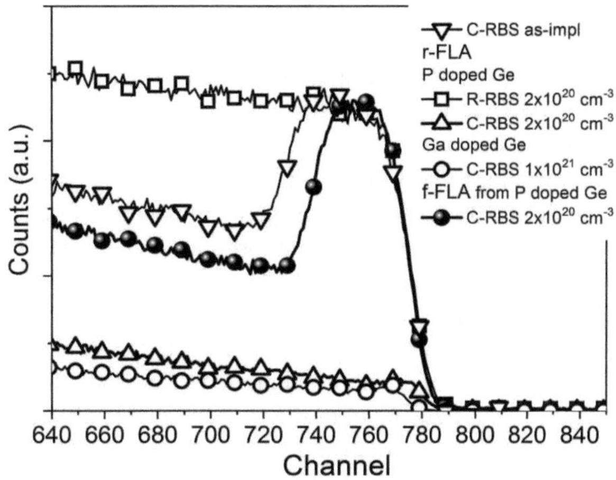

Fig. 4.19 Random (R-RBS) and channelled (C-RBS) spectra obtained from P and Ga implanted Ge followed by r-FLA

the implanted regions ends up with a lower, monocrystalline layer and an upper, polycrystalline layer (Fig. 4.18a). However, in case of r-FLA, the temperature at the surface is now lower than at the interface which prolongs the incubation time. This allows the fabrication of much thicker, completely monocrystalline layers by FLA, where the crystalline/amorphous interface moves towards the surface with an average speed of up to 10 m/s (Figure 4.18b).

The existence of a polycrystalline Ge layer limits both the maximum electron concentration and the electron mobility in the P doped sample. In the polycrystalline layer the defect density is much higher than in the single crystalline Ge while the grain boundaries affect the carrier mobility. In order to avoid the formation of polycrystalline layer during FLA the temperature gradient within the implanted layer must be inverted. Thus, Prucnal et al. performed the FLA from the rear side (Fig. 4.18b) which got hot during the initial stage of annealing and the heat is transferred through the wafer towards the implanted layer. The crystalline/amorphous interface is hotter than the surface of the implanted Ge. In such a case recrystallization starts from the crystalline/amorphous interface and seed nucleation is suppressed. Importantly, the recrystallization takes place via explosive SPE where the c/a interface moves faster than the diffusion of both n- and p-type dopants. Therefore, the dopants are trapped at the lattice site of Ge and doping concentrations much above the solid solubility limit in Ge are possible.

Figure 4.19 shows the RBS spectra collected in random and channelling direction (RBS R/C) and obtained from an ion implanted Ge wafer followed by r-FLA. The RBS/C spectrum of a virgin, un-implanted sample is shown as well. The RBS/C technique is very sensitive to point defects and atoms located at interstitial position. By comparing the RBS/R and RBS/C spectra it is possible to conclude about the crystal quality and the location of the implanted atoms in the matrix. The substitutional fraction, estimated by the comparison of random and aligned spectra, can be calculated according to:

$$f_S = [1 - \chi(X)]/[1 - \chi_R(X)] \quad (4.2)$$

where $\chi_R(Y)$ is the minimum yield in the fraction of dechannelled particles calculated in the detection region of the matrix elements and $\chi(X)$ is the minimum yield in the aligned spectrum of the peak of implanted element. For the single crystalline matrix, the minimum yield in the aligned spectrum should be below 5%, and for the ideal case f_s is 100%. After P implantation with a fluence higher than 1×10^{15} cm^{-2} the implanted Ge becomes amorphous, which can be easily visualized by RBS/C where the yield in the aligned spectrum is at the same level as that for RBS/R. Surprisingly, the yield of the aligned spectrum recorded from the sample implanted and annealed from the front side is at the same level like that of not-annealed Ge. Only the width of the non-channelled layer is smaller which can be due to either a not fully recrystallized layer or the existence of a polycrystalline film at the surface of the implanted Ge. μ-Raman and X-ray diffraction (XRD) spectroscopy (not shown here) confirm the formation of a polycrystalline layer at the surface and the only partial epitaxial regrowth of implanted Ge during front side FLA. The yield of the RBS/C spectra obtained from both P and Ga implanted Ge followed by r-FLA is at the same level as for virgin Ge with $\chi(\text{Ge}) = 4.6\%$. This in turn confirms the full epitaxial recrystallization of implanted Ge. Unfortunately, the substitutional fraction of Ga and P cannot be calculated from the RBS spectrum due to the resolution limitation of the RBS techniques for elements lighter than the matrix. However, most of the Ga and P atoms are assumed to be in a substitutional position due to the fact that the yield for all channelling spectra obtained after r-FLA (from virgin and implanted Ge) are at the same level.

The microstructural properties of implanted Ge followed by FLA from both the rear and front side were investigated by TEM in cross-section. Figure 4.20a, b show the bright field STEM image of Ga implanted Ge followed by r-FLA for 20 ms and EDX mapping for Ge, Ga and Si, respectively. According to these TEM investigations Ge is single crystalline after implantation and r-FLA. In the presented figure the Ga concentration is four times higher than the solid solubility limit. Nevertheless, no segregation of Ga was found. If any Ga clusters are formed their diameter must be smaller than 1 nm which is the resolution limit for Ga in Ge in our TEM system. The obtained images together with EDX mapping are in good agreement with RBS data fully supporting the assumption that after r-FLA the implanted Ge is single crystalline and the implanted elements are incorporated into the lattice site of Ge. The electrical properties of Ga doped Ge are discussed in Sect. 4.2.3.4.

4.2.3.3 n-Type Germanium: Optical Properties

Intrinsic and relaxed Ge is an indirect bandgap semiconductor with a bandgap of 661 meV and a direct transition at the Γ point of 800 meV at RT. The separation between the direct and the indirect transition is only 139 meV. Therefore, it is possible to convert Ge from an indirect to a quasi-direct semiconductor by: (i) strain engineering, (ii) ultra-high n-type doping or (iii) alloying with Sn. Figure 4.21 shows the

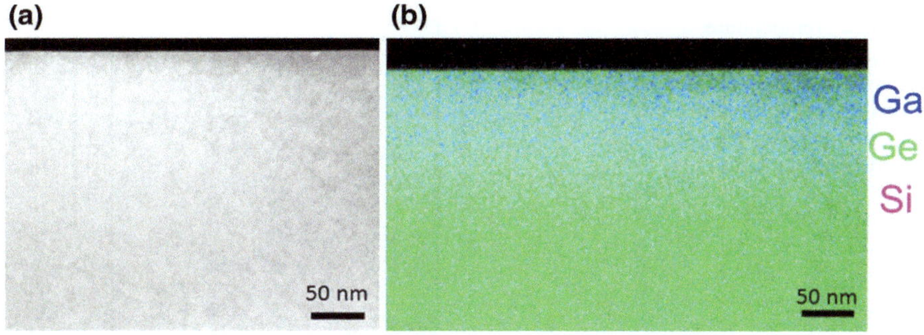

Fig. 4.20 Bright field STEM image of Ga implanted Ge with a fluence of 2×10^{16} cm^{-2} followed by r-FLA for 20 ms (**a**) and EDX image presenting the distribution of Ga, Ge and Si (**b**)

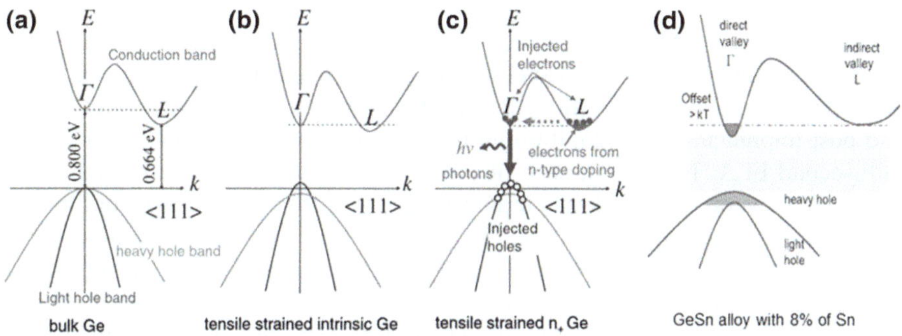

Fig. 4.21 Schematic band structure of bulk intrinsic Ge (**a**), of tensile strained i-Ge (**b**), of heavily n-type doped and tensile strained Ge (**c**) and of GeSn with 8% of Sn (**d**). After [122]

schematic band structure in intrinsic bulk Ge (i-Ge), in tensile strained Ge, in heavily n-type doped Ge and in a GeSn alloy with 8% of Sn [122]. Using only one of the presented methods the achievement of a direct bandgap in Ge is extremely difficult. The minimum biaxial tensile strain required for the direct bandgap in Ge is above 1.9% [123]. This is realized by the growth of Ge on foreign substrates with a lattice parameter much higher than the lattice parameter of Ge, e.g. on InGaAs. But such layers are metastable, and post-growth annealing, e.g. during p-n junction formation or metallization, releases parts of the strain.

The direct bandgap of Ge can be also formed by ultra-high n-type doping. It was shown that at RT, for an electron concentration higher than 8×10^{19} cm^{-3}, the L valley is filled with electrons up to the level which is energetically equal to the energy of the bottom of the Γ valley. Then the thermal energy is sufficient to fill the Γ valley with electrons. Because the radiative recombination via the direct transition has a 10^4 times higher probability than the indirect one, the highly n-type doped Ge becomes a quasi-direct semiconductor. The direct bandgap of Ge can be also formed by alloying Ge with Sn. The minimum Sn concentration required for the formation of a direct bandgap GeSn alloy is in the range of 8%, but can be significantly reduced

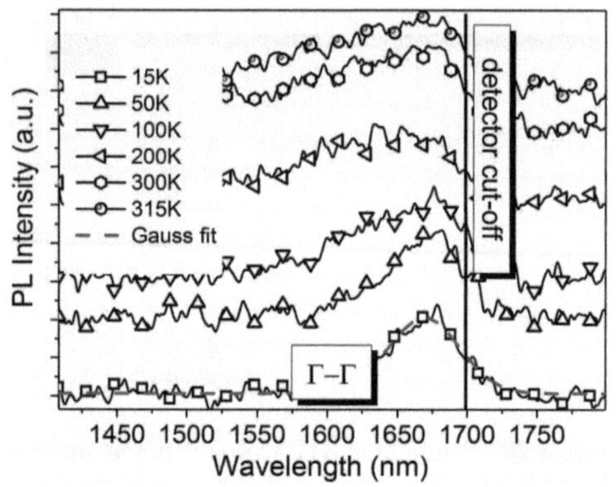

Fig. 4.22 Temperature dependent PL spectra obtained from P-doped Ge after r-FLA. The effective carrier concentration is above 9×10^{19} cm^{-3}. After [124]

(below 6%) by a combination of at least two of the discussed mechanisms, e.g. biaxial tensile strain and Sn alloying. Moreover, the proper bandgap engineering in Ge via Sn-doped and strained GeSn alloys can also be realized using ion beam implantation and post-implantation non-equilibrium thermal processing like nanosecond LA or millisecond FLA. In general, each of the presented methods alone is challenging because of several stability limits if annealing is performed in thermal equilibrium: stable n-type doping in Ge is limited to 3×10^{19} cm^{-3}, the solubility limit for Sn in Ge is below 1% and stable biaxial tensile strain in Ge can be easily controlled up to 0.2%. Therefore, the formation of a direct bandgap Ge needs the use of strongly non-equilibrium processing. It is realised mainly by MBE growth and in situ doping, but also by ion implantation followed by ms-range FLA. Figure 4.22 shows the temperature-dependent photoluminescence (PL) spectra obtained from a P doped Ge layer followed by r-FLA [124]. The low temperature PL is dominated by the Γ-Γ transition which is a direct evidence for the conversion of Ge from an indirect to a direct bandgap semiconductor. The PL spectra recorded above 150 K are partially cut due to the detection limit of the detector.

The ultra-doped n-type Ge can also be used for plasmonics operating in the near and mid-infrared region. To this day the most commonly used materials for plasmonics are metal nanoparticles whose main optical response is in the visible spectra range. But semiconductors offer the possibility to tune the plasma frequency edge via controlling the doping layer. According to theory, the plasma frequency edge is proportional to the square root of the carrier concentration n and inversely proportional to the square root of the effective mass m^* [105, 112]. The plasma frequency ω_p for a given doping level and particular material can be represented according to:

$$\omega_p \approx \sqrt{\frac{ne^2}{m^* \varepsilon_\infty \varepsilon_0}} \tag{4.3}$$

where ε_∞ and ε_0 are the dielectric constant (for ultra-doped n-type Ge $\varepsilon \approx 16$) and the permittivity of free space, respectively. e is the elementary charge.

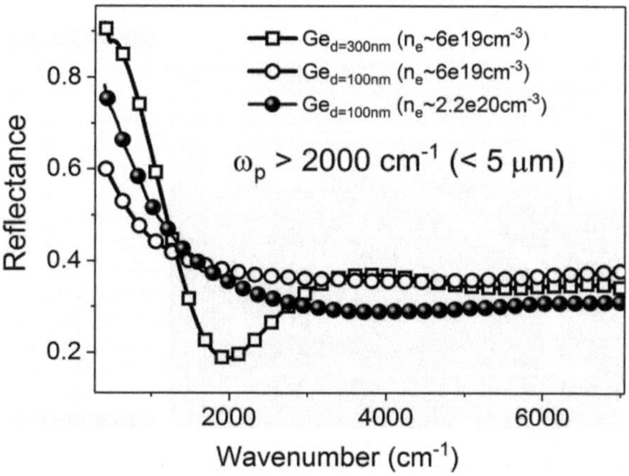

Fig. 4.23 RT reflectivity measurements from ultra-doped n-type Ge thin films after r-FLA 61.5 J cm^{-1} with an active carrier concentrations at 2 K of around 6×10^{19} cm^{-3} and 2.2×10^{20} cm^{-3}. The layer thickness of ultra-doped Ge is around 100 nm or 300 nm

Figure 4.23 shows the reflectance spectra as a function of the wavenumber of the incident light for different doping levels. Ge wafers were implanted with P ions and subsequently annealed by r-FLA for 3 ms. The electron concentration was 6×10^{19} cm^{-3} and 2.2×10^{20} cm^{-3} while the thickness of the doped layer was either 100 or 300 nm. For a given n and m^* the edge of the plasma frequency ω can be easily calculated using (4.3) and is confirmed experimentally for the 300 nm thick sample where the electrical field build up by carriers is strong enough to reflect most of electromagnetic radiation in the mid-infrared below $\omega = 2000$ cm^{-1}. The 100 nm thick doped Ge is, even with a doping concentration above 2×10^{20} cm^{-3}, semi-transparent for near infrared (NIR) radiation.

4.2.3.4 Superconductivity in p-Type Germanium

Since the first observation of superconductivity in 1911 hundreds of different materials have been found which exhibit superconducting states like metals: Hg or Pb, alloys like GeNb, NbTi, ceramics like YBCO, MgB_2, and other organic and inorganic superconductors. Science 2004 group IV semiconductors join the field of superconductivity. Ekimov et al. have discovered superconductivity in B doped diamond [125] and later Bustard et al. have shown that heavily doped p-type Si is also superconducting with a critical temperature below 1 K [126]. For the first time the superconductivity in heavily doped p-type Ge using ion implantation was presented by Herrmannsdörfer et al. [127]. A Ge wafer was implanted with Ga ions with a fluence of 4×10^{16} cm^{-2} and an energy of 100 keV which corresponds to a peak concentration of about 6% (Fig. 4.24a). After ion implantation samples were annealed by FLA for 3 ms or by RTA in the temperature range from 700 to 940 °C for 60 s. It was shown that after both FLA and RTA p-type Ge is superconducting with a critical temperature in the range of 0.2 and 1.2 K. The critical magnetic fields in superconducting Ge:Ga are in the range of 0.3 T which is one of the highest values ever

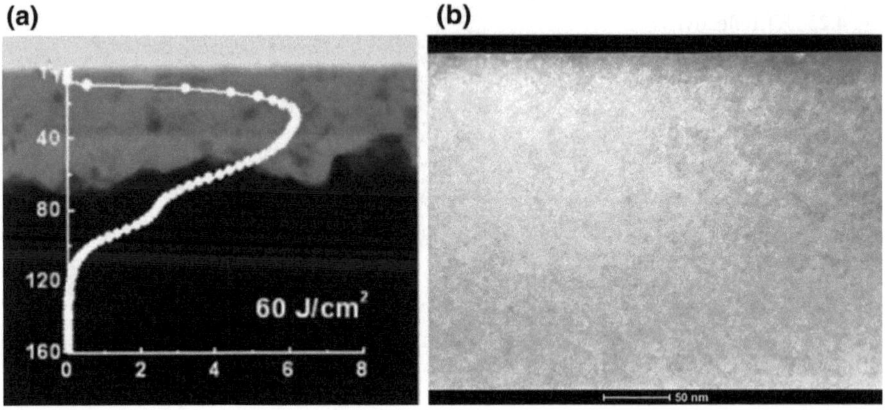

Fig. 4.24 Cross-section TEM images obtained from Ga implanted Ge followed by front side FLA (**a**) (after [129]) and r-FLA (**b**). The Ga fluence was 4×10^{16} cm^{-2}. The inset in (**a**) shows the depth distribution of Ga after FLA. Reprinted from [129], with the permission of AIP Publishing

observed in superconducting semiconductors. The maximum carrier concentration in Ga doped Ge is in the range of 1×10^{21} cm^{-3} which is two times higher than the solid solubility limit of Ga in Ge. Moreover, the structural properties of FLA treated Ge:Ga samples exhibit the formation of a polycrystalline layer like in P implanted Ge [128]. Usually, dopants in a polycrystalline matrix whose concentration is higher than the solid solubility limit have the tendency to decorate the grain boundaries. Using high resolution cross section TEM the formation of Ga clusters was not confirmed due to the resolution limits of the system. If we assume that Ga clusters can be formed the interpretation of the superconductivity in Ge offers space for at least two explanations: (i) the superconductivity is due to the formation of Cooper pairs in diluted Ge or (ii) due to Ga clusters with a diameter below 1 nm.

Detailed studies using atom probe microscopy confirm the formation of Ga clusters and agglomerates, but the origin of superconductivity in Ga doped Ge is still unclear [130]. Recently, single crystalline Ge:Ga with a carrier concentration above 1×10^{21} cm^{-3} was fabricated by using r-FLA (Fig. 4.24b). Such samples exhibit superconductivity with a critical temperature of up to 0.45 K (Fig. 4.25).

In summary, the maximum carrier concentration in Ga implanted Ge is independent of the annealing type, but after r-FLA the material is single crystalline without Ga clusters. Therefore, the superconductivity in such samples originates from ultra-doped Ge while after front side FLA Ga clusters can affect the critical temperature. After front side FLA or RTA the critical temperature can be higher than 1 K which is close to T_c of gallium clusters incorporated into a glass substrate.

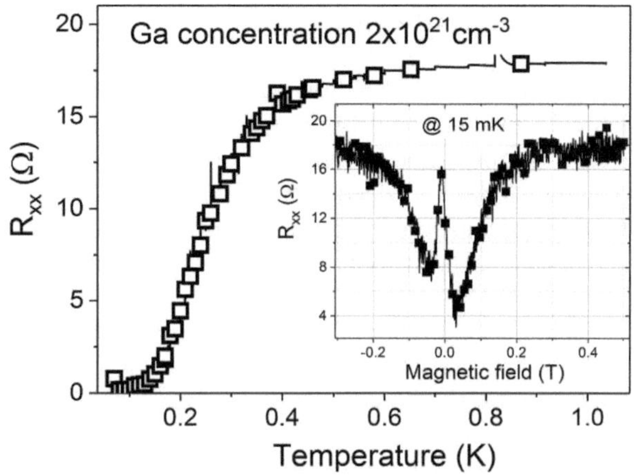

Fig. 4.25 Temperature dependence of the resistance in Ge:Ga for a Ga concentration of 2×10^{21} cm^{-3}. The inset shows the change of the resistance as a function of the magnetic field

4.2.4 Doping of Silicon Carbide

4.2.4.1 Introduction

In this world of downsizing and cost reduction leakage is one of the biggest enemies of microelectronics. As it behaves indirectly proportional to the bandgap of a semiconductor wide bandgap materials like silicon carbide (SiC) are favoured, especially for the application in high-voltage electronics. On the other hand, with regard to forward bias large bandgaps account for a high threshold voltage that needs to be overcome for conduction to set in via a p-n junction. This in turn slows down operation with respect to the use of SiC in high frequency technology. Therefore a balance needs to be found which may be done by doping wide bandgap semiconductors thus creating additional (donor/acceptor) levels within the "forbidden zone".

For robust performance microelectronic devices need to be able to dissipate latent heat. That means a high thermal conductivity of the substrate is desired among others and again this is helped by wide bandgap semiconductors. Further the large bandgap is the reason for SiC exhibiting a high electrical breakdown field which raises its importance for high temperature, power and frequency applications [131–133].

Boron and Al are used for p-type doping of SiC, while n-type doping is brought about by nitrogen. Thereby lateral diffusion is just as important and crucial for advanced microelectronic manufacturing as vertical diffusion. However, annealing implantation induced defects and electrical activation after thermal treatment have been found difficult. Conventional annealing is limited due to the risk of Si sublimation at elevated temperatures above 1750 °C. The reason for this lies in the smaller heat of sublimation for Si compared to the heat of reaction between Si and carbon forming SiC [134]. Challenges and solutions regarding doping of SiC will be looked at more closely in the following sections.

4.2.4.2 p-Type Doping

Boron and Al act as shallow acceptors, but only if they occupy Si sites in the SiC lattice. This is more easily achieved for Al than for boron. It is true for both though that a high acceptor concentration accounts for a comparatively low hole concentration due to the relatively deep energy levels created by p-doping in SiC [135]. However, finding suitable implantation and annealing parameters for optimum electrical dopant activation has been found challenging as contradictory results have been given by literature [131].

Hole concentration and hole mobility in 6H-SiC were studied for FLA (up to 2000 °C) and FA (below 1750 °C to avoid Si sublimation) at various implantation temperatures and dopant concentrations such that a buried layer of 500 nm thickness was implanted with boron about 100 nm below the SiC surface. To save this buried implantation zone from thermal degradation FLA was carried out on the wafer backside. The results for both kinds of dopants seem to oppose each other on the first sight. With increasing boron concentration (Fig. 4.26), the number of holes per cubic centimetre decreases thus rising its mobility, while the opposite is true for Al indicating that in the latter case the dopants are more successfully incorporated on lattice sites. Moreover, a significant increase was found after FLA in contrast to FA, especially for high dopant concentrations. The mobility rise for boron doped samples is especially steep for FLA annealed samples and occurs at a threshold concentration of about 3×10^{20} cm^{-3}.

Further, it was found that the hole concentration as a function of temperature behaves similarly for both FLA and FA and, moreover, independent of the boron

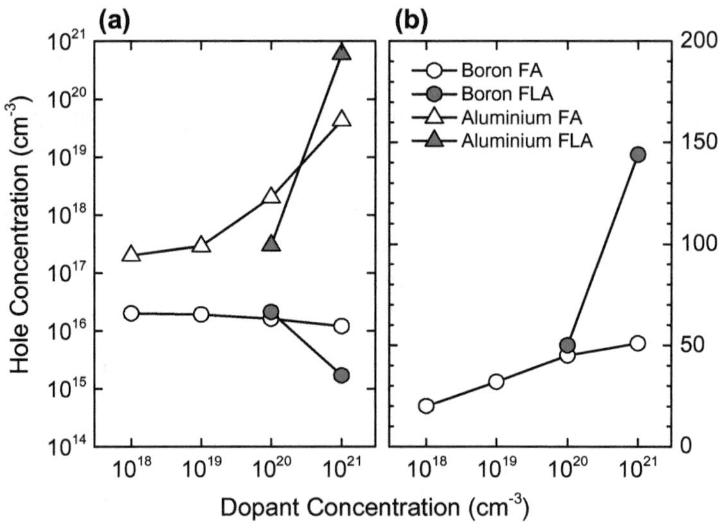

Fig. 4.26 Hole concentration (**a**) and mobility (**b**) in 6H-SiC as a function of concentration, annealing procedure and temperature. Data taken from [137]

concentration. This made the authors also assume similar processes, namely Ostwald ripening, dopant segregation and outdiffusion of boron, behind the results which would indeed reduce the effective dopant concentration thus reducing the hole concentration. Further studies have shown that outdiffusion is negligible for FLA. For high dopant concentrations Ostwald ripening is dominant for both FLA and FA, although the electrical properties after FLA (20 ms FWHM at 2000 °C) show slightly larger hole concentrations for boron doping than after FA. For large dopant concentrations above 5×10^{20} cm^{-3} this difference may be more pronounced [131, 132, 136–138].

The rise of hole concentration as a function of Al concentration, however, underlines the higher solubility of Al in the SiC lattice occupying the Si lattice sites more readily than boron as mentioned before. The hole concentration can be further increased by temperatures above the point of Si sublimation using FLA (Fig. 4.27). For high dopant concentrations the semiconductor almost shows metallic-like behaviour following FLA. This translates into a hole concentration which becomes independent of the measurement temperature for dopant concentrations presumably larger than 5×10^{20} cm^{-3}. This effect has been described as a "freezing-in of the enhanced solubility of Al in SiC" and cannot be found after FA where thermal conduction is the dominant process. The described behaviour after FLA for large Al concentrations is attributed to its low diffusion coefficient in contrast to boron in combination with the short pulse of the flash providing a maximum in temperature (here: 2000 °C) with an abrupt drop at the end of the flash allowing for instantaneous radiation cooling. It is the small diffusion coefficient of Al which preserves its high solubility upon cooling which is why this phenomenon cannot be found for boron (Fig. 4.26), which exhibits a smaller solubility while the diffusion coefficient is larger. Therefore, annealing above a certain critical temperature is inefficient for boron with respect to hole concentration though it is larger after FLA than after FA [131, 137, 138].

The hole mobility decreases with temperature for lower Al concentrations though independent of the kind of annealing. However, overall the mobility is a bit smaller for the FLA annealed sample which may be brought about by quenching-induced defects [138]. Only for higher concentrations the decrease is a lot smoother due to enhanced dopant interaction (Fig. 4.27) [135–137].

The influence of thermal treatment on diffusion was looked at, but no significant broadening of the implanted layer was found after FLA. Slow positron annihilation spectroscopy has shown that defects caused by p-type doping are minimised by increasing substrate temperature during ion implantation and subsequent annealing. In fact, the annealing technique itself does not significantly alter defect creation or rather annihilation. The electrical activation of the dopants in terms of sample resistivity as a function of temperature and dopant concentration has also been studied. While keeping the diffusion time to a minimum FLA allows for comparable conductivity results with respect to FA [131, 132, 136, 138, 139].

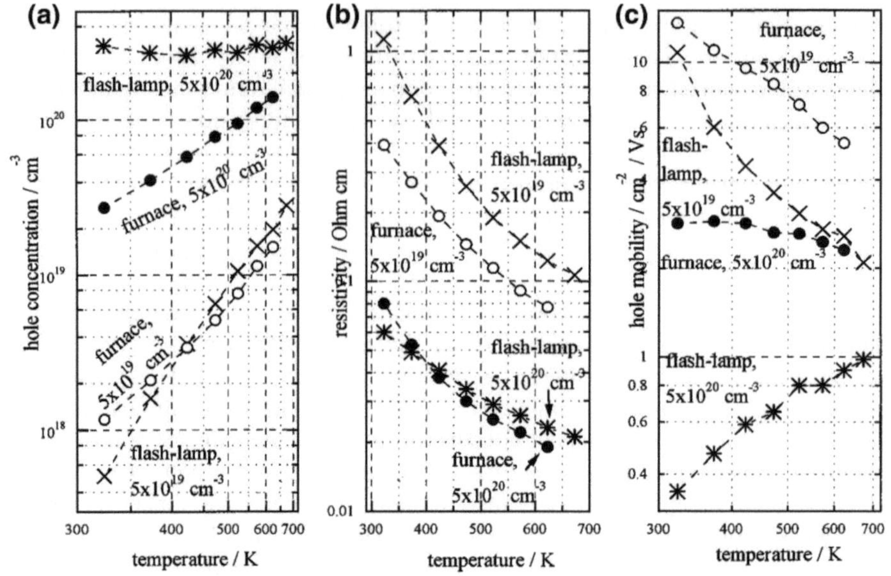

Fig. 4.27 Hole concentration (**a**), resistivity (**b**), and hole mobility (**c**) as function of temperature for Al-implanted 6H-SiC after FLA or FA for two different Al concentrations of 5×10^{19} and 5×10^{20} cm^{-3}. Reprinted from [138], with the permission of AIP Publishing

4.2.4.3 n-Type Doping

N-type doping in 6H-SiC proceeds via two steps. First the energetically lower hexagonal sites of the 6H modification are occupied, followed by the cubic lattice sites. The donor energy levels were found at 79 meV and at 128 meV, respectively, while after FLA only the first remains indicating mainly hexagonal lattice sites being occupied. The reason for this may be found in the short annealing time during FLA that only allows for enough time to occupy the hexagonal lattice sites. Therefore, no significant improvement of the electrical activation was found during FLA despite enhanced annealing temperatures. However, it could be shown that FLA followed by conventional annealing leads to an increase in hole concentration [136, 137].

4.2.4.4 Others

Conventionally, for the epitaxial growth of cubic SiC by FLA the underlying Si substrate is melted which, however, leads to an enormous interface roughness and a high defect density therein (Fig. 4.28a) due to the large lattice mismatch of 21%, which is not the case for its hexagonal polymorphs. Upon resolidification this roughness expands towards the surface and creates something that is often called “buckling”. FLA can already help without further precautions to minimise the defect density by

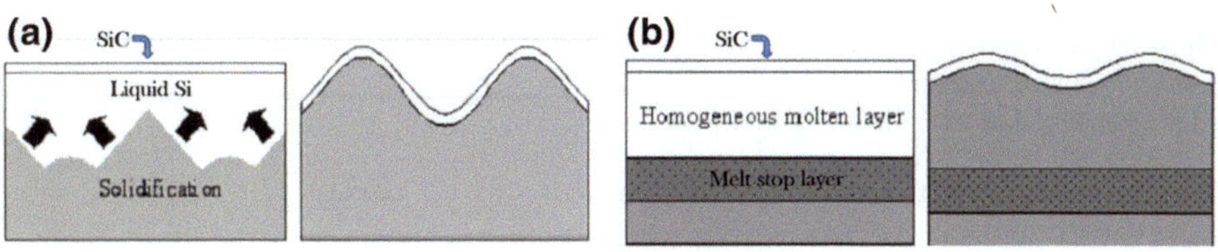

Fig. 4.28 Solidification of Si beneath the Si/SiC interface without (**a**) and with (**b**) melt stop. Reprinted from [143], with the permission of AIP Publishing

choosing the annealing properties such that the maximum radiant flux is deposited at the 3C-SiC/Si interface. The thus created locally confined melted layer partially dissolves the SiC which—using the annealed SiC surface as a seed—recrystallizes with a minimum of defects. This method has become known under the acronym FLASiC and will be explained in full detail in Sect. 4.3.3. The defect ingrowth can be restricted that way, but the undesired surface roughness due to stress release is not impeded. Therefore, an upgrade of the just explained FLASiC process was developed, called inverse)-FLASiC which again will be explained at length in Sect. 4.3.3 [140–143].

An alternative to the i-FLASiC process is the introduction of a so-called "melt stop". If a high dose of carbon is implanted into the Si beneath the interface 3C-SiC/Si as shown in Fig. 4.28b, the melting point of this thin layer is raised such that it acts as a melt stop to the undoped substrate beneath. Thereby a homogeneously molten Si layer is created upon FLA between the SiC surface and the melt stop because the lateral growth of recrystallisation seeds (shown as pyramids in Fig. 4.28a) is limited through restricted mass transport of Si atoms to the interface. Therefore, a relatively smooth surface can resolidify from this homogeneously molten layer and buckling is diminished. This technique has also been successfully applied to heterostructures. Similarly, this technique may work for semiconductor doping if the melt stop is put beneath the implanted layer to control diffusion during annealing [143, 144].

Taking it from the dissolution of SiC into the Si melt at the interface [145] improved the formation of SiC-on-insulator. The samples were prepared by depositing a 35 nm thick 3C-SiC layer on Si. After FLA they found an upper highly defected region while beneath the defect density was significantly reduced. So they wet oxidised the surface thus creating a thick SiO_2 layer (Fig. 4.29). Simultaneously, a buried silicon oxide layer was created beneath the less defected SiC through channels that had been formed by the wet oxidation. Afterwards the SiO_2 cap was removed by etching and a thicker SiC layer could be grown onto the less defected SiC thin layer giving the desired SiC-on-insulator -structure with minimal defect density [145].

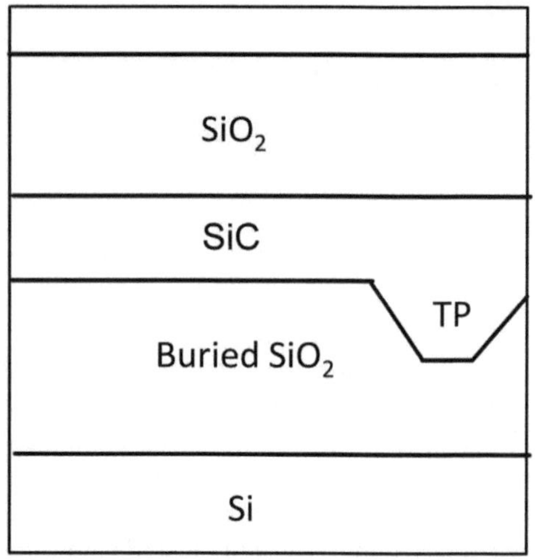

Fig. 4.29 Sketch of a SiC-on-insulator structure after annealing and wet oxidation, but before the SiO_2 cap is removed; TP refers to trapezoidal protrusions—the regions with low defect density as described in the text [145]

4.2.5 Doping of III–V Compound Semiconductors

4.2.5.1 Constraints of Annealing

Most of the III–V compound semiconductors have a direct bandgap which makes them attractive for light emitters operating in the spectral range from the UV to the mid-infrared. The bandgap engineering relays on the composition change within ternary or quaternary alloys. The controllable doping of these semiconductors is crucial for optoelectronic and high speed electronic devices. The III–V compound semiconductors can be doped in situ during the growth or ex situ using conventional ion implantation or diffusion methods. Light emitting diodes and lasers must be fabricated from nearly defect-free semiconductors. Therefore, only epitaxial layers in situ doped are considered for light emitters. The doping during epitaxial growth is limited resulting in layers of p-type or n-type material. This defines the circuit design and only vertical devices can be fabricated. The application of ex situ doping using ion implantation is limited due to defect formation during the doping process. Unfortunately, many defects are temperature-stable and cannot be completely removed during post-implantation annealing like vacancies which have the tendency to form clusters during annealing. Therefore, the post-implantation annealing of III–V compound semiconductors is very challenging. The annealing temperature of the implanted layer must be as high as possible to reconstruct the crystal structure and to activate dopants. On the opposite side, the annealing time must be reduced to a minimum to avoid dopant diffusion. But due to the thermal instability of binary semiconductors the maximum annealing temperature at normal conditions must be much lower than the melting temperature which reduces the probability to remove the implantation damage. For instance, the melting temperature of GaAs at normal pressure is about 1200 °C but using conventional FA the maximum temperature at which GaAs remains still stable is about 700 °C. The annealing of GaAs above 700 °C causes the decomposition of GaAs; As evaporates from the surface and a Ga-rich surface layer is formed. The maximum annealing temperature can be significantly increased by applying a capping layer and reducing the annealing time. Using RTA and a protective capping layer GaAs can be annealed for a few seconds at a temperature as high as 1000–1050 °C. If the annealing time is reduced to the millisecond time scale the annealing temperature can reach 1150 °C without surface decomposition [146, 147].

Using ns-pulses for annealing the doped layer of the III–V compound semiconductors can be even molten. Thus, similar to group IV semiconductors, LPE of ion implanted layers in III–V compound semiconductors is achieved by using nanosecond PLM [148]. During solidification of the molten layer, the implanted elements are incorporated into the lattice of the III–V compound semiconductor. Using ion implantation and LPE the activated dopant concentration can be a few orders of magnitude bigger than the solid solubility limit. This was successfully presented for Mn doped binary compound semiconductors like GaAs, InP, GaP and InAs [148]. The situation is much more complicated in the case of binary semiconductors with a very high melting point like GaN (about 2500 °C). At normal conditions GaN cannot be molten because N sublimates before the melting temperature is reached. Hence, the activation of dopants in GaN using PLM and LPE is not possible. Recently, GaN became a basic semiconductor for solid-state lightening, high power electronics and the next generation of the mobile networks "5G". Bulk GaN wafers are extremely expensive which is why the commercially used GaN is epitaxially grown on Si (high power electronics) or on sapphire (light emitting diodes). Although ion implantation is destructive for the crystal structure and post-implantation annealing may not remove all defects created during doping, ion implantation can be used for high power electronics and high-speed electronics where vertical devices are required.

4.2.5.2 Doping and Bandgap Engineering in Gallium Arsenide

The optoelectronic properties of III–V compound semiconductors can be easily tuned by changing the composition e.g. by alloying with N. This alloying causes a reduction of the fundamental bandgap, increases the carrier mobility and reduces the carrier effective mass. For example, the incorporation of only 1% of N into GaAs shrinks the effective bandgap of GaAs by 120–180 meV [149]. The strong change of the bandgap in III–V-N is due to the local perturbation of the crystal lattice potential when As atoms are replaced by smaller atoms with a much higher electronegativity like N [149]. The change of the bandgap with increasing N concentration is defined by the bowing parameter (see Fig. 4.30) and can be explained using the band anticrossing model. This model is composed of two simple models: (i) extended states of the GaAs matrix and (ii) localized states of N in GaAs. It explains the change of the bandgap in GaAsN based on the conduction band splitting. N in GaAs forms an energy level above the conduction band minimum (about 250 meV). The conduction band splitting is caused by an N-induced perturbation which increases with increasing

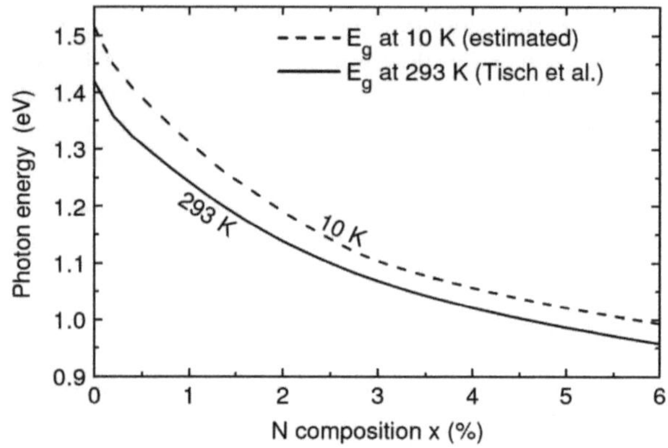

Fig. 4.30 The bandgap change in GaAsN as a function of the N concentration [150]. © Juha Toivonen

N concentration. The solid solubility of N in GaAs is extremely low. Even by using MBE, it is challenging to grow $GaAs_{1-x}N_x$ with an N content higher than 1%

GaAsN alloys with N concentrations much higher than the solid solubility can be achieved by using ion implantation and non-equilibrium thermal annealing. Gao et al. have investigated the bandgap engineering in GaAsN made by ion implantation followed by FLA for 3 ms [151]. The peak temperature obtained at the surface of implanted GaAs was in the order of 1100 °C which is much higher than the decomposition temperature of GaAs (about 700 °C). The annealing of GaAs at such high temperatures was only possible due to the use of a SiO_2 capping layer and non-equilibrium thermal annealing. For comparison ns-range PLM followed by RTA was used as well. The highest incorporation efficiency of N into the lattice of GaAs was achieved after double-flash FLA for 3 ms. The $GaAs_{1-x}N_x$ layers with $x = 0.38\%$ and 0.76% exhibit RT PL at about 890 and 910 nm, respectively. According to the band anti-crossing model the observed PL emission corresponds to the effective incorporation of N in the range of 50%. Surprisingly, the sample only annealed by the laser did not show PL at all. Samples implanted with N and annealed first with laser and later by RTA shows very week PL at about 880 nm. The temperature dependent PL reveals additional PL peaks in the range of 1100–1600 nm (see Fig. 4.31a). The NIR PL is due to the radiative transition between deep donor and acceptor states within the bandgap of GaAs. Usually such a PL emission is observed in n-type GaAs and can be enhanced by ion implantation [152]. Figure 4.31b shows the change of the bandgap as a function of temperature for virgin GaAs and GaAsN (0.76% of N). In both cases the tendency is the same (bandgap reduction with increasing temperature) which additionally confirms that the observed PL in N-implanted GaAs annealed by FLA is due to the modified bandgap in GaAs. The optical emissions at 0.8, 1.3 and 1.5 μm match the transmission window in optical glass fibres. Therefore, such a PL emission is interesting not only from the point of view of fundamental research but also from that of potential applications in telecommunication.

The nature of the 1.3 μm emission from GaAs was investigated by Prucnal et al. [146] using temperature-dependent PL spectroscopy and positron annihilation spec-

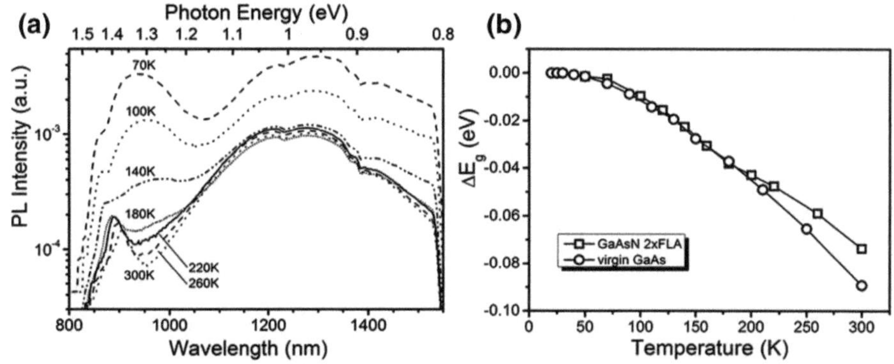

Fig. 4.31 Temperature dependent PL observed from N implanted GaAs followed by double flash FLA for 3 ms (**a**) and the change of the near-band edge emission peak as a function of temperature obtained from virgin and N (0.76%) implanted GaAs (**b**)

troscopy. The latter technique is sensitive to open volume defects being either neutral or negatively charged. Positively charged defects like As or Ga vacancies cannot be detected by this way. The main defect-related emissions from GaAs originate from Ga and/or As vacancies/interstitials complexes [153]. Unfortunately, there is no available technique, neither during the growth nor by using post-growth treatments like ion implantation, irradiation etc., which allows the full control over the type and density of defects arising in GaAs. Interestingly, the 1.3 μm emission in GaAs is observed from intrinsic or n-type GaAs. The p-type doping quenches this PL emission completely. Prucnal et al. have used ion implantation and post-implantation FLA to modify the electrical and microstructural properties of intrinsic GaAs. Already the virgin, intrinsic GaAs wafer exhibits weak NIR PL emission at RT. The NIR PL intensity can be significantly enhanced by applying non-equilibrium annealing like FLA. On the one hand FLA is too short to activate the surface decomposition but long and powerful enough to form and activate some point defects. This may lead to an enhancement of the 1.3 μm emission. In contrast to virgin GaAs, after FLA the defect related emission is enhanced by a factor of more than 200. Simultaneously, the near band edge emission is quenched. This can be either due to the degradation of the structural properties of the annealed sample or due to the energy transfer from excited electrons to the defect centre. In the case of the N implanted sample both the near band edge emission and the defect related luminescence are more intense in comparison with the virgin wafer (see Fig. 4.32).

After doping of GaAs with Mn ions which are much heavier than N ions the near-band edge emission is well distinguishable, but the defect related NIR luminescence is fully vanished. Mn is a deep acceptor with an energy level located about 110 meV above the valence band minimum. Therefore, Mn ions incorporated into As lattice sites of GaAs are deep acceptors. A Mn ion located at an interstitial position or replacing Ga is a double or single deep donor, respectively. Mn ions which do not replace As act as compensation centres for holes. Thus, properly fabricated, Mn-doped GaAs

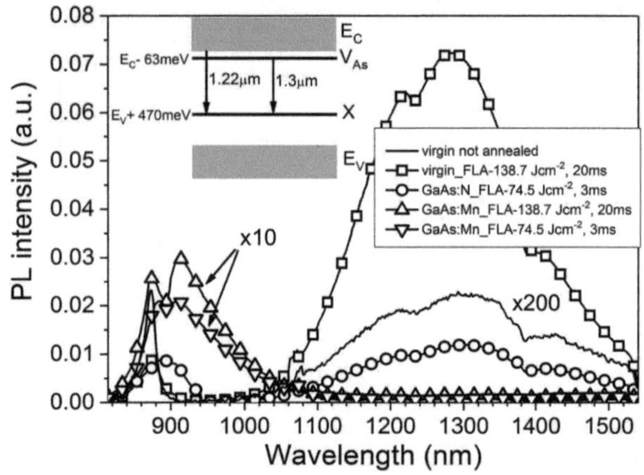

Fig. 4.32 PL spectra obtained from virgin, not annealed GaAs, from virgin GaAs after FLA and from ion implanted GaAs annealed either for 3 ms or for 20 ms. The peak temperature was fixed at about 1100 °C. The inset shows the schematics of the NIR optical transitions in GaAs under green light excitation. After [146]

is p-type. Prucnal et al. have proposed that the NIR PL is due to a radiative transition between the conduction band and some deep acceptor levels (X-centres) which are created during the growth or during post growth irradiation [146]. The absence of the NIR PL in Mn doped GaAs can be due to the deactivation of X-centres in GaAs via deep acceptors. The NIR PL emission is composed of two peaks located at 1.22 and 1.3 μm. If we assume that the 1.3 μm emission is due to the radiative recombination of electrons in the conduction band and holes localised at *X*-centres, the 1.22 μm PL must originate from a deep donor-acceptor pair emission. In the case of irradiated GaAs or GaAs annealed at high temperature the most common donor defect is the As-vacancy. As-vacancies can have an ionization state of 1+, 2+ or 3+. The corresponding energy levels are located at about 30, 60 and 140 meV. Considering the bandgap of GaAs and the PL emission peaks the energy level of the X-centres must be located at about 0.47 eV above the valence band maximum. The schematic representation of the energy levels and possible optical transitions are shown in the inset of Fig. 4.32. The existence of the X-centres was also confirmed by positron annihilation spectroscopy. According to this both the ion implanted sample and the FLA annealed samples contain only one type of defects which can trap positrons. In the case of N implanted GaAs the depth distribution of X-centres overlaps the depth distribution of N. In conclusion, the NIR PL emission originates from the radiative recombination between electrons in either the conduction band or at As-vacancies and holes at the X-centres.

The NIR PL emission can be easily enhanced by the proper choice of the implanted element. Gao et al. have investigated the influence of the ion implantation process on the intensity of the NIR PL emission and on the temperature-dependent PL stability of GaAs wafers. After ion implantation samples were annealed by FLA [152]. In order to verify the influence of different ions on the NIR PL emission the virgin, not processed wafer and unimplanted, but annealed samples were tested as well. Due to the fact that the NIR PL is absent in p-type GaAs only dopants which are isoelectronic to Ga or As were chosen. Figure 4.33 shows the PL spectra obtained from the virgin, not processed wafer, from the unimplanted wafer after FLA and

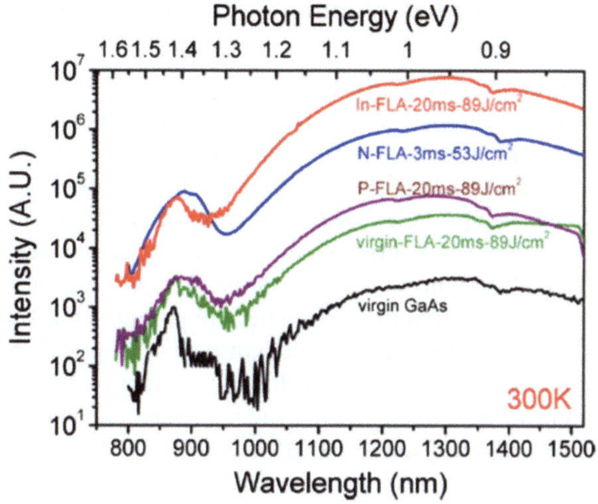

Fig. 4.33 PL spectra observed from intrinsic GaAs wafers after ion implantation and FLA. Reprinted from [152], with the permission of AIP Publishing

from samples implanted with P, N or In followed by FLA for either 20 or 3 ms. The annealing time was always optimised for the maximum NIR PL emission. The GaAs wafers used here were standard, intrinsic, (100) oriented wafers grown by the Czochralski method. According to the PL already virgin GaAs wafers contains both As vacancies and X-centres which are optically active. After FLA without ion implantation the PL emission is tenfold enhanced. The peak temperature at the surface during FLA was in the range of 900 °C. According to a previous study annealing for 20 ms is sufficient to remove some of the point defects but is too low to activate the decomposition of GaAs and the formation of As vacancies. This suggests that during FLA the defect concentration responsible for the NIR PL emission is not affected but some of the non-radiative de-excitation channels are removed.

After ion implantation and FLA the NIR PL emission is significantly enhanced. The biggest enhancement of the NIR PL was achieved after In implantation followed by FLA for 20 ms. The NIR PL is almost three orders of magnitude stronger than that observed from virgin GaAs wafer. Moreover, the temperature dependent PL reveals that the NIR PL is quite temperature-stable. If temperature rises from 20 to 300 K the In-implanted GaAs exhibits a PL intensity reduction by only 67% which is unusual for defect related emissions. In general, the strongest NIR PL was recorded after FLA for 20 ms, which can be related with the reduction of point defects during annealing. One exception is N where the strongest PL was observed after FLA for 3 ms. This can be due to the very low solubility of N in GaAs and the tendency of N to form bubbles during longer annealing. The intense PL emission observed at RT opens the possibility to apply ion irradiated GaAs in optoelectronics.

4.2.6 Diluted Magnetic Semiconductors

Conventional nanoelectronics is based on the charge manipulation where the reaction time of the transistors is limited by the charge mobility. The read-out/access time of electronic circuits can be significantly reduced if the basic functions of electronic devices will be controlled by the manipulation of electron spins instead of charges. To realize spintronic (short for spin electronics) circuits magnetic semiconductors are required. The first magnetic semiconductors were realized by the incorporation of small amounts of magnetic atoms into a II–VI semiconductor host and into oxides [154]. The doping of magnetic atoms into semiconductors converts the host into a magnetic material without significant deterioration of their optical and electrical properties. Due to the fact that the magnetic atoms are at a substitutional position in the semiconductor matrix the ferromagnetic materials are called diluted magnetic semiconductors/oxides (DMS).

4.2.6.1 Manganese Doping of III–V Compound Semiconductors

The most studied DMS is probably (Ga, Mn)As which is widely considered as the prototype of DMS with a Mn content in the range of 1–15%. In 2000, Tomasz Dietl et al. published a theoretical paper in Science where they explained the origin of ferromagnetism in Mn doped III–V compound semiconductors and theoretically predicted the Curie temperature (T_C) for different (III, Mn)V DMS [155]. Figure 4.34 shows the theoretically predicted T_C for different semiconductors containing 5% of Mn. According to the theoretical calculation T_C can exceed RT for (Ga, Mn)N and (Zn, Mn)O. Unfortunately, to this day DMS operating at RT have not been realized yet. (Ga, Mn)N has a Tc below 30 K and the magnetism in Mn doped ZnO has not yet been fully clarified due to the existence of magnetic defect centers in virgin ZnO. The most optimistic material is (Ga, Mn)As with T_C slightly below 200 K. However, the basic phenomena discovered in DMS can provide fundamental insights into ferromagnetism in other types of materials like ferromagnetic metals or semimetals which are highly relevant for spintronic applications [156].

As mentioned before, Mn doped III–V compound semiconductors are the most studied DMS. To realize the DMS Mn atoms must replace the group III atom in the crystal structure. Being in the proper substitutional position Mn is a deep acceptor (in GaAs, GaP, InP, GaN) with an activation energy much above 100 meV or a shallow acceptor like in InAs or InSb. Figure 4.35 shows the schematic representation of potential Mn locations in GaAs (a) and the localisation of the acceptor level with respect to the valence band minimum in different III–V compound semiconductors doped with Mn. Mn being in the interstitial position is a double donor while Mn which replaces group V atoms in the crystal is a single donor. This means that each Mn atom in the interstitial position introduces two electrons to the matrix. Moreover, most of the point defects in III–V compound semiconductors are shallow or deep donors. Since the ferromagnetism in (III, Mn)V compound semiconductors

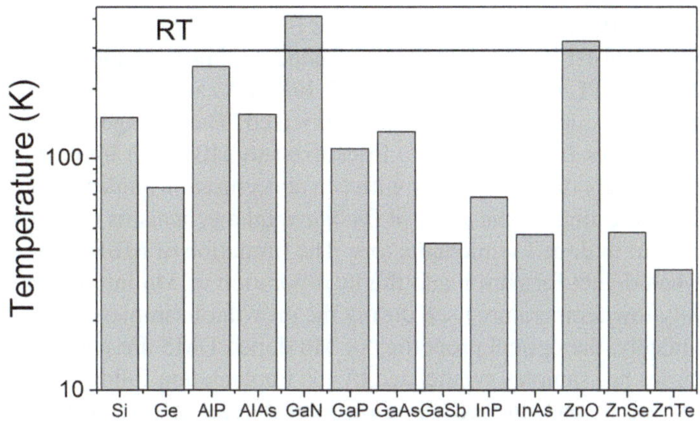

Fig. 4.34 Theoretically predicted T_C in different DMS containing 5% of Mn. After [155]

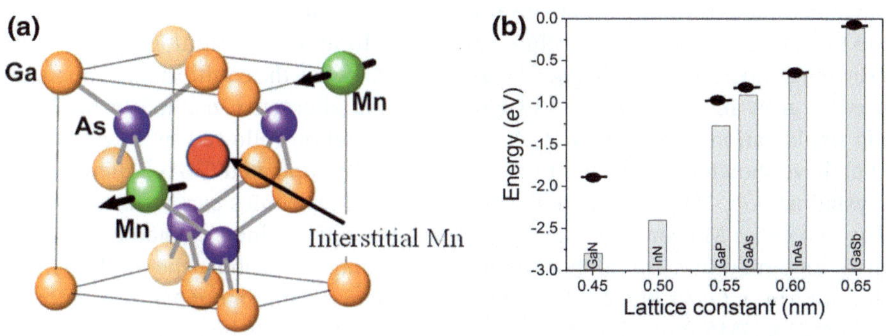

Fig. 4.35 Schematic representation of Mn doped GaAs with Mn atoms in an interstitial or substitutional position (**a**) and experimentally determined position of the acceptor level caused by Mn doping in various III–V compound semiconductors (**b**). After [159]

is hole-mediated, any electrons injected into the ferromagnetic system significantly decreases T_C [157].

The model of hole-mediated ferromagnetism was proposed for the first time by Zener in 1958 to explain the ferromagnetism in metals [158]. Later Dietl adopted the Zener model to explain the ferromagnetism in Mn doped III–V and II–VI semiconductors [155, 159]. In order to fabricate DMS the minimum Mn concentration is in the range of 1% which is more than four orders of magnitude above the solid solubility limit. The corresponding hole concentration is above 5×10^{19} cm^{-3}. Therefore, DMS can be fabricated only by using strong non-equilibrium processing. Nowadays two techniques are available; MBE and ion implantation followed by sub-second thermal annealing. By using MBE different arsenic based DMS and (Ga, Mn)N were produced with T_C strongly affected by the crystalline structure and the Mn content. Ion implantation followed by sub-second thermal annealing, namely nanosecond pulsed laser annealing (PLA) and millisecond range FLA, can be used to fabricated DMS based on different III–V compound semiconductors [148]. Using ion implantation followed by 30 ns PLA, DMS based on GaP, InP, InAs and GaAs with a T_C close to the predicted theoretical value have been fabricated. The average Mn concentration in the matrix varies from 0.5 to 6%. In contrast to MBE, ion implantation allows to fabricate strongly diluted systems with Mn concentrations much below 1%. Such systems are interesting for the study of fundamental mechanisms responsible for the ferromagnetism in doped semiconductors. The limitation of MBE is due to competition between defect formation and the incorporation of Mn into the lattice due to the relatively low temperature used during the growth of samples.

Unfortunately, the optical properties of Mn doped DMS are poor. Neither MBE grown samples nor samples synthetized by ion implantation followed by PLA show PL. This is due to both the existence of point defects in the ferromagnetic layer and the formation of a degenerated semiconductor which limits the PL efficiency. Independent on the fabrication method about 10% of the introduced Mn atoms are in the interstitial position. They create efficient non-radiative decay channels which kill the PL. Moreover, to ensure a high amount of Mn in the substitutional position the growing temperature for Mn doped GaAs is much below the equilibrium temperature used to grow high quality GaAs ($\ll$600 °C). An alternative method to MBE and ion implantation followed by PLA is ion implantation followed by FLA. Prucnal et al. have shown that utilising FLA to anneal a Mn implanted GaAs wafer, a single crystalline (Ga, Mn)As layer with superior optical properties can be formed [147]. The maximum Mn concentration which can be effectively incorporated into GaAs is in the range of 0.6%. This is about three orders of magnitude above the solid solubility limit of Mn in GaAs but still two times lower than the minimum Mn concentration required for DMS formation. Nevertheless, the fabricated film is single crystalline and paramagnetic, whereas the bandgap decreases with increasing the Mn concentration.

Figure 4.36 shows a high resolution TEM image of a Mn implanted and FLA annealed GaAs sample with 0.6% of Mn. The Mn depth distributions in the as-implanted (green curve) and the annealed sample (red curve) are included in the cross-section image presented in Fig. 4.36a. Moreover, the magnifications of three regions are presented in Fig. 4.36b–d. The region close to the surface (A) is very sensitive to the formation of point defects during ion implantation and annealing. During ion implantation the surface region is often selectively sputtered which leads to the formation of a Ga-rich layer. In addition, during high temperature annealing As has the tendency to evaporate first via sublimation. In both cases a Ga rich layer can be formed. Although the peak temperature is close 1100 °C during FLA, a decomposition of the surface region was not observed. Region B corresponds to the area with a maximum Mn concentration. During the FLA process Mn slightly out-diffuses leading to a box-like distribution of Mn and the formation of a homogeneously doped layer with a thickness of about 200 nm. Due to the high Mn content the formation of secondary phases like MnAs can be expected. However, we did not observe such clusters at least within the resolution limit of our experimental setup. It is known that MnAs is ferromagnetic with a T_C about 300 K [160]. Therefore the formation of secondary phases like MnAs can be easily verified by measuring the

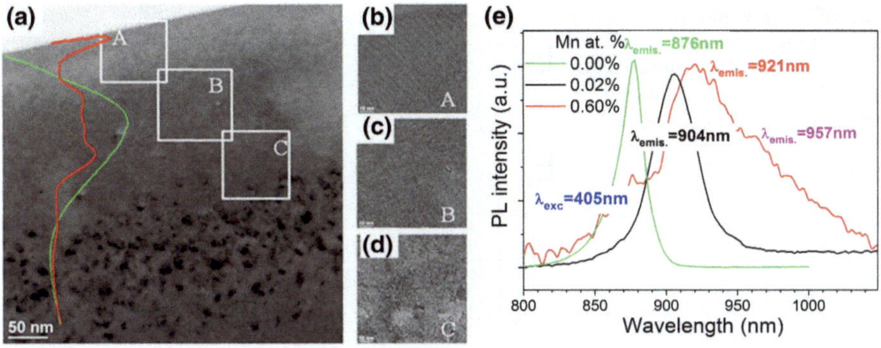

Fig. 4.36 High-resolution cross-section TEM image of a Mn implanted and FLA annealed GaAs wafer (**a**) and magnifications of three regions marked as A, B and C (**b–d**) respectively. The distribution of Mn in the as-implanted and annealed sample is marked with a green and a red line, respectively. **e** displays the RT PL obtained from virgin GaAs and wafers containing 0.02 or 0.6% of Mn after FLA for 3 ms. Samples were excited with a 405 nm laser

magnetic properties of the fabricated layer. Since our system is only paramagnetic the formation of MnAs can be excluded. Finally, region C with its high density of defects corresponds to the region of EOR defects typically observed in ion implanted and annealed semiconductors. The influence of the Mn doping on the bandgap of GaAs was investigated using RT PL. It is worth noticing that MBE grown (Ga, Mn)As and samples annealed by PLA do not emit light.

Samples containing different Mn concentrations and made by ion implantation followed by FLA for 3 ms were excited with a 405 nm laser. The virgin GaAs exhibits a PL emission peak at 875 nm due to the radiative recombination of electrons and holes localised in the conduction and valence band, respectively. The Mn doped samples show that the PL peak is significantly shifted towards longer wavelengths. The shift of the PL peak is due to the change of the valence band edge towards the impurity band formed in GaAs by Mn atoms. The decrease of the bandgap in Mn doped GaAs is a direct evidence that holes introduced into the system are localised in the valence band but not within the impurity band (see Fig. 4.37). Hence the presented PL data fully support the Zener model which is used to explain the ferromagnetism in Mn doped GaAs and which is in contrast to the impurity band model [147, 148].

4.2.6.2 Magnetic Materials Made by Ion Implantation and FLA

Due to the strong segregation of transition metals in group III–V compound semiconductors and the very low solubility limit the use of conventional annealing is limited. Even the utilization of millisecond range FLA cannot produce ferromagnetic bulk or thin film DMS. But ferromagnetic *nanostructures* can be fabricated by utilizing the segregation of Mn and the formation of secondary phases in III–V compound semiconductors. Khalid et al. have shown that using Mn implantation into GaAs and

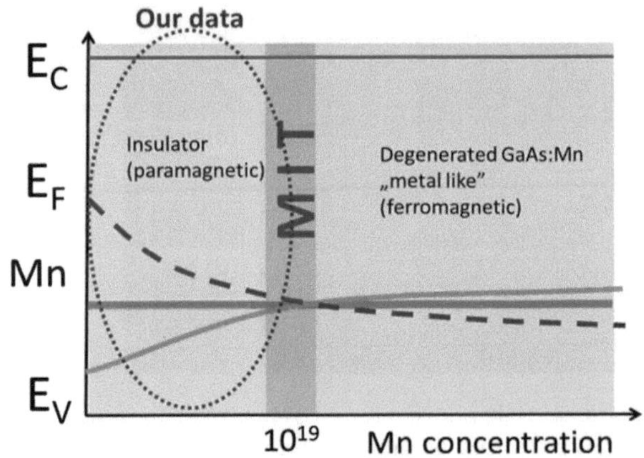

Fig. 4.37 The bandgap model of GaAs heavily doped with Mn. The region of the FLA annealed samples is marked by a dashed ellipse. The IMT is also shown

InP, MnAs and MnP magnetic nanostructures can be formed during ms-range FLA [160]. Thereby, T_C is higher than RT which is promising for potential applications in spintronics. The formation of ferromagnetic semiconductors is not limited to the group II–VI and III–V compound semiconductors, but also group IV semiconductors doped with Mn can become ferromagnetic if the proper phase is formed. Further details about Mn-doped NWs are given in Sect. 4.4.3.

4.3 Crystallization

4.3.1 Non-equilibrium Crystallization Modes

4.3.1.1 Fundamentals

Non-equilibrium crystallisation can also be termed explosive (re)crystallisation (EC) and describes the process of spontaneous reordering of an amorphous state into a (poly)crystalline state driven and self-catalysed by the latent heat of the *ignition* which may be a laser or a flash pulse [161–163]. The reordering velocity v of the process—in the order of several meters per second—depends critically on temperature which is why it can also be used as a mean for temperature measurement [164]:

$$v \propto e^{-E_A/k_B T} \tag{4.4}$$

where (v = reordering velocity; E_A = activation energy; T = temperature; k_B = Boltzmann constant) [161, 165].

A frequently studied workpiece is amorphous Silicon (a-Si) of several micrometres in thickness on a glass substrate. During the process of (re)crystallisation microstructures are created with grains of various orientation and size. The relationship of which to the specific kind of process will be discussed later. The (re)crystallisation is stopped again if the proceeding front cools down below a threshold temperature [166].

In the year 1986 already, Geiler et al. postulated four processes by which (re)crystallisation can take place keeping in mind that the samples were treated in an RTA regime and that ignition was carried out by a laser pulse instead of a flash. The factors governing the distinction are phase and free energy and can be categorized as follows illustrated by Fig. 4.38:

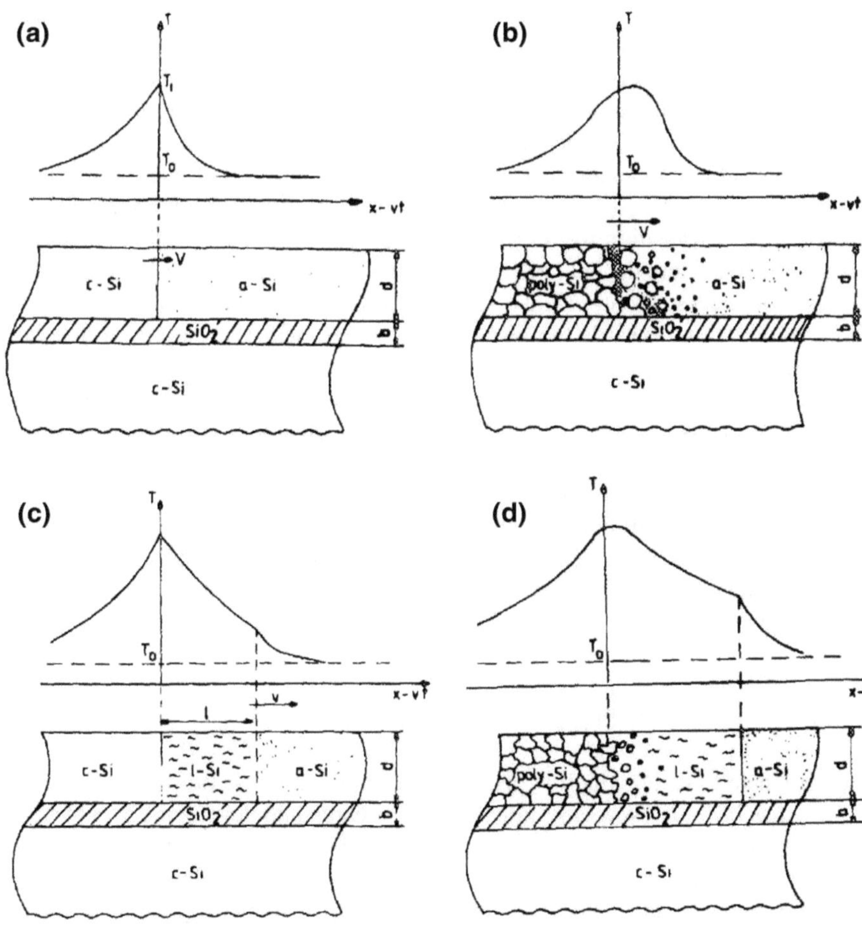

Fig. 4.38 Sketch of the various (re)crystallization modes as proposed by Geiler in 1986: **a** ESPE, **b** ESPN, **c** ELPE, **d** ELPN. More details can be found in the text. Reprinted from [161], with the permission of AIP Publishing

(a) ESPE—explosive solid-phase epitaxy
(b) ESPN—explosive solid-phase nucleation
(c) ELPE—explosive liquid-phase epitaxy
(d) ELPN—explosive liquid-phase nucleation

Epitaxy occurs solely when the (re)crystallisation is brought about by direct contact of the crystalline and the amorphous state when the first can serve as a seed to the latter [167]. Nucleation, however, includes the formation of nuclei *within* the metastable amorphous phase and thus gives rise to polycrystalline Si instead. If the phase transformation proceeds via the metastable liquid state in the case that the temperature is in between the melting point of the amorphous phase and the crystalline state (RTA) or even above (FLA and LA), the whole process is termed liquid phase epitaxy/nucleation instead [161].

These crystallization mechanisms are governed by the ignition of explosive crystallization at the Si film edges and the homogeneous heating of interior a-Si. Crystallization starts from the edges, because there a-Si receives more flash lamp light than the interior due to its disclosed cross section and due to multiple reflections off the chamber wall. Once crystallization takes place at the edges, thermal emission due to the phase transition from metastable a-Si with its larger building enthalpy to stable c-Si occurs, and the heat emitted diffuses into neighbouring a-Si areas, which induces further crystallization and results in a self-catalysed lateral movement of the crystallization front towards the inside. This process will last on until a threshold temperature is underrun.

It has been found that both explosive crystallization via the solid and the liquid phase are probably involved during FLA-(re)crystallisation. However, it has also become clear that these processes are not equal and which of them is favoured or rather dominates the regrowth is a complex interplay of the flash parameters regarding energy density (i.e. melt duration among others) and the temporal profile, the film thickness and the assembly of the a-Si as well as the hereby occurring stress inside the amorphous material. These factors in turn not only determine the (re)crystallisation route, but also its velocity and the microstructures formed [168–170].

4.3.1.2 Process Modelling

Lead zirconate titanate, which is favoured for its ferroelectric properties in storage memory, has become especially important to research and development in terms of large-scale *ferroelectric random access memory* (FeRAM). A technique is needed that protects the underlying temperature-sensitive substrate while transforming the material on top into its crystalline phase to give a ferroelectric capacitor. This technique had been found in FLA. With a comparatively low energy density of 27 J cm^{-2} the amorphous lead zirconate titanate could be crystallised without significant transient heat dissipation [171].

Despite these yet important, albeit less abundant applications in electronics an important group of applications reaches into (re)crystallization of a-Si on glass for displays. FLA is again the tool of choice for inexpensive, still efficient area-spread processing. Further, the glass-transition temperature around 600 °C must not be overcome which can easily be handled by surface thermal treatment as it is provided by FLA. There are different means to deposit the a-Si onto the glass substrate, like chemical vapour deposition (CVD), sputtering and electron-beam (EB)-evaporation, which will become important when having a closer look at the (re)crystallization modes during FLA. Further, in contrast to RTA, a-Si of only a few micrometres in thickness may be flashed without causing damage to the substrate. However, FLA may impose a similar temperature profile like RTA throughout a thin Si layer via thermal conduction, but it needs a much shorter annealing time.

The impact of FLA to (re)crystallisation of a-Si on glass has been closely looked at through an optical, solely thermal and thermo-stress model to understand the kinetics behind and to be able to predict the favoured process by calculating the transient temperature profile in the substrate and the stress that is thus imposed on the same one. Despite the required components for FLA the model also considers preheating of the substrate (here: glass). Before a-Si deposition by low-pressure CVD a protective silicon dioxide layer was applied.

The optical model considers a stack of materials consisting of an inert atmosphere, thin melted films (and possible further thin films such as the silicon dioxide capping layer) and finally the glass substrate. Required optical properties are taken from literature. The thermal model is based on the differential equation for thermal conduction taking into account the building enthalpy of the (re)crystallised Si as well as an external source, namely the energy impact by flash lamp exposure and building on this basis, the thermo-stress model assumes quasi-static behaviour of the wafer, which means that the wafer deforms instantaneously upon temperature rise. The mechanical response of the wafer is calculated according to the *finite element method* (FEM) [168, 172]. A comprehensive review on literature regarding temperature modelling may be found in Sect. 3.2.

4.3.1.3 Thin Film Thickness

Following the model from Smith et al. [168] and McMahon et al. [172], which was introduced in the preceding section, a direct relationship between a-Si film thickness and the threshold energy density has been found. The model allows to predict for which film thickness which (re)crystallisation mode is favoured and thus to adjust the annealing parameters accordingly. Figure 4.39 shows the threshold energy density that is required for the (re)crystallisation to proceed through the solid (SPC) or liquid (super lateral growth) phase as a function of film thickness. Super lateral growth can be described by ELPE according to the definition given by Geiler et al. [161]. The distinction will become clear in the following subsection when the impact of the energy density of the incident flash will be looked at more closely.

The researchers found that the threshold energy density for (re)crystallisation in fact decreases for increasing film thickness (first before approaching a kind of saturation) which may be explained by the glass substrate serving as a heat sink.

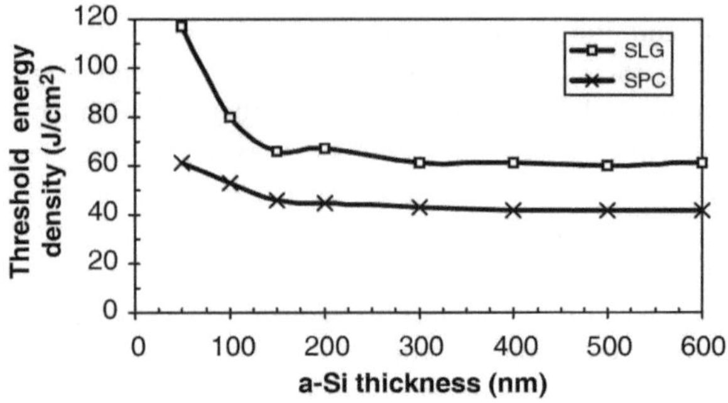

Fig. 4.39 Threshold energy density for (re)crystallisation of the amorphous phase of Si as a function of a-Si thickness. Blank boxes represent (re)crystallisation via the liquid phase (ELPE) and crosses show ESPN (SPC/nucleation). Reprinted from [168], Copyright (2005), with permission from Elsevier

Thicker layers, however, although they require higher energy densities for melting, absorb more flash light, i.e. the conversion rate of input and available energy is higher, and they can efficiently catalyse their own (re)crystallisation by the latent heat gained through this process itself. Therefore, a decrease in threshold energy density is found for the first 200 nm of a-Si, while it merely changes for larger film thicknesses. As super lateral growth involves (re)crystallisation via the liquid state it requires more energy than SPC for any given film thickness of amorphous material, but the trend can be seen for both (Fig. 4.39).

The dependence of (re)crystallization on film thickness has also been found by Ohdaira et al. Studying polycrystalline Si with a thin film thickness between 1.3 and 4.5 μm the results indicated that for a thin film thickness larger than 2 μm the (re)crystallization is governed by what the authors call *intermittent explosive crystallisation.* Close to the crystallisation front both the solid and the liquid phase take part in the regrowth. Further away from the moving a-Si front, however, regrowth is governed by ESPN only. As the temperature to transfer the material into the liquid phase in these distant areas may not be sufficient crystallisation occurs intermittently. Ohdaira et al. found that this process dominated the overall regrowth for a thin film thickness of more than 2 μm. If, however, the initial temperature was high enough ESPN could proceed continuously as the whole process is then evolving much faster through thermal diffusion. For a-Si films thinner than 2 μm the process of ESPN might be replaced by ESPE which was suggested by the results as there was no indication of lateral movement found [166].

4.3.1.4 Energy Density

We have already looked at the influence of energy density on (re)crystallization in the context of its relationship to film thickness (Fig. 4.39). Another group, who aimed to proof that FLA was an alternative to excimer LA holding the balance to RTA as an inexpensive mean for area annealing, made similar findings. Indeed, it could be shown that FLA-initiated (re)crystallization proceeds via the solid or the liquid phase and a direct reference to the amount of energy deposited by the flash could be established. Pecz et al. found that SPC dominates for low and liquid phase regrowth for high energy densities, whereas the transition is described by super lateral growth. This process is bound to a very narrow energy regime where not yet complete melting has been attained and regrowth starts from small solid nucleation seeds of a-Si. Ohdaira et al. refers to it by ELPE on solid nucleated grains [166, 168, 169, 173].

Through FLA regrowth of a-Si microstructures are formed spontaneously whose lateral dimensions are also a function of flash pulse length and thus energy density deposited by the flash. These microstructures show differently sized grains depending on the dominant recrystallisation process (solid phase route or partial melting) which will be looked at more closely in the following section [166].

4.3.1.5 (Re)Crystallization Velocity and Structural Consequences

Ohdaira et al. found different kinds of periodic microstructures with about 1 μm spacing or less on recrystallised a-Si and a-Ge, respectively. They consisted of grains of different sizes. Fine grains—as the researchers assumed—originate from ESPN only and large grains of several hundred nanometres in size were related to partial melting of a-Si. The mechanism involved corresponds to what has been explained before by super lateral growth. Solid a-Si distant to the propagation front may not homogeneously melt due to the comparatively large latent heat of a-Si. The remaining solid a-Si undergoes ESPN and the thus created c-Si-nuclei can grow larger by a process similar to ELPE where the a-Si-melt can epitaxially recrystallize on the c-Si-nuclei/grains due to the "significant change in surface morphology" [165]. The results are similar for FLA on a-Ge (Fig. 4.40) despite the observed differences in microstructure spacing, which may be explained by the particular thermal diffusivity of the two materials. Further results gained by Ferro et al. suggest that the cooling rate determines strongly crystal orientation and thus lattice mismatch and strain [110, 165, 174].

The group around Ohdaira also investigated the lateral crystallisation velocity to study the processes involved. In order to do so they used several single pulses with frequencies of 1–10 kHz to generate quasi-long flash pulses with time spreads between 5 and 7 ms. To find the underlying pattern the frequency of the single pulses was raised stepwise. In doing so they could investigate the (re)crystallisation online in order to judge when or whether the regrowth proceeds via the liquid or solid state or via both.

Earlier the group had shown that FLA-initiated (re)crystallisation takes the ESPN route with partial melting (ELPE). This however, contradicted Geilers findings that it indeed proceeds via ELPE well knowing that his experiments on explosive recrystallization were initiated by a quantum-switched Nd-laser during RTA instead of a flash pulse. While aiming to solve this contradiction the researchers found instead that crystal regrowth is a sensitive function of various process parameters, one of

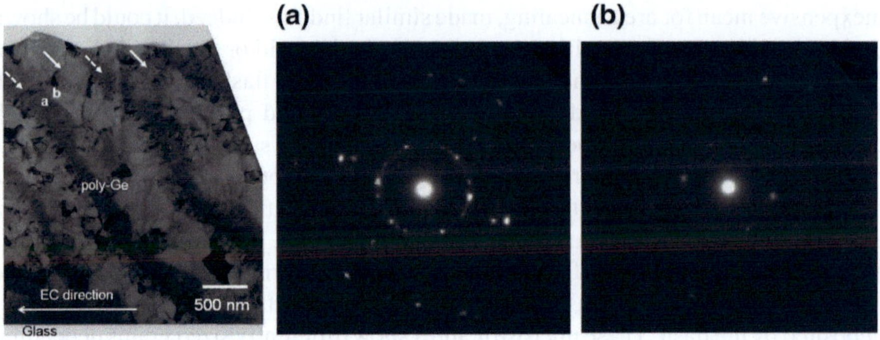

Fig. 4.40 Images from cross-sectional TEM (left) and EB diffraction patterns to show the existence of fine grain—regions (dashed arrows, **a**, ESPN) and large grain—regions (solid arrow, **b**, ESPN & ELPE/SLG). c = a-Si. Reprinted from [177], Copyright (2012), with permission from Elsevier

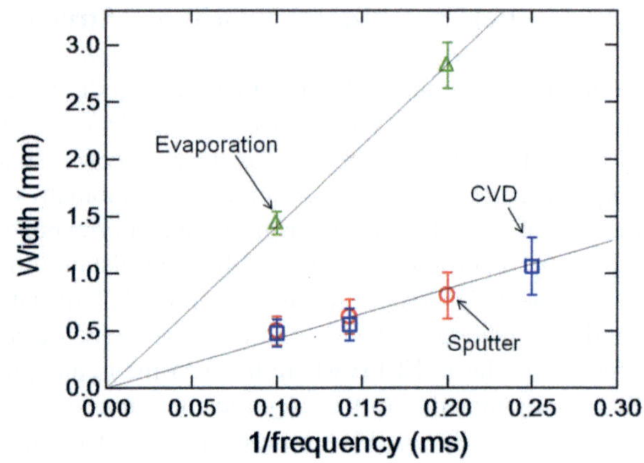

Fig. 4.41 The width of microstructures formed from different mechanisms of (re)crystallization according to the mean of a-Si film deposition as a function of pulse duration. Reproduced from [175] with permission of Electrochemical Society, Inc.

which being the way to deposit a-Si in the first place. They found (re)crystallisation velocities of 5 m/s for catalysed CVD and sputtered a-Si, while velocities of 15 m/s were gained when the a-Si was deposited by EB evaporation (Fig. 4.41). The first indicates a mixture of ESPN and ELPE whereas the latter corresponds to ELPE only. However, for sufficiently thick a-Si films made by EB evaporation ESPN may become the dominant part. The influence of the deposition process on crystal regrowth will be further looked at more closely in the following [175, 176].

Ohdaira et al. also studied 3 μm thick amorphous Ge films sputtered on a quartz glass substrate to find similar results to those of a-Si. Only the interval between the microstructures was found to differ due to the different thermal properties. Again judging from the (re)crystallisation velocities the regrowth involves both, ESPN and ELPE [177].

4.3.1.6 Other Influences on the Recrystallization of Amorphous Silicon

As indicated in the preceding section the mean of preparation of a-Si itself influences the process of (re)crystallisation. Scientists found by the study of microstructures and dopant profiles after regrowth that the latter is governed by (re)crystallisation from the liquid phase for samples prepared by EB-evaporation only, while for CVD-deposited or sputtered a-Si solid phase nucleation also takes over an important role in the process which coincides with the results from the other groups introduced in the preceding sections. One possible explanation for this difference according to the researchers may be the tensile stress in EB-evaporated a-Si precursors. CVD deposition and sputtering rather leads to compressive stress. Similar conclusions could be drawn when the crystallized poly-Si film was observed. The grains after crystallisation of EB-evaporated a-Si were found to be in a certain orientation while regrowth of catalysed CVD or sputtered a-Si has been found to yield randomly oriented grains. Moreover, Raman spectra reveal a much higher crystallinity for EB evaporated a-Si. Figure 4.42 confirms this assumption showing a sharp peak with smaller FWHM for EB-evaporated recrystallized samples. The authors assume that the shift of the Raman peak is due to the tensile stress which is larger in EB evaporated a-Si. However, this fact may be an advantage for the development of highly efficient solar cells although investigations into sputtered and CVD a-Si and thus into different levels of hydrogen-termination and defect generation have not revealed significant influence on the mechanism of (re)crystallization [170, 178–180].

In case of catalytic CVD it is known that both tensile and compressive strain affect (re)crystallisation, which can be controlled by process temperatures. But it was found that they exhibit different consequences on the (re)crystallisation of a-Si, i.e. tensile stress leads to larger grains than compressive stress. However, these findings could not be reproduced for EB evaporated a-Si which indicates that there are other critical factors determining the process. Moreover, tensile stress is assumed to suppress ESPN for EB evaporated a-Si which is yet again supporting the findings of the preceding subsection [179, 181].

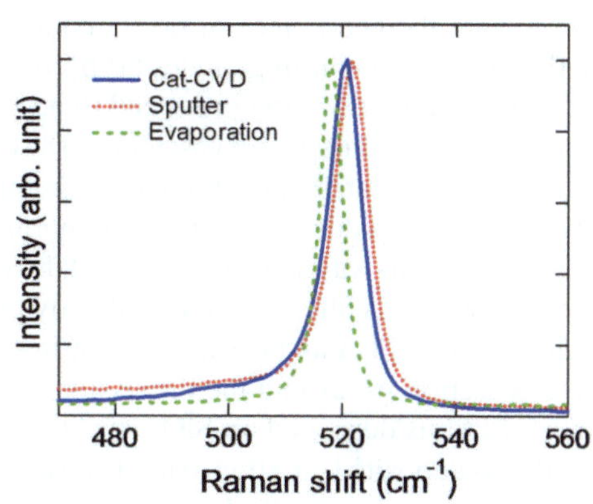

Fig. 4.42 Raman spectrum of a flash-lamp (re)crystallised a-Si film on glass. Reproduced from [175] with permission of Electrochemical Society, Inc.

The microstructures formed during (re)crystallisation have also been investigated for the application of low-reflective display panels. In fact, these microstructures rise quite significantly the absorptivity of the regrown Si and subsequent wet etching may further support this effect. Altogether after recrystallization (60 s annealing) the reflectance could be drastically reduced for both with and without an anti-reflection coating made of 100 nm thick indium tin oxide (ITO) [182].

4.3.1.7 Defect Annealing

Simultaneous defect annealing has been shown to be successful during FLA-assisted (re)crystallization. Almost no ingrain defects had been found. While excimer LA holds the same advantage with respect to both defect removal and transient thermal diffusion, FLA shows its superiority being an inexpensive mean suited for large substrates [169].

"Hydrogen content" is a delicate subject in photovoltaics (PV), which is due to the fact that a low hydrogen content may prevent the wafer from delamination or crack formation on the surface during FLA, but cannot help to prevent low lifetimes of minority carriers which are caught by deep level acceptors/donors such as voids being created by (re)crystallization. Hydrogen, however, may bind free dangling bonds and thus prolong minority carrier lifetime. This was the reason for the Ohdaira group to use FLA in a water vapour atmosphere under high pressure. The defect density could be thus reduced by one order of magnitude for EB evaporated 3 μm thick polycrystalline Si films [183–185].

4.3.2 Thin Films of Amorphous Silicon

There is a broad palette of applications for thin Si films ranging from PV over various microelectronic devices to sensor applications. Both amorphous and polycrystalline thin films are of interest for thin film PV, and thin film transistors (TFTs) made of poly-Si are the heart piece of pixel and driver circuitries for liquid-crystal displays and organic light emitting diodes [186]. In addition, the ability to deposite SiO_2 and Si layers in an alternating order and to process them allows the extension of device density without further downscaling [187]. This so-called 3D integration benefits from shortened interconnect lengths accompanied by a reduced resistive capacitive (RC) delay and reduced power consumption [188, 189]. In most cases Si is deposited in an amorphous way and has to be crystallized by a thermal treatment. The maximum thermal budget is often limited by already prepared devices which is why short time annealing like FLA and LA have an intrinsic advantage compared to other methods. In this section we extend the discussion of the previous one and focus on FLA and thin a-Si films deposited on SiO_2 or glass carriers.

If Si is deposited on a transparent carrier, and if the film thickness is small enough, a significant part of the flash light might not be absorbed. Figure 4.43a compares the

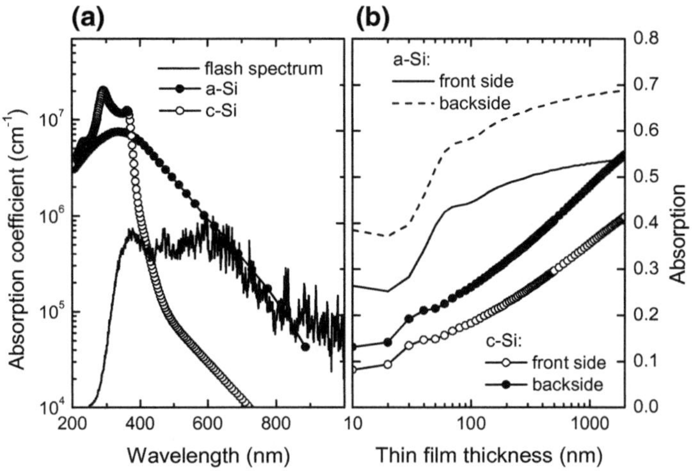

Fig. 4.43 **a** Absorption coefficient of a-Si (solid circle) and c-Si (open circle) in comparison with a typical flash lamp spectrum (solid line, arbitrary intensity). **b** Absorption in a thin amorphous (solid line) or crystalline (open circle) Si film on glass as a function of film thickness. If flashed from the backside, the absorption in the thin amorphous (dashed line) or crystalline (solid circle) Si film is significantly higher

absorption coefficient of a-Si and c-Si with regard to a typical lamp spectrum. Both materials are highly absorbing in the UV region, but there is a significant difference in the visible: the absorption coefficient of c-Si is more than one order of magnitude smaller than that of a-Si. In the NIR their absorption coefficient is low and absorption in thin layers is usually negligible. One exception is a-Si and layer thicknesses in the μm range which may absorb most of the NIR radiation between 700 and 900 nm which is not reflected. As a result, FLA on thin a-Si films is reasonable, whereas for c-Si it becomes more challenging. However, this turns into an advantage if larger carriers are flashed stripe-wise: an area already flashed and being polycrystalline is much less affected by a second flash than areas which are still amorphous. The situation is visualized in Fig. 4.43b showing the absorbed fraction of light (with respect to the flash lamp spectrum of Fig. 4.43a) for a thin a-Si or c-Si film on quartz as a function of thickness, not considering chamber wall reflection. The slight oscillations on the rising edge of the curves are due to interference effects. As seen, the final absorption of a thin a-Si film is 2–3 times higher than that of a c-Si film for thicknesses in the range of a few 10 to a few 100 of nm. If applicable, FLA from the backside will increase absorption for thin layers as the quartz carrier acts as an anti-reflection coating. In fact, the required energy density needed to crystallize a ca. 2.1 μm thick film of a-Si on glass was significantly reduced when FLA was applied from the backside [190]. However, the use of highly reflective chamber walls will generally increase absorption and will lower the differences between a-Si and c-Si as well as between front and backside FLA.

Many application profit from larger grain sizes as grain boundaries usually impair the device performance. Especially for TFTs it would be advantageous to prevent grain boundaries in the channel region and to process the whole transistor in a single grain [186]. Spontaneous nucleation of crystalline seeds in an amorphous matrix already occurs at fairly low temperatures, namely 400–500 °C in case of Si. Thus, full recrystallization can be achieved at these temperatures, provided that the annealing time is long enough. As the activation energy for nucleation in the case of Si (~5.1 eV) is larger than that for grain growth (~3.3 eV) [191], only a very low number of seeds will nucleate which leads to the formation of large grains. However, such long annealing times are impractical in most cases. Annealing times can be made shorter in parallel with an equivalent increase of the annealing temperature, but this shifts the equilibrium between nucleation and grain growth towards nucleation with the consequence of smaller and smaller grain sizes.

This is also the case for thin a-Si layers with thicknesses of a few 10 or a few 100 of nm which are annealed by FLA and where crystallization entirely takes place in the solid phase. Using relatively short flash pulse of 400 μs, grain sizes of few tens of nanometres were achieved for a 50 nm thick film of a-Si deposited by plasma-enhanced chemical vapour deposition (PECVD) on large-area glass panels [173]. This work was continued with more detailed investigations devoted to the crystallinity as a function of preheating temperature [192]. For a given FLA pulse time, a moderate increase of the energy density E_D leads to a moderate increase of grain size, namely from 40 to 60 nm for $E_D = 50$ to 87 J cm^{-2} for 20 ms FLA and a 50 nm thick a-Si film [169]. As shown by the authors, the use of a double-flash was able to increase the grain size to 100 nm.

However, situation changes if temperatures exceed the melting temperature of a-Si and enter the region of explosive SPC or super lateral growth (see previous section). In the first case a-Si is overheated due to the large heat ramps, such as 10^5–10^6 Ks^{-1} for FLA, and the released heat of crystallization additionally drives crystal growths. As a consequence, crystallization velocity is strongly increased leading to larger grains. In case of super lateral growth [193] the major part of a-Si will melt, nucleation is limited to the remaining solid material, and the combination of only a few seeds and high crystallization velocity also leads to larger grains. Grain size will at maximum in a situation where the a-Si layer is nearly entirely molten and only a very small number of seeds remain at the Si-SiO_2 interface. However, if a-Si is entirely molten, nucleation will take place in the entire liquid, and grain size will drop again. Finally, for thin a-Si films exposed to FLA the degree of crystallization as deduced from Raman measurements seems to be higher in case of LPC than in case of SPC [27].

The occurrence of this pattern during FLA—small grain sizes for SPC, large grain sizes in the region with a partly molten a-Si film, and again small grain sizes for LPC—is already known from the 80s [194]. In addition, the same behaviour can be found in case of LA [195]. Figure 4.44 shows an example of ultra-short FLA with 40 μs where the transition from SPC to a regime with a partly molten a-Si film occurs at $E_D = 1.8$ J cm^{-2} [196]. If taking advantage of explosive SPC or super lateral growth, grain sizes in the μm range [169, 197] or even above 10 μm [198–200] can

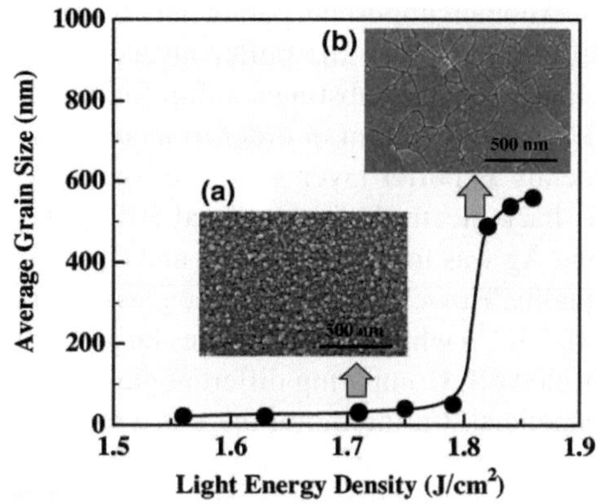

Fig. 4.44 Average grain size as a function of energy density for a 40 μs flash pulse. Reproduced by permission of The Electrochemical Society [196]

be achieved. Another idea is the use of pre-defined seeds to increase grain size and to grow grains at defined locations. Defined thickness variations of the underlying SiO_2 [186] or the a-Si film known as grain filters [201] can serve as seeds. Although in both cases annealing was performed by an excimer laser, these techniques should work for FLA as well. Furthermore, structured absorbing layers like a metal mirror [187] can also be used to create a local, but defined variation of temperature.

There are numerous publications from the group of Ohdaira et al. which investigated the crystallization of μm-thick a-Si films during FLA. One part of these investigations compares the impact of different depositions methods on the structure of the final, polycrystalline films. As already mentioned in Sect. 3.5, a higher H content can lead to delamination of the film during FLA [202]. Thermal desorption spectroscopy reveals that H is quite stable in a-Si and poly-Si films and will not significantly desorb below 600 and 900 °C, respectively. In case of poly-Si H is suspected to accumulate at grain boundaries and at other defects not fully annealed during FLA [184]. Another work compared the crystallization behaviour of 2–4 μm thick a-Si films on 60 nm Cr on quartz substrates deposited by catalytic CVD, sputtering or EB evaporation [175]. As already mentioned in the previous section, the deposition by catalytic CVD and sputtering leads to small and somewhat larger compressive stress, respectively, whereas films made by EB evaporation feature tensile stress. In case of catalytic CVD the film stress can also be tuned by the deposition temperature [203]. Stress tends to relax during FLA which leads to different surface morphologies depending on the kind of stress. In case of compressive film stress periodic surface structures with a period in the order of 1 μm evolved which were assumed to be formed due to the emergence of a particular explosive crystallization mode which alternatively repeats two kinds of crystallization mechanisms [165]. In contrast to this tensile stress led to flat surfaces which, however, are decorated by cracks. It has to be noted that the flash pulse with a pulse length of 5–7 ms as used in [175] was composed of μs subpulses with a repetition rate in the kHz range [176] which additionally promotes the formation of periodic structures.

Another important parameter to consider is the underlying material, either the substrate itself or the buffer layer between substrate and the a-Si film. In case of glass or quartz substrates a thin SiO_2 buffer layer is often deposited in advance of the actual a-Si film in order to increase adhesion. In [204] the suitability of several metals as buffer layer was investigated with the motivation to use the metal layer as back electrode for potential PV applications. Whereas the melting point of Al and Ag was too low, both Mo and Cr led to crystallization on quartz with satisfying quality. However, on soda lime glass with a thermal expansion coefficient of $9.35 \times 10^{-6}\ K^{-1}$ (which is ca. 17 times larger than that of quartz glass) Si films on Mo were peeled off. Comparing different glass types it was found that a higher strain point is beneficial. Furthermore, there is a study about the crystallization behaviour of thin a-Si films on a SiC surface [174]. It also underlines to the importance of a capping layer on top of the a-Si film in case of LPC: if not done, Si recrystallizes in larger droplets instead of a closed layer. If a sputtered a-Si film serves as an underlying buffer to an a-Si film deposited by catalytic CVD, the recrystallization temperature und thus the required energy density during FLA is reduced [205].

Finally, there are few publications which study the electrical properties of thin polycrystalline Si films crystallized by FLA. Choi et al. [192] used laser-crystallized thin Si layers, implanted them with P and recrystallized the layer again by FLA. By optimizing the FLA conditions they were able to achieve carrier concentrations and sheet resistances of up to $8 \times 10^{19}\ cm^{-3}$ and below 700 Ω/sq, respectively. In case of 2.3 μm thick a-Si layers, in situ doped with B or P during catalytic CVD, carrier concentrations above $10^{20}\ cm^{-3}$ and a specific resistivity below 0.01 Ωcm were achieved [206]. Saxena and Jang [199] published transistor data from devices processed in such layers and documented how drain current and especially mobility become worse with increasing number of grain boundaries in the channel of the TFT. Recently, TFTs with large grains were fabricated by FLA [200] which feature a high degree of alignment resulting in channel mobility values of 380 $cm^2/(Vs)$ and 143 $cm^2/(Vs)$ for electrons and holes, respectively. In case of 4.5 μm thick a-Si layers the minority carrier lifetime was measured after different post-FLA annealing steps [207]. Whereas high pressure water vapour annealing achieves minority carrier lifetimes around 10 μs and terminates defects according to electron spin resonance spectroscopy [185], FA can reach values in excess of 100 μs [207].

4.3.3 Silicon Compound Semiconductors

4.3.3.1 Silicon Germanium

Conventionally, SiGe is grown on Si substrates by means of MBE or CVD. For small film thicknesses, SiGe grows with the smaller lattice constant of Si which introduces additional stress into the lattice and thus a higher carrier mobility can be found than for relaxed SiGe or bulk Si. The relation of Ge content to Si determines the degree of lattice mismatch until a critical sheet thickness is achieved at which SiGe relaxes, but also causes undesired defects and dislocations. This process has been successfully replaced by Ge implantation and subsequent thermal annealing [208, 209].

It is a main focus to control Ge implantation to a certain depth in the Si bulk wafer. One possibility to restrain the width of this "dopant" layer is to decrease the annealing time as it is done with FLA. Supportively, the material of interest can be modified in such a way that it becomes the place where dopants preferably gather. This may be achieved by selective melting. The solubility of dopants is higher in the melt. Therefore, a melted layer would be such a preferred place. Further this approach helps defect reduction.

In order to specify, however, the transient distribution of the implanted species in the bulk a method is required that allows for selective melting of the substrate. It has been found that at moderate energy densities of the flash pulse depending on the pulse length (e.g. around 130 J/cm^2 for 20 ms) and for sufficiently high implantation energies and doses Ge lowers significantly the melting point of Si by tens of degrees. The thickness of this Si:Ge layer is controlled by the energy density and the length of the flash pulse. It has been found that the implantation profile shows a plateau-like shape for elevated energy densities which can only be explained if the dopants can diffuse in the liquid phase. The confinement of the melted zone to the Ge-rich buried layer equally hinders the ingrowth of defects while a flash of 3 ms shows less defects in comparison to a flash of 20 ms. Literature claims a defect reduction by more than three orders of magnitude [210–212].

The subsequent cool down and resolidification of the melted SiGe layer leads to liquid phase epitaxial regrowth. This procedure may be likewise used for the fabrication of strained Si due to the lattice mismatch between Si and SiGe if the recrystallized SiGe was further cooled down. This in turn rises the carrier mobility for the same reason as SiGe shows larger values [208, 210–212].

For the first time scientists have demonstrated the suitability of FLA for annealing high-performance SiGe heterojunction bipolar transistors. These transistors are mainly used for high frequency applications in electronics and digital technology. The base exhibits a gradual Ge dopant profile which changes the energy bandgap equally (Fig. 4.45). This in turn creates an additional electric field aiding the current between the emitter and the collector and thus allowing for higher transit frequencies [213, 214]. The results have shown that FLA is a true alternative to hitherto existing spike anneals. Neither out-diffusion of Boron nor a broadening of the implantation profile was found. Quite contrary, diffusion and sheet resistance could be reduced and the base width could be narrowed for higher transit frequencies. Further, the results suggest a lowering of the external base resistance [214].

4.3.3.2 Silicon Carbide

SiC is a wide bandgap semiconductor and therefore especially suitable for high power and high frequency operation. Its only isotropic polytype 3C-SiC is typically grown by heteroepitaxy on Si. Its preferred use over the other SiC polytopes is manifold and comprises its high electron mobility and saturation velocity as well as its large

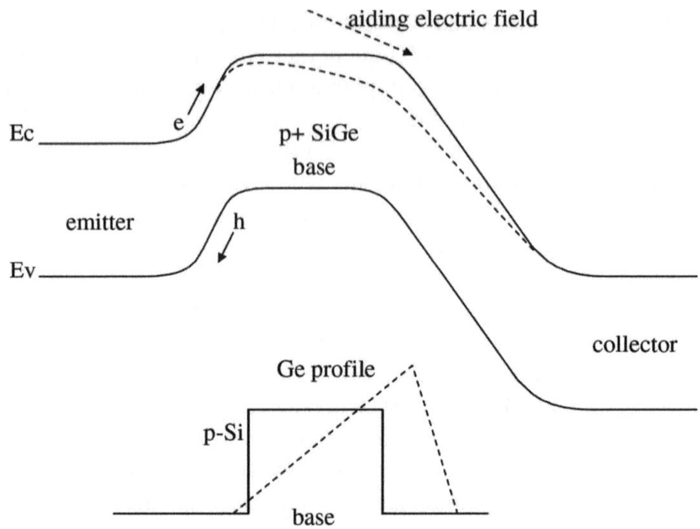

Fig. 4.45 Band diagram for a SiGe heterojunction bipolar transistor (n-p-n) for a constant (solid graph) and a graded Ge profile (dashed graph) in the base; E_c = conduction band, E_v = valence band, e = electrons, h = holes. Reprinted from [213], Copyright (2005), with permission from Elsevier

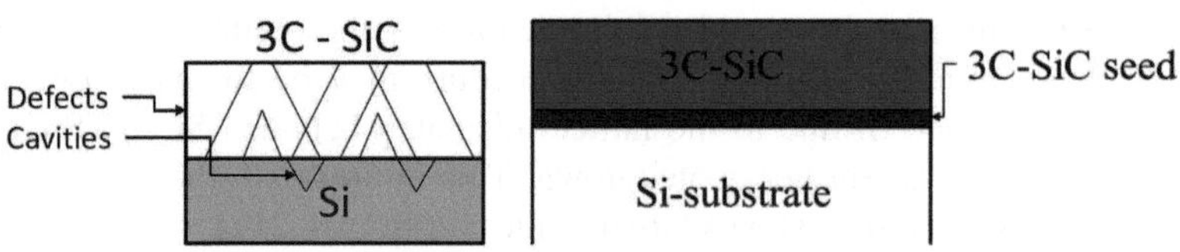

Fig. 4.46 Initial structure for FLASiC and i-FLASiC: Conventional heteroepitaxy of SiC on Si. A thin SiC (seed) layer is first grown onto the Si surface (left), defects are created by the large lattice mismatch. A much thicker SiC layer is subsequently grown (right) still exhibiting a high defect density [142]

thermodynamic stability. However, this procedure usually produces a high defect density in the vicinity of the Si/SiC interface. The large lattice mismatch between Si and SiC (around 20%) as well as the difference in thermal expansion coefficients during deposition (up to more than 20%) lead to crystal defects, such as dislocations and stacking faults (Fig. 4.46). The mismatch is also the origin for the need of a step-wise growth on Si-substrates, but the overgrowth of the afore created defective thin seed crystal by a much thicker SiC cap in the sequence of this process additionally helps defect propagation. In fact, depending on its thickness due to annihilation effects the latter contains still a comparatively high defect density. Therefore the growth of thick SiC layers is considered [140, 142, 215–217].

FLA, however, has shown to limit defect growth and defect density in general. The process known as *flash lamp annealing supported deposition of SiC*, namely FLASiC, is shown in Fig. 4.47 and can be described as follows. A thin layer of

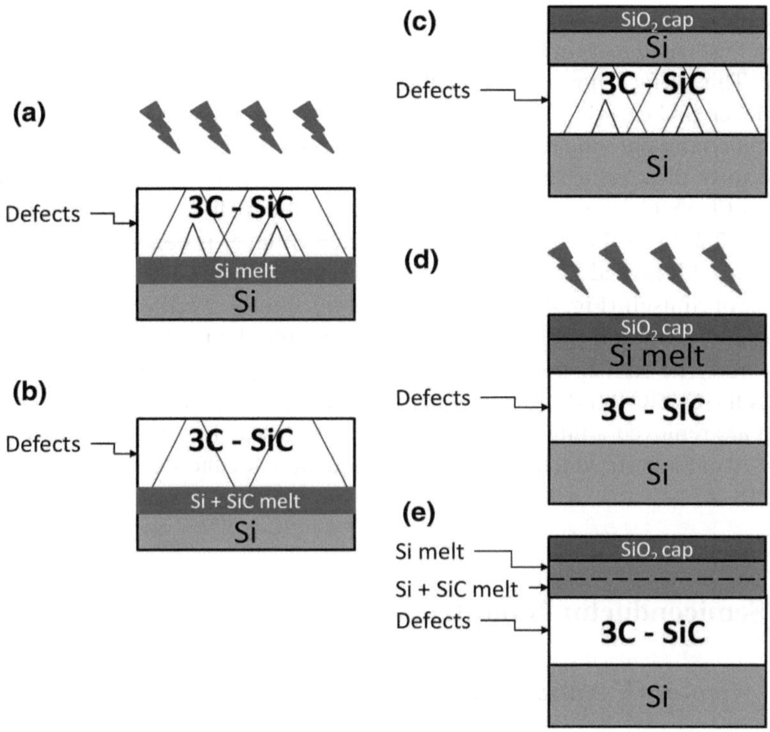

Fig. 4.47 Schematic sketch showing the sequence of FLA-supported deposition of 3C-SiC (left) and the sequence of inverse FLA-supported deposition of 3C-SiC to overcome buckling and low crystallinity at the surface (right); Left: FLASiC—**a** Si melts during flash lamp annealing, **b** partial dissolution of defected SiC into the Si melt, annealing of non-melted SiC; Right: i-FLASiC—**c** Deposition of catalytic layer system, **d** melting of Si catalytic layer, **e** partial dissolution of SiC into the Si melt. After [142]

3C-SiC is grown onto a Si substrate. In a second step FLA with a sufficiently high energy density is performed to initiate melting at the Si/SiC interface (Fig. 4.47a). As the melting point of Si is much lower than that of SiC the upper part of the bulk Si in the vicinity of the Si-SiC interface—here a kind of a melt stop—will melt first and 3C-SiC near the interface dissolves into the molten Si while the SiC further towards the surface is annealed (Fig. 4.47b). But as only a small layer of the Si substrate is melted the defect ingrowth is limited. Indeed, it has been found that defect density could almost be cut in half in comparison to samples without FLA [140, 142, 216, 217].

Upon cooling the Si/SiC melt resolidifies via LPE and depending on the annealing parameters and the growth rate the resulting SiC exhibits a different thickness. The Si below and the annealed SiC above the melted layer serve as seeds to the recrystallizing Si/SiC melt. Although the propagation of defects is limited due to the locally confined melting of the Si substrate, the melting itself leads to a buckling of the

Si/SiC interface. The consequences can also be seen at the surface, even though a lot smaller. Further, defects may be hindered to spread into the bulk, but not towards the surface. This surface, however, could have served as a seed to the subsequent growth of a thicker SiC cap (Fig. 4.46). Instead, it needs to be etched to grow anew SiC homoepitaxially onto the resolidified thin SiC seed with a much lower defect density. To solve these issues all the same an enhancement of FLASiC was introduced: i(nverse)-FLASiC. The idea behind is similar to the melt stop approach described in Sect. 4.2.4. Instead of the thicker SiC another Si layer is deposited on top of the thin SiC seed followed by a cap whose melting temperature needs to be well larger than that of Si itself (Fig. 4.47c). Again only this thin Si overlayer is melted during FLA (Fig. 4.47d) with partial dissolution of SiC into the melt (Fig. 4.47e), but the bulk is not. The further processes resemble those explained above, but in inverse order as the Si/SiC interface is now above the SiC seed. Once the Si overlayer and the cap are removed a thin SiC layer with both a low defect density and a smooth surface is uncovered which gives best prerequisites for the further growth of SiC [142, 218–221].

4.4 Semiconductor Nanostructures

4.4.1 Group IV Nanocluster

After the discovery of quantum size effects in microcrystalline Si:H [222] and of red PL from porous Si [223, 224], comprehensive research activities devoted to Si nanostructures, especially Si nanoclusters (NCs), have been initiated. Due to the superior quality of the Si-SiO_2 interface most publications deal with Si NCs in SiO_2. There is a plurality of potential applications, including the use of Si NCs for luminescent materials and in memory applications. Indeed, quantum confinement allows the tailoring of the bandgap, and, even more importantly, the luminescence efficiency is strongly enhanced. The latter effect is due to the fact that with decreasing NC size both the chance to find a NC without a luminescence quenching defect increases and that optical selection rules are weakened by the Heisenberg uncertainty principle. Due to their much larger absorption cross section, Si NCs in a SiO_2 matrix were also found to pump nearby Er^{3+} ions very efficiently [225, 226]. If placed between the channel and the control gate of a FET, Si NCs can also serve as a non-volatile memory [227, 228]. There is also a larger bandwidth of fabrication methods. One of them is the deposition of a SiO_x layer or a superlattice made of thin layers of SiO_2 and SiO_x, followed by high temperature annealing [229]. Alternatively, Si NCs can be formed by ion implantation of Si followed by Ostwald ripening. However, in both cases high thermal budgets are needed. On one side, Si has a very low diffusion constant in SiO_2 in the order of 10^{15} cm^2 s^{-1} for a temperature of 1400 °C [54], even if the increase of diffusion by defects is considered. On the other side, diffusion is needed to grow NCs. As a consequence, there are only few studies using FLA in the context of Si NCs. The following paragraph is focused on these studies, whereas for all other details we refer to the numerous reviews in literature [230–232].

In the search for maximum PL and electroluminescence (EL) output of SiO_2 layers implanted with group IV elements, it came out that the luminescence is mainly due to oxygen deficiency centres, namely to the neutral oxygen vacancy [233]. These defects arise in the bulk SiO_2 or in the vicinity of Si NCs, and maximum luminescence output was achieved if either no or only small NCs were present. Thus, FLA can potentially be used to anneal out unwanted defects and to *prevent* the growths of NCs. However, if the annealing temperature is high enough, even FLA can induce the formation of Si NCs. In case of Si implantation into a 500 nm thick SiO_2 layer on Si first Si NCs can be traced by TEM for RTA 1200 °C, 1 s, and FLA 1350 °C, 20 ms [234]. This investigation was later expanded to nanosecond LA showing that this time scale is definitely too short to grow NCs [235]. However, LA is sufficient to crystallize already existing, amorphous Si clusters. Another study revealed that FLA alone is not sufficient to completely eliminate dangling bonds at the Si NC interface [236]. Recently, FLA was used to fabricate nanocrystalline Si on quartz with a NC size in the order of 9 nm by the decomposition of SiO_x deposited by reactive sputtering [237].

Ge features a significant higher diffusivity in SiO_2 than Si [233], which is why the growth of Ge NCs can be initiated by lower thermal budgets. Nevertheless, in most cases FA or RTA were used to perform this, and only in two studies investigating the coupling between Er^{3+} and Ge NCs FLA was used to form the Ge NCs [238, 239]. In the 90s it was discovered that Ge-implanted SiO_2 exhibit strong EL in the blue-violet spectral region caused by Ge-related oxygen deficiency centres [240] which are located in the bulk or in the surface region of a Ge NC. As a consequence, the maximum number of luminescence centres is expected for either no Ge NCs or an ensemble of very small Ge NCs with a large surface-to-volume ratio. In contrast to Si NCs, Er^{3+} ions in the close vicinity of Ge NCs efficiently pump the Ge-related oxygen deficiency centres. This is called the inverse energy transfer [241] and is assumed to be due to the low interface quality of Ge NCs with a large number of Ge-related oxygen deficiency centres in combination with their relatively short luminescence decay time [242]. In this framework a 200 nm thick SiO_2 layer was implanted with Ge leading to a Ge peak concentration in the order of 7–8 at.%. The following FLA at 1050 °C for 20 ms and a preheating of 700 °C led to small Ge NCs with a size in the order of 2–4 nm. Afterwards, the structure was implanted by Er (peak concentration 0.5 at. %), followed by FLA for 20 ms. This time FLA was used to recrystallize the Ge NCs without further growth and to activate the Er^{3+} ions. In fact, the Ge-related EL increases with the increase of temperature of the second FLA step from 800 to 1100 °C, whereas the Er-related EL at 1550 nm is strongly reduced compared to the case without Ge NCs [238]. The final result after both FLA steps is shown in Fig. 4.48a. With respect to microstructure, the direct comparison between FA and FLA [239] reveals that with increasing FLA temperature the NC density increases, whereas the NC size remains more or less constant. In case of FA large Ge NCs evolve in the central part of the SiO_2 layer whose size gradually increase with increasing FA temperature (Fig. 4.48b). In this example the structure was annealed at 950 °C

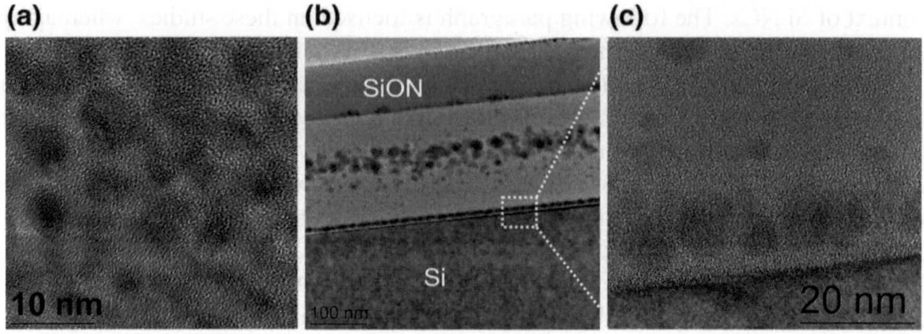

Fig. 4.48 Ge NC evolution in Er-doped SiO_2 after FLA (**a**) and FA (**b**). The image in (**c**) displays an enlarged section of the Si-SiO_2 interface region of image (**b**). More annealing details can be found in the text. (**a**) and (**b**) are reprinted from [239], with the permission of AIP Publishing

for 60 min and 1100 °C for 30 min in nitrogen after the Ge and Er implantation, respectively. Furthermore, there is a significant Ge diffusion towards the interfaces of the SiO_2 layer where Ge forms another band of NCs very close to the interfaces (Fig. 4.48c). In addition, a phase transition to Er_2O_3 or $Er_2Ge_2O_7$ may occur during annealing.

Finally, FLA can induce the growths of NCs in dielectric matrices if the corresponding diffusion constants are high enough. Another possibility is the crystallization of clusters which are amorphous due to fabrication or because of a previous ion implantation. In a recent example a superlattice made of thin, alternating layers of amorphous Ge and ZrO_2 were deposited by RF-sputtering [243]. During FLA at 1100 °C for 3 ms and a preheating of 200 °C, *crystalline* Ge nanoparticles in an *amorphous* ZrO_2 matrix were formed. Whereas Ge most probably crystallizes from the liquid phase, the short time scale inhibits SPC of ZrO_2. The combination of crystalline Ge nanoparticles with an amorphous ZrO_2 matrix showed a large hysteresis in capacitance-voltage measurements and thus a high potential for non-volatile memory applications.

4.4.2 Integration of III–V Nanocrystals into Silicon

4.4.2.1 Methods

In the last years, the unbroken and continuous striving for ever smaller and faster devices has brought miniaturization to its physical limits, and further improvements are only expected for new architectures like nanostructures and new materials. III–V compound semiconductors became an attractive alternative to Si due to their outstanding properties. Their electron mobilities are up to 10 times higher than that of Si, and III–V transistors exhibited current-gain and power-gain cut-off frequencies

Table 4.1 Properties of Si, Ge and selected III–V semiconductors. All data are for 300 K and can be found in [246] if not otherwise stated

	Si	Ge	InAs	InP	GaAs
Electron affinity (eV)	4.05	4.0[a]	4.9[a]	4.38[a]	4.07
Lattice constant (Å)	5.43102	5.64613	6.0584	5.8686	5.6533
Melting point (°C)	1414	937[a]	942[a]	1060[a]	1240
Bandgap (eV)	1.12	0.66	0.36 (d)	1.35	1.42 (d)
Electron mobility @ 300 K ($cm^2\ V^{-1}\ s^{-1}$)	1450	3900	33,000	4600	8000
Hole mobility @ 300 K ($cm^2\ V^{-1}\ s^{-1}$)	500	1900	460	150	400
Linear coefficient of thermal expansion ($10^{-6}\ K^{-1}$)	2.59	5.9[a]	4.52[a]	4.6[a]	5.75

(d) direct bandgap
[a]Extracted from [245]

of several 100s of GHz up to 1 THz [244]. In contrast to Si, many III–V compounds have a direct bandgap which opens a broad palette of optoelectronic applications from integrated light sources to light detection. Other applications can be found in the field of memory devices or in PV. Table 4.1 compares some material parameters of group IV and III–V semiconductors which are of interest with respect to this section.

On the other hand, Si substrates are robust and highly cost-effective, and SiO_2 as the natural oxide of Si provides an excellent electrical and thermal insulation. Therefore, III–V integration into Si is an attractive possibility to combine the advantages of both material systems. However, there are a couple of serious challenges to meet. As seen in Table 4.1, there is a substantial lattice mismatch between Si and most of the III–V compounds which leads to defect creation and an appropriate deterioration of the device quality. A mismatch in thermal expansion coefficients may lead to similar problems if the wafer passes through several annealing cycles. Finally, the replacement of Si requires the development of new material systems which are aligned with the corresponding III–V compounds such as high-quality gate insulators or low-resistivity contacts [247].

In the past years, there have been great efforts to develop reliable and cost-effective processes for the III–V integration into Si. A common method is epitaxy, namely MBE or metalorganic vapour-phase epitaxy, in which the III–V compound is grown layer by layer from the gas phase. However, such layers typically suffer from a large density of dislocations, many antiphase domains, and autodoping effects [248]. Several strategies have been developed to improve the layer properties. The use of buffer layers and several deposition steps can confine the defect-rich zone and ensure a surface layer with lower defect density. Another approach is the epitaxial lateral overgrowth in which a lithographically patterned SiO_2 mask is used to stop the propagation of dislocations towards the surface [249]. A method called aspect ratio trapping uses lithographically fabricated trenches in the substrate material to grow III–V layers with a reduced defect density [250, 251]. There are also several methods to grow III–V NWs on Si [252], including the VLS growth and the selective area growth. An alternative for the integration of III–V compounds into Si is wafer bonding which offers the possibility to process large wafer sizes as far as correspondingly large III–V bulk substrates are available. For further details we refer to [247] about the advantages and disadvantages of wafer bonding and to [253] for general aspects of III–V integration.

Another possibility is the use of ion implantation followed by an annealing step. In this scenario the elements of the compound semiconductor are introduced by ion implantation into Si or a thin dielectric film on Si. The following step of annealing can initiate the diffusion of the incorporated elements, the nucleation and the growth of III–V NCs via Ostwald ripening. Thereby, the NC size distribution and the NC density is the combined result of the implantation fluences, the annealing temperature and the annealing time. White et al. were able to create several II–VI and III–V compound semiconductor NCs in SiO_2, Al_2O_3, and Si by ion implantation [254]. The sequence of the implanted ions can also influence the formation of NCs as shown in case of InAs in Si [255]. If As is implanted first, the resulting InAs NCs have a smaller size and a more uniform size distribution than in the opposite case. The combination of sequential ion implantation and a subsequent thermal treatment was also investigated by the group of Komarov et al. [256, 257]. Although ion implantation offers an excellent control over the distribution of the implanted species, it also leads to defect creation and eventually to amorphization. To avoid this, some authors use hot implantation [255–258] and getter layers [257]. However, hot implantation suffers a high loss of implanted species [258, 259] or experience strong diffusion leading to a bimodal depth profile and a bimodal size distribution [260]. Finally, the formation of InGaAs NCs by sequential ion distribution and FA was recently demonstrated [261].

4.4.2.2 Bulk Silicon Substrates

Whereas FA and RTA mainly form small, spherical NCs, FLA can extend the variety of shapes. A comparative study [262, 263] implanted As and In into a Si wafer capped with a thin SiO_2 layer, whereas the implantation energies were chosen in such a way that the projected range of implantation R_P was in Si, but just below the Si-SiO_2 interface (Fig. 4.49a). As calculated by TRIM [6], the peak concentration of both As and In is in the order of 7 at.% (Fig. 4.49e). During FLA (Fig. 4.49b), different InAs NCs shapes will evolve depending on the specific annealing conditions. After RTA at 850 °C for 1 min, randomly oriented and spherical InAs NCs with a size in the order of 15 nm were formed. However, after FLA at 1200 °C for 20 ms, pyramidal InAs NCs with a size in the order of 40–50 nm appear in Si at the Si-SiO_2 interface which are formed between ⟨111⟩-oriented twins (Fig. 4.49d). Such a NC is visible in Fig. 4.50a together with some small InAs NCs with an average diameter of 7 nm in the SiO_2 layer. The material to form these small NCs originates from the tails of the implantation profiles which extend into the SiO_2 layer (Fig. 4.49e). The lattice constant of the pyramidal InAs NCs is slightly less than that for bulk InAs indicating a compressive strain. However, with increasing annealing temperature

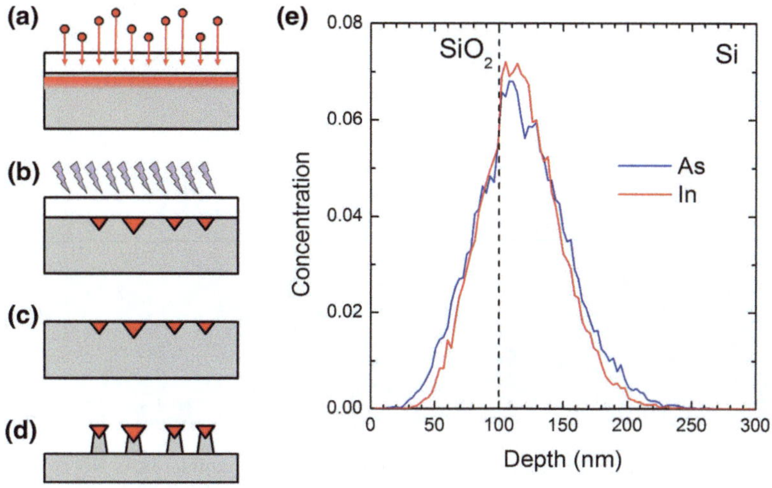

Fig. 4.49 Basic processing scheme for the fabrication of an InAs-Si heterojunction: ion implantation (**a**), FLA (**b**), removal of the SiO_2 capping layer (**c**) and selective wet etching (**d**). The diagram displays the as-implanted profile of As and In as calculated by TRIM [6] (**e**)

both the NC size and the lattice constant was found to increase slightly which in turn leads to a decrease of the strain. To give a number, the increase of the FLA annealing temperature from 1100to 1200 °C reduces the strain from 0.107 to 0.043 [263]. The main difference between the RTA and FLA is, besides the time scale, that the NCs are formed by SPC and LPC, respectively. This issue is more comprehensively discussed at the end of this section.

The InAs pyramids can also be used to fabricate an InAs-Si heterostructure. To do so, the capping SiO_2 layer was removed (Fig. 4.49c) and the InAs pyramids served as mask for selective wet chemical etching of Si (Fig. 4.49d) [264]. As a result, a vertical hetero-junction with a Si column and an InAs NC on top as shown in Fig. 4.50b has evolved. As measured by EDX (not shown), the concentration ratio between In and As inside the pyramid is almost 1:1. In order to investigate the electrical properties of the InAs-Si heterostructure a 180 nm thick Au layer was sputtered on the backside, whereas the InAs NC was contacted via a conductive atomic force microscopy tip (see the inset of Fig. 4.51b). Kelvin probe force microscopy measurements reveal that the InAs NC is highly n-type and that the Fermi level in the centre of the InAs NC is approximately 100 meV above the conduction band edge. This roughly corresponds to a doping concentration in the order of 10^{18} cm^{-3} [265] and is most probably caused by the incorporation of Si during the formation of the InAs NC (see the end of this section for more details). As the Si substrate was p-type, the InAs-Si heterostructure exhibits a type II alignment as sketched in Fig. 4.51a and forms a p-n junction. In fact, the current-voltage curve as shown in Fig. 4.51b exhibits the typical diode behaviour with an ideality factor of 4.6. This relatively high number can be caused by defects at the heterojunction which act as recombination centres and favour Poole-Frenkel

Fig. 4.50 Bright field high resolution TEM micrographs of a pyramidal InAs NC at the Si-SiO_2 interface after preheating at 600 °C for 1 min followed by FLA at 1200 °C for 20 ms (**a**) and an InAs-Si heterostructure after SiO_2 removal and selective wet chemical etching (b). The structure in (**b**) was annealed at 1300 °C for 20 ms, and was once again capped with a SiO_2 layer for TEM preparation. More details can be found in [262–264]

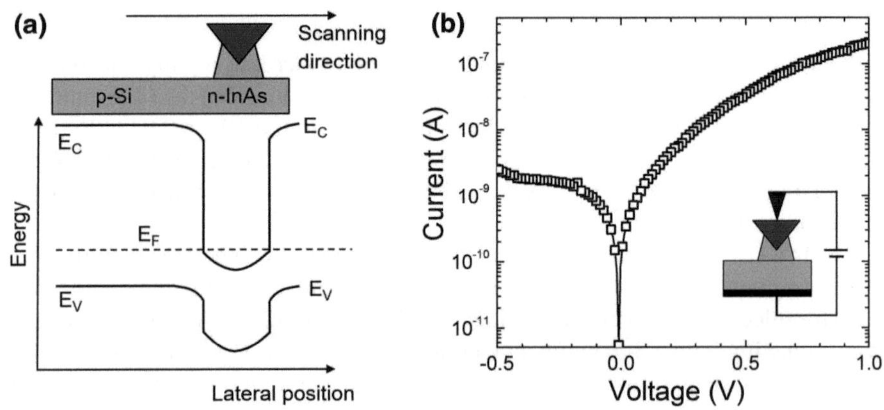

Fig. 4.51 Schematic band structure of the InAs-Si heterostructure as deduced from Kelvin probe force microscopy measurements (**a**) and current-voltage characteristics of the corresponding np heterojunction (**b**). The inset in (**b**) displays a basic sketch of the current-voltage measurement setup. After [264]

conduction [266]. The average p-n junction area between the InAs NC and the Si column is assumed to be in the order of 700 nm^2.

The combination of ion implantation and FLA also works in case of other III–V compounds. In a recent study, the evolution of InP NCs with increasing thermal budget during FLA was traced [267]. In this case, P and In was implanted into a Si wafer with a 60 nm thick SiO_2 capping layer in such a way that the profile maximums are approximately 20 nm below the Si-SiO_2 interface. After annealing at 1100 °C

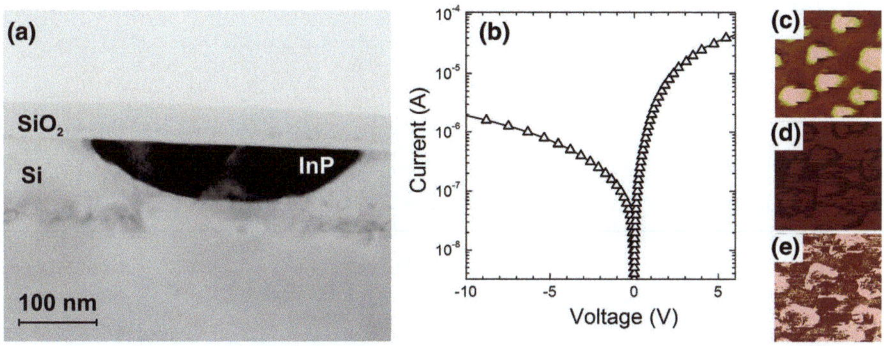

Fig. 4.52 Bright field high resolution TEM micrographs of an InP NC after FLA with 1200 °C and preheating of 700 °C, 2 min (**a**), current-voltage characteristic of an InP-Si heterostructure after wet selective etching (**b**) as well as the topography as measured by atomic force microscopy (**c**). The current response received in atomic force microscopy is given for the reverse (**d**) and foreward direction (**e**). After [267]

for 20 ms and a preheating of 600 °C, 1 min, triangular InP NCs with an average size of 60 nm are formed. The NCs are polycrystalline and contain amorphous InP or even In inclusions. If the FLA temperature is increased to 1300 °C, zinc-blende hemispherical single crystalline (111)-oriented InP NCs will be formed which, however, are characterized by tensile stress and are still decorated with In nanoclusters in some cases. A further increase of the thermal budget of the preheating (600 °C, 3 min) leads to fully relaxed, (100)-oriented InP NCs with sizes up to 300 nm. At least in the area investigated by TEM no In inclusions has been found. Finally, FLA at 1200 °C with a preheating of 700 °C, 2 min, leads to the formation of 400–500 nm large InAs NCs as shown in Fig. 4.52a. The NC shown in this figure is composed of a (100)-oriented crystal in the middle and (111)-oriented crystals on the edge. Whereas the lattice constants for the central NC match that of bulk InAs, the (111)-oriented NCs exhibit tensile strain.

Similar to the case of InAs, InP-Si heterostructures were fabricated by removing the SiO_2 layer, oxidizing the surface by an oxygen plasma, and selectively etching the oxide on top of the InP NCs [268]. Once again, the incorporation of Si during FLA leads to n-type doping of InP, and a p-n heterojunction is formed. The corresponding current-voltage characteristic in Fig. 4.52b shows the typical behaviour of a diode, but with a relatively large leakage current in reverse direction. The InP NCs are visible as bright spots in the atomic force microscopy image in Fig. 4.52c, whereas the current response in forward and reverse direction is given in Fig. 4.52d, e, respectively. As seen, no significant current flow was detected except in forward direction and at those locations which correspond to the locations of InP NCs. To investigate the suitability of this material for PV applications, a 15 nm thick, semitransparent Au film was deposited on top of the InP NCs. Under halogen lamp illumination, an open circuit voltage of about 0.39 V was measured [268].

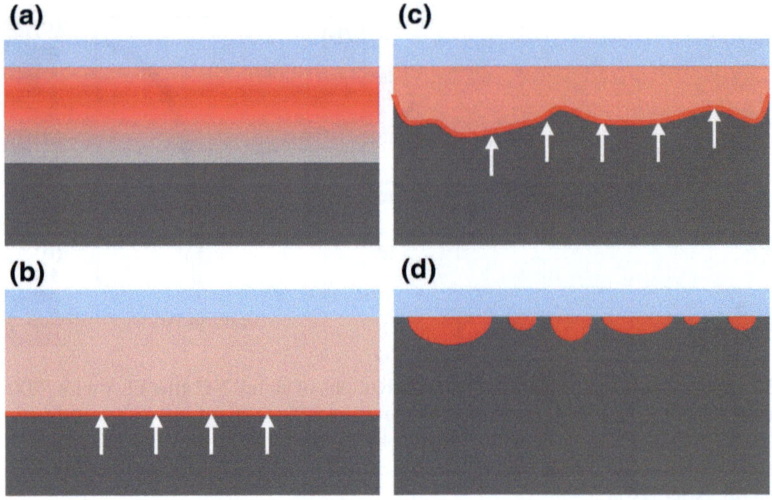

Fig. 4.53 Growth mechanism for III–V NC in bulk substrates

Finally, the formation of GaAs NCs in Si was proven [269], and the results are similar to those of InAs and InP. Due to the higher melting temperature of GaAs higher thermal budgets during FLA are needed to initiate NC formation. Depending on the FLA conditions, randomly oriented, (100) oriented and (111) oriented GaAs NCs are formed. The incorporation of Si into GaAs during FLA is also more complex. Si incorporation into (100) GaAs grown by MBE leads mainly to n-type doping with Si preferably located on a Ga lattice site [270]. However, in case of (110) or (111) GaAs, a significant part of Si also occupies As lattice sites which results in a much weaker n-type doping. In the present case of GaAs NCs fabricated by ion implantation and FLA, Raman investigations reveal the existence of Si_{As} acceptors and neutral complexes formed by donor–acceptor pairs Si_{Ga}–Si_{As}. Thus, GaAs NCs are assumed to have a lower effective carrier concentration due to charge compensation.

The main mechanism of III–V NC formation in bulk Si during FLA is LPE as depicted in Fig. 4.53. After ion implantation, the layer stack as shown in Fig. 4.53a consists of the Si substrate (dark grey), an amorphized Si layer (bright grey) with a Gaussian-like distribution of the implanted group-III and group-V atoms (red), and a SiO_2 capping layer (blue). During FLA the temperature is high enough to melt the amorphous layer without melting the crystalline substrate (Fig. 4.53b) as the melting point of crystalline Si is about 200 K higher than that of a-Si [271, 272]. In addition, the implanted species will lead to a further reduction of the melting point. The diffusion constants of impurities in the liquid phase are so large (Table 4.2) that the concentration will be constant over the implanted layer.

Immediately after the FLA pulse, the melt cools down and recrystallization of Si starts from the interface between amorphous and crystalline Si which acts as an epitaxial pattern. As the crystallization front moves towards the surface (Fig. 4.53c)

Table 4.2 Diffusion coefficients D_S and D_L in the solid and liquid phase of Si, respectively, as well as the segregation coefficient k. D_S refers to a temperature close to the melting point of Si of 1414 °C

Element	D_S (cm^2 s^{-1}) [54]	D_L (cm^2 s^{-1}) [18]	k [54]
Ga	1.0×10^{-10}	4.8×10^{-4}	~0.01
In	2.6×10^{-10}	6.9×10^{-4}	0.0004
P	3.4×10^{-7}	5.1×10^{-4}	0.35
As	6.0×10^{-12}	3.3×10^{-4}	0.3

the remaining melt is enriched with the implanted species according to their segregation coefficient (Table 4.2). In and Ga have a very low segregation coefficient, and consequently only a tiny fraction of them is incorporated into the regrown Si. In contrast to this, the segregation coefficient of P and As is quite large, and up to one third of them can be incorporated into Si. This leads (i) to strong n-type doping of the regrown Si and (ii) to an increasing imbalance between group III and group V elements in the melt. There will be regions where crystallization is faster than in others making the crystallization front more uneven and rough. Crystallization stops when the crystallization front approaches the SiO_2 capping layer, and the shrinking molten layer decays into individual pockets which later evolve to III–V NCs (Fig. 4.53d). With increasing enrichment the melt transforms from a Si melt with impurities to a compound semiconductor melt with Si impurities, accompanied by an adequate decrease of the solidification temperature. If the temperature drops below the solidification temperature of the corresponding III–V compound, a Si-doped III–V NC is formed. According to Sect. 3.2.4 the cooling process consists of a fast phase in the ms range governed by thermal conduction followed by a slow phase of several seconds governed by convective heat transfer and thermal radiation. It is assumed that InAs, GaAs and InP NC are mainly formed in this early phase due to its relatively high melting temperature. However, the melting point of In (157 °C) and Ga (30 °C) is so low that elemental inclusions can remain in the liquid phase for seconds. As mentioned above, the formation of InAs NC was accompanied by the formation of elemental In clusters. Thus, the required excess of In can be explained by the stronger enrichment of the melt with In compared to As due to the different segregation coefficient, and the relatively large size can be the result of the long time span available for diffusion between the flash pulse and the solidification of the In cluster.

4.4.2.3 Germanium and Silicon-on-Insulator Substrates

The formation of III–V NCs by FLA is not limited to Si. A recent study compares the formation of InAs NCs in Si and Ge using a preheating of 470 °C for 3 min, a 20 ms flash pulse and energy densities in the range 9–18 J cm^{-2} and 46–93 J cm^{-2} for Ge and Si, respectively [273]. In both cases InAs NCs were formed, but their morphology

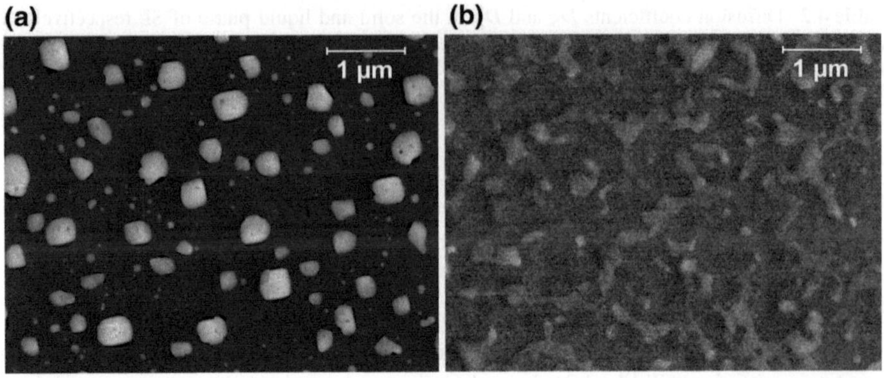

Fig. 4.54 SEM images of InAs NCs in bulk Si (**a**) and Ge (**b**). FLA annealing was performed using a preheating of 470 °C for 3 min and a 20 ms flash with an energy density of 93 J cm^{-2} (**a**) or 13 J cm^{-2} (**b**). Contrast, brightness and sharpness have been modified for optimal illustration. After [275]

and crystal quality differ significantly. In case of Si, round-shaped NCs at the SiO_2 interface with sizes up to 500 nm were formed (Fig. 4.54a). In case of Ge the NC shape is much more variable including dendritic structures that connect individual precipitates (Fig. 4.54b). Moreover, some structures can get sizes of several µm, and smaller InAs NCs can also be found in a depth of 30–120 nm below the Si-SiO_2 interface. The lattice mismatch between InAs and Ge with ca. 7% is below the limit of pseudomorphic growth [274] which allows the formation of InAs NCs with the same crystal orientation as the Ge matrix. However, the preheating temperature is sufficient to already start recrystallization prior to the flash, and the solidification temperatures of Ge and InAs are very close to each other. Thus, in contrast to the mechanism in Fig. 4.53, Ge and InAs may solidify simultaneously and develop a different shape and size distribution. From a user-relevant point of view the formation of III–V NC in Si seems to be more promising because of the better control of NC size and shape.

Another option to tailor the NC formation process is the change to SOI wafers. Such investigations were performed with the intention to fabricate InAs NCs [276] or $In_xGa_{1-x}As$ NCs [277] in the device layer. The original device layer was thinned down by thermal oxidation from 90 to about 60–65 nm, whereas the resulting 50–60 nm thick thermal oxide served as a protection against sputtering during ion implantation again. The implantation energies were chosen in such a way that the projected range of implantation was in the middle of the device layer. The implantation fluences were sufficient by far to amorphize the device layer which melts during the FLA treatment. During crystallization from the liquid phase large block-like InAs NCs can be formed if the implantation fluence is sufficiently large. Figure 4.55a–c displays HAADF-STEM images from SOI structurers implanted with 1, 2 and 3 $\times$ 10^{16} ions/cm^2 which correspond to an atomic concentration of 3, 6 and 9% for each element, respectively [276]. For the lowest implantation fluence, the InAs NCs have a more irregular shape and are mainly located at one of the Si-SiO_2 interfaces.

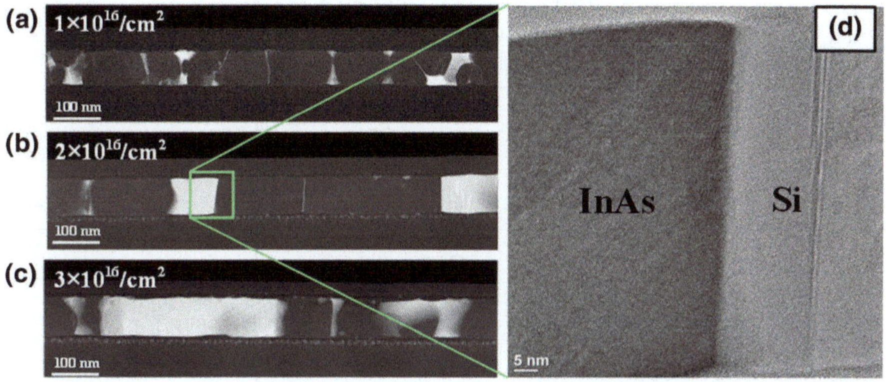

Fig. 4.55 HAADF-STEM images of SOI structures implanted with various fluences and subjected to FLA for 20 ms, an energy density of 82 J cm^{-2} and a preheating of 600 °C for 3 min (**a–c**). **d** displays a high resolution TEM image of a selected Si-InAs interface for a sample implanted with the medium fluence. Reprinted from [276], with the permission of AIP Publishing

For medium and higher implantation fluences large InAs NCs of high crystal quality evolve which have vertical sizes equal to the device layer thickness and a high-quality interface to the surrounding Si matrix (Fig. 4.55d).

The formation process is similar as in case of bulk Si. When the implanted Si layer is liquid, the impurities will evenly distribute in the layer, but now there is no epitaxial pattern for regrowth. As a consequence, Si crystallizes first with random orientation, and the grains grow until they are limited by the Si-SiO_2 interfaces. Further growth occurs only in a lateral direction which also allows the InAs NC to crystallize over the whole layer thickness, provided that a sufficient amount of In and As atoms are available. The situation becomes a little bit more complex if ternary III–V NCs like $In_xGa_{1-x}As$ should be formed. In a recent investigation [277] As with a fluence of 3 $\times$ 10^{16} ions/cm^2 was implanted in the device layer of a SOI wafer, followed by the implantation of Ga and In where the accumulated fluence of Ga and In was equal to that of As. After FLA the formation of $In_xGa_{1-x}As$ was evidenced and investigated by Raman and XRD measurements. In case of ternary compounds the peak positions of the corresponding phonon modes and XRD reflections reveal a systematic deviation from the nominal composition according to the In:As implantation ratio (Fig. 4.56). Although the tracing of the $In_xGa_{1-x}As$ composition from measurements has many imperfections and uncertainties, the majority of the formed NCs seems to be Ga-rich compared to the nominal value. In fact, the bimodal peak shape in case of XRD suggests the existence of a distribution of NC populations with different compositions which is, for convenience, represented by a dominating Ga-rich composition and a minor In-rich composition. As the melting temperature of $In_xGa_{1-x}As$ continuously decreases with increasing In content, $In_xGa_{1-x}As$ NCs with a Ga-rich composition are assumed to crystallizes first during the cooling phase after FLA. This shifts the In:Ga ratio in the cooling melt to the favour of In and leads to the formation of

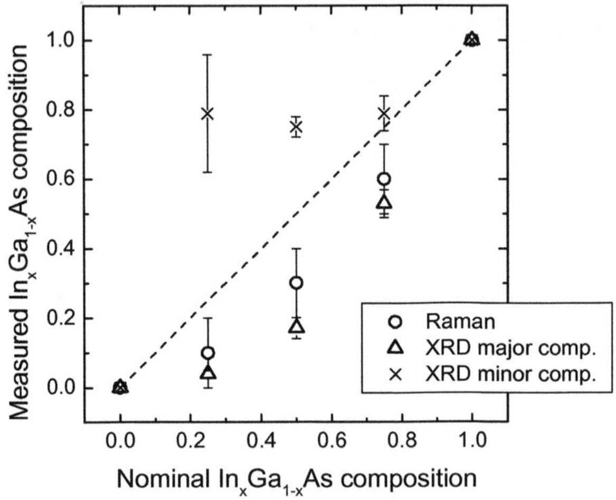

Fig. 4.56 Composition of $In_xGa_{1-x}As$ NCs as deduced from Raman (open circles) and XRD measurements as a function of the nominal composition defined by the In:As implantation ratio. Due to the bimodal peak shape in case of XRD, the existence of at least one dominating, Ga-rich component (open triangles) and a minor, In-rich component (crosses) can be postulated. The dashed line is a guide for the eye. More details can be found in [277]

In-rich $In_xGa_{1-x}As$ NCs or elemental In clusters in low concentrations at the end of the cooling phase. Finally, the formation of GaAs and InP NCs in a SiO_2-Si-SiO_2 stack deposited by PECVD was demonstrated [278], although this process was accompanied by unwanted void formation probably caused by the high amount of hydrogen introduced during the PECVD deposition.

4.4.2.4 III–V Nanocrystals at Defined Positions

With regards to potential applications the fabrication process should be able to form III–V NCs at locally defined positions. Thus, the next step is the introduction of a masked layer which is thick enough to protect the underlying layers from being implanted [276]. This masked layer has two advantages: at first, NCs can only be formed at places where the masked layer has openings, and secondly, the neighbouring areas which are protected against implantation can serve as an epitaxial pattern for recrystallization. For better understanding, Fig. 4.57a–c traces the main processing steps. From top to down, the structure in Fig. 4.57a comprise the masked layer (orange), the SiO_2 cladding layer (white), the Si device layer (grey), the SiO_2 separation layer and the Si substrate. The masked layer contains implantation windows at the desired positions of NCs. If implanted (Fig. 4.57b), the ions will be incorporated either in the Si device layer or in the masked layer, whereas the first one is completely amorphized. After the removal of the masked layer, the structure is subjected to FLA (Fig. 4.57c) which leads to the melting of the implanted areas. In the following cool-

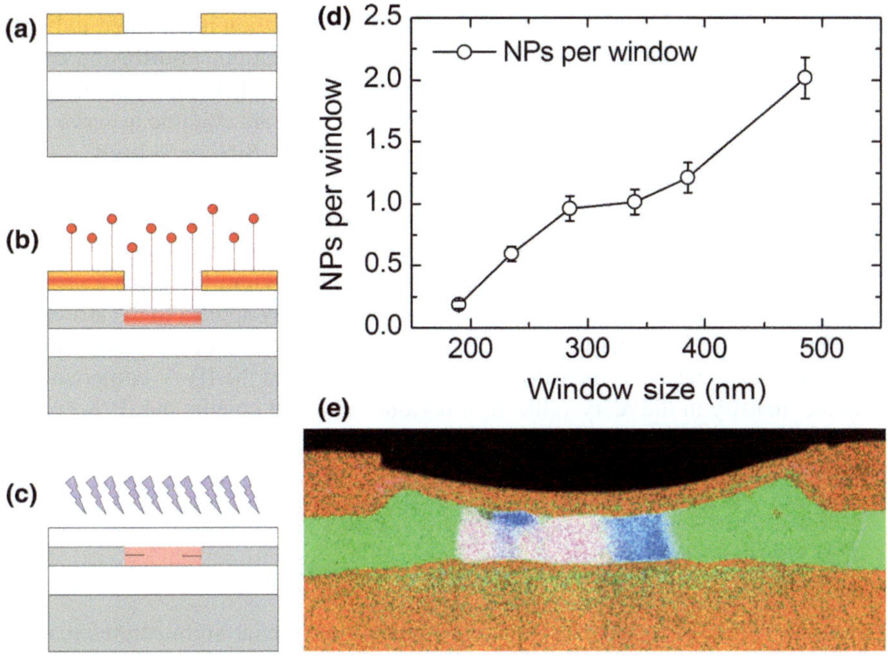

Fig. 4.57 Basic processing scheme for the fabrication of III–V NCs at locally defined positions: SOI wafer with masked layer for implantation (**a**), ion implantation (**b**) and FLA (**c**). Average number of nanoparticles as a function of implantation window size (**d**). The solid line is a guide for the eye. TEM image of a selected NP combined with EDX data. The colours mark different materials, namely SiO_2 (beige), Si (green), InAs (blue) and NiSi (pink). Partly adapted from [279]

ing phase Si starts to recrystallize from the sidewalls (black arrows in Fig. 4.57c), and one or more NCs are formed in the central region of the implantation window.

In a recent investigation [279] As and In were implanted into such a structure with a masked Ni layer patterned by EB lithography. The implantation windows were quadratic in shape, had edge lengths between 190 and 460 nm and were typically arranged in arrays. After Ni removal and FLA using a 20 ms pulse, nanoparticles are formed in the area of the implantation window. A statistical analysis over several hundreds of implantation windows (Fig. 4.57d) reveals that the average number of nanoparticles increases with increasing window size and that there is a small plateau of about one nanoparticle per window for windows with edge lengths between 285 and 340 nm. However, there is a statistical variation of the number of NPs, of their size, their shape and their exact position within the implantation window. In addition, Si seems to recrystallize in a polycrystalline way according to TEM investigations (not shown). Therefore, the aspiration to create InAs NCs at defined positions is only partly fulfilled. Figure 4.57e displays a larger NP in an implantation window with an edge length of 460 nm. The image combines TEM and EDX data and marks the different materials with different colours. As seen, the NP is composed of InAs and unintentionally formed NiSi. The NiSi formation is assumed to be the result of Ni sputtering and re-deposition during implantation, in-diffusion during implantation, and diffusion and phase formation during FLA. As sputtering is a general phenomenon, only materials being intrinsic to Si are suitable to serve as an implantation mask layer. Potential materials include SiO_2, SiN, or Si itself.

In summary, III–V NCs can be integrated in Si by using ion implantation into bulk Si or SOI structures followed by annealing. Compared to FA or RTA, FLA has the advantage to offer LPC and to suppress diffusion in the solid phase. Although the results are very interesting from a scientific point of view, there are a series of obstacles with regard to potential applications. This mainly applies to the aforementioned statistical variations in size, shape and position of the NCs. In addition, the crystal quality in terms of interface quality between Si and the III–V compounds or of defect density in the recrystallized Si regions is not yet convincing. However, as seen in the next section, the situation improves for semiconductor NWs.

4.4.3 Nanowires

The downscaling of transistor size using conventional planar architectures reached the technology limits already a decade ago, mainly due to the current leakage and current tunnelling through the channel. In 2004 Intel has presented a full commercial silicon processing scheme employing a 45 nm MOSFET technology. Only few years later (2010), the channel length was reduced to 32 nm. However, further progress in nanoelectronics was possible by the implementation of non-planar architectures, namely 3D transistors. The conventional planar MOSFET device was replaced by the tri-gate fin field effect transistor (FinFET) structure where the gate is wrapped around the channel from three sides. Using FinFETs the operation voltage was reduced to 0.7 V and below, the gate delay was reduced to less than 1 ps, and heat dissipation was significantly improved. Nowadays, the channel length in FinFETs is reduced to 12 nm and the next generation of nanolectronics is going to use FinFETs with a channel length below 10 nm. The reduction of the channel length in FinFETs and the increase of the density of transistors per chip generate a significant increase of the production costs. Therefore, alternative technologies and solutions are considered for the next generation of nanoelectronics which may be essential for continuing the progress in nanoscale devices. The miniaturization of the FinFETs requires ultra-high doping levels beyond 5×10^{19} cm^{-3} which affects the carrier mobility. To keep the subthreshold slope at 60 mV/dec or even below the FinFET structure should be replaced by gate-all-around FETs where the electronic devices are made of NWs fabricated either by a top-down approach or by a bottom-up method. The top-down approach relays on EB lithography and etching while the bottom-up refers to growth methods like VLS growth, MBE, CVD etc. Due to the technological problems the in situ doping of grown NWs is challenging. This especially applies to the formation of n- and p-type regions within the same NW. Using the top-down approach the conductivity type and the doping level of fabricated NWs are defined by the substrate.

Therefore, ion implantation followed by thermal annealing is the main technique used for the doping at the nanoscale.

4.4.3.1 Ion Implantation of Nanowires

The doping of NWs can be realized in situ during the growth or ex situ using either ion implantation or diffusion. The in situ doping of semiconducting NWs is challenging and mainly limited by the solubility of dopants in the catalyst (metallic droplet). First dopants must be diluted within the metallic catalyst and will be later transferred from the liquid catalyst to the NW. Schmid et al. have shown that using a Au-catalyst the electron concentration in phosphorous doped Si NWs can be as high as 1.5×10^{20} cm^{-3} for a gas phase ratio of $SiH_4/PH_3 = 1.5 \times 10^{-2}$ [280]. They have shown that the doping of Si NWs is a self-limiting process and that a further increase of the gas ratio doesn't increase the effective carrier concentration. In the case of Ge NWs the in situ doping allows to obtain an electron concentration in the range of 10^{19} cm^{-3} mainly due to the limited solubility of P in Ge, while p-type Ge NWs shows an effective hole concentration above 10^{20} cm^{-3} [281]. The in situ doping of III–V compound semiconducting NWs is even more challenging than the doping of Si. The n-type GaAs NWs hardly reaches a doping level of 10^{18} cm^{-3}. A similar doping level is obtained for p-type NWs [282]. Moreover, the formation of lateral p-n junction within a single NW is very difficult due to the different activation energy and different thermal diffusivity for p- and n-type dopants in the host. The formation of radial junctions where the shell layer is grown with conditions different from those of the core and without the catalyst is simpler. The alternative to in situ doping is ion implantation of NWs where the type of doping as well as the junction depth and shape are determined by the lithography steps and the ion implantation parameters. By using ion implantation and post-implantation annealing both a radial and a lateral junction can be fabricated [283]. Unfortunately, ion implantation is a destructive technique introducing a lot of different defects together with the dopants. These defects must be annealed-out without affecting the dopant distribution. Moreover, during ion implantation strain is accumulated within the NWs leading to the bending of the NWs towards the ion beam direction (Fig. 4.58).

The strain accumulation in implanted NWs depends on the ion species, the ion energy (projected range R_p) and the initial position of the NWs with respect to the ion beam [284]. Figure 4.58 shows a typical case when as-grown NWs are implanted with high fluence. Using conventional growing parameters, NWs grow along the (111) direction which means that in case of an (100) substrate NWs will be sloped to the sample surface at 37.5°. However, in case of an (111) oriented substrate NWs will grow perpendicular to the sample surface. In order to have homogeneous doping along the NWs such a sample must be tilted with respect to the incident ion beam. But the final effect will be the same. Figure 4.58i shows the initial stage of ion implantation. R_p is designed to be slightly below the middle of the NWs. Incorporation of ions into the NWs matrix causes the expansion of the lattice parameter of the implanted part of the NW in both directions while the unimplanted part remains unchanged.

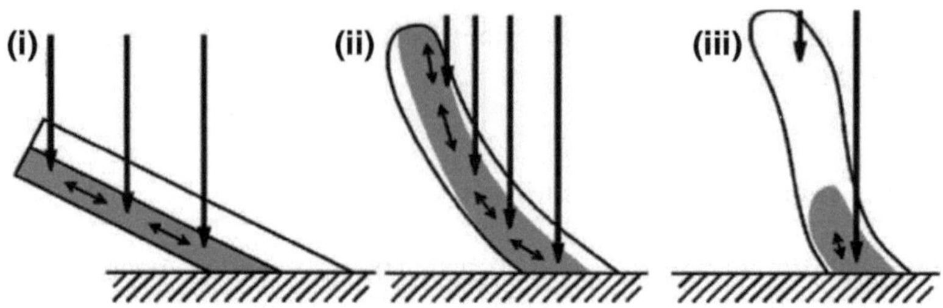

Fig. 4.58 Schematic representation of strain accumulation in a NW during high fluence ion implantation: (**i**) initial stage of ion implantation, (**ii**) bending of NWs for middle ion fluences (**iii**) NWs after high fluence ion implantation. The strain accumulation in the NW during ion implantation is marked by the grey areas. Reprinted from [284], with permission from Elsevier

The expansion along the NW causes an elongation of one side of the NW leading to its bending.

The shape change of the NWs during implantation introduces inhomogeneity both to the implantation profile and to the strain accumulation (Fig. 4.58ii). As a final result NWs implanted with a high ion fluence (above 1×10^{15} cm^{-2}) cannot preserve their initial shape due to the continuous plastic deformation caused by ion beam irradiation. The change of the NW shape also causes a non-homogeneous distribution of implanted elements along the NW (Fig. 4.58iii). Figure 4.59a shows a scanning electron microscopy (SEM) image of as-grown Si NWs on an (100) oriented Si substrate recorded at 30° from the normal to the sample surface. Si NWs are growing randomly along the (111) direction with an average diameter in the range of 100 nm. Figure 4.59b shows the as-implanted Si NWs implanted with Mn ions to a fluence of 1×10^{16} cm^{-2}. R_p was designed to be in the middle of the NWs (50 nm). The shape of Si NWs is very similar to the schematic representation shown in Fig. 4.58iii. During ion implantation of heavy elements like As and In with a fluence above 10^{16} cm^{-2} the NWs are deformed (Fig. 4.59c). Due to the plastic deformation, sputtering and the diffusion of Si atoms at the NW surface towards the beam direction are finally connected with each other [283]. Only NWs which are aligned parallel to the ion beam remains unaffected. Figure 4.59d shows Si NWs made by reactive ion etching and implanted with As^+ ions to a fluence of 5×10^{15} cm^{-2}. The NWs stood perpendicular to the Si surface and the sample was tilted by 20° with respect to the ion beam. It is clearly seen that the base of the NW remains unchanged whereas the top of the NW starts to bend towards the ion beam. Therefore the doping of NWs using ion implantation is challenging and many aspects must be taken into account designing the process flow for NW doping. The bending of NWs can be avoided either by implanting horizontal NWs (the beam is perpendicular to the NW axis) or rotating the NW during ion implantation around the ion beam axis. Jochannes et al. have shown that the rotation of NWs during implantation supresses the bending of NWs and allows an homogeneous doping [283].

The implantation of horizontal NWs was successfully applied to Si NWs with chalcogen atoms. Chalcogens in Si are fast diffusers and the solid solubility is limited

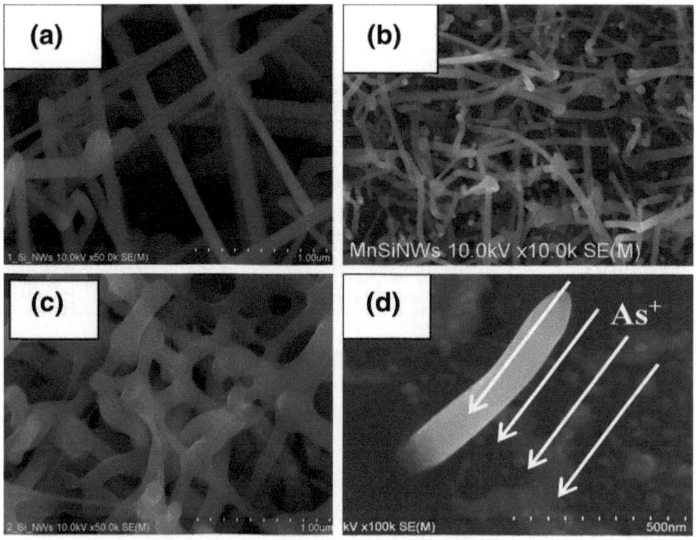

Fig. 4.59 SEM images of as-grown Si NWs (**a**) and as-implanted NWs with a total ion fluence of 1×10^{16} cm^{-2} (**b**) and 8×10^{16} cm^{-2} (**c**). Si NWs presented in (**b**) were implanted with Mn ions while Si NWs shown in (**c**) were implanted with As and In ions with the same fluence for each element (4×10^{16} cm^{-2}). **d** shows the bending of a single Si NW towards the ion beam. With permission of A. Lugstein, TU Vienna

to the level of 10^{16} cm^{-3}. However, chalcogen concentrations exceeding this limit by four orders of magnitude can be achieved by using strongly non-equilibrium processing, e.g. ion implantation followed by ms-range FLA (Sect. 4.2.2). Berencen et al. have shown that Si NWs grown by the VLS method can be doped with Se up to 2.3% without Se segregation and without a significant change of the NW shape [285]. The proper design of the implantation parameters allows the amorphization of only the top part of the NW (R_p was smaller than the NW diameter) while the bottom of the NW remains crystalline (see Fig. 4.60b). In the following step of FLA the crystalline part of the NWs is intended to be used as a template for the recrystallization of the implanted part. Simultaneously with the recrystallization process of the implanted NW the Se atoms are incorporated into the lattice of the Si crystal. Figure 4.60c shows the EDX mapping of the Si (green), O (red) and Se (blue) distribution in an as-implanted NW obtained in cross-section geometry. According to simulations the Se atoms are located in the upper part of the Si NW. Interestingly, after FLA for 3 ms the Se distribution is almost unchanged. The fraction of Se in substitutional sites is in the range of 70% [285].

The main difference between NWs implanted with low (Fig. 4.60a) and high (Fig. 4.60b) fluences is related with the cross-sectional shape of the Si NW. The deformation of the shape of the NWs is due to the effective mass transport during ion implantation. The plastic deformation of nano-objects via ion beams dominates over the sputtering process [283]. Interestingly, the plastic deformation of irradiated NWs

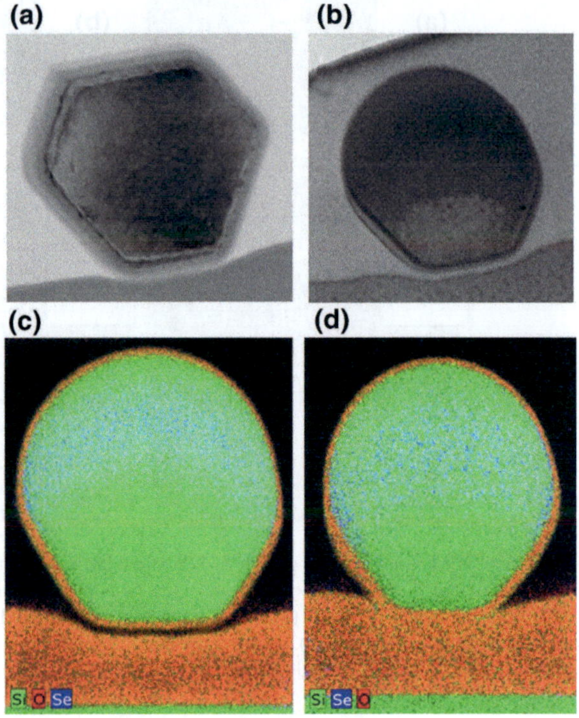

Fig. 4.60 The bright field cross-section TEM images of P implanted Si NWs with a fluence of $2 \times 10^{14}\ cm^{-2}$ (**a**) and Se hyperdoped Si NW with a fluence of $2 \times 10^{16}\ cm^{-2}$ (**b**). **c** and **d** show the EDX mapping of the Si, O and Se distribution in the cross-section geometry obtained from as-implanted and annealed hyperdoped Si NWs, respectively

is independent on the orientation of the NWs with the respect to the incident beam but it is strongly affected by the kind of ion, the ion fluence and the ion energy. As already discussed (Sect. 4.2.2) the hyperdoping of Si NWs with chalcogens strongly reduces the effective bandgap of Si. This makes Se hyperdoped Si attractive for short wavelength IR optoelectronics, especially photodetectors.

Figure 4.61 shows the current-voltage characteristics obtained from a Se hyperdoped Si NW and a virgin Si NW with and without illumination. The illumination was performed using a NIR laser operating at 1.55 μm (0.80 eV) which is much smaller than the bandgap of Si (1.12 eV). Therefore the current-voltage curves obtained from virgin Si NWs overlaps each other, i.e. there is no photon absorption. In contrast to virgin Si NWs the Se hyperdoping causes a significant enhancement of the NW conductivity. Moreover, the conductivity of Se hyperdoped NWs further increases under the 1.55 μm illumination. This is the direct prove of the effective bandgap reduction upon the Se hyperdoping.

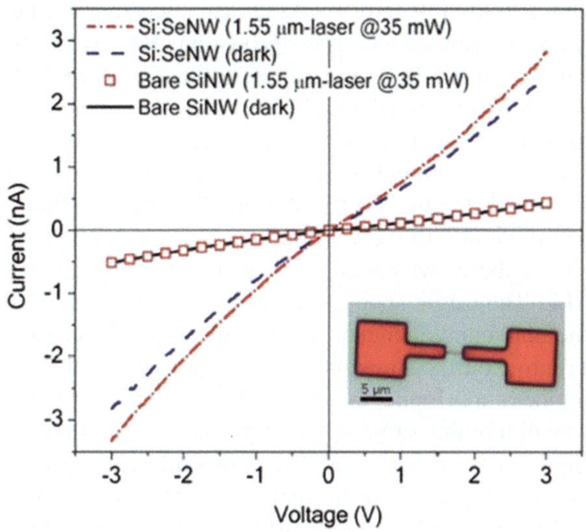

Fig. 4.61 RT current-voltage characteristic under dark conditions and under 1.55-μm illumination of a single bare NW and a 1 × 10^{16} Se/cm^2-hyperdoped NW, with diameters of approximately 130 nm. The inset shows an optical microscope image of an NW and its contacts created by EB lithography. Reprinted from [285], with permission from Wiley

4.4.3.2 Liquid Phase Processing of Nanowires

To reduce the surface sputtering and roughening via energetic ions the Si NWs were coated with 30 nm of SiO_2. After ion implantation Si NWs were annealed using FLA for 20 ms with different energy densities. By proper adjusting the flash energy the temperature within the NW can be fully controlled. The implanted NW can be recrystallized either via SPE or LPE. If the temperature during ms-range annealing is lower than the melting point of NWs the recrystallization takes place via SPE. The diffusion of implanted elements is limited to few nanometres which allows the precise control of the dopant distribution like in Se hyperdoped Si. This phenomenon was proven to exist both in thin films and in NWs. But annealing of implanted semiconductors at temperature higher than the melting point allows the recrystallization via LPE. It was shown that by using ns-range PLA the impurity concentration in the semiconductor can easily exceed the solubility limit by four orders of magnitude and higher [148]. Importantly, the dopant diffusion coefficient in the liquid phase is up to seven orders of magnitude higher than in the solid phase.

Moreover, the diffusion coefficient in the liquid phase is very similar for most elements in different semiconductors. The advantage of LPE compared to SPE was successfully exploited to integrate III–V compound semiconductors with Si and especially into Si NWs [286]. In some aspects the integration of III–V materials into NWs is an advancement of the research activities discussed in the previous section. Due to the much higher diffusivity of dopants in liquid Si, during ms-range annealing the implanted elements can segregate within the single Si NW forming either clusters or separated segments. This phenomenon was used to fabricate III–V compound semiconductors in Si NWs. The Si NWs grown by the VLS technique were implanted with In and As ions to a fluence of 3 × 10^{16} cm^{-2} for each element. The total concentration of implanted elements was in the range of 15%. The as-grown NWs refers to samples with NWs grown on (100) Si substrates which mean that NWs are tilted by about 36° to the sample surface, whereas transferred NWs describe samples with NWs dispersed on a Si substrate. Ion implantation was performed either into as-grown NWs (Fig. 4.59a) or they were transferred onto a virtual substrate before ion implantation. In both cases the ion fluence is the same but the ion distribution can be slightly different. The transferred NWs are more homogeneously implanted then the as-grown due to the bending of as-grown NWs during ion implantation Fig. 4.62a shows the SEM image of In and As implanted, transferred NWs after ms-range SPE. The bright spots covering the entire length of the Si NW are clusters made of the implanted elements. These clusters are InAs quantum dots formed during FLA. Figure 4.62b, c show the SEM images obtained from implanted NWs after FLA at temperatures higher than the melting point of a-Si ($T = 1150\,°C$). Here the additional function of the SiO_2 capping layer is to prevent the change of the NWs shape during liquid phase processing. The annealing temperature ensures the melting of Si and is much higher than the melting point of InAs (942 °C). During the liquid phase of Si the In and As atoms diffuse and segregate. Due to the different melting point Si solidifies faster than the InAs. Si being in the supercooled liquid state pushes dopants being present at concentrations above the solubility limit to the front of the solidification interface. The solidification always starts from the coldest place, the ends of the NW, towards the middle. As a consequence, Si NWs recrystallize along the (111) orientation, followed by the crystallization of InAs segments leading to the formation of a hybrid Si-InAs-Si NW.

Figure 4.62d, e shows the electron energy loss spectroscopy (EELS) plasmon mapping of Si and O taken after FLA from the NW presented in Fig. 4.62c. The EELS image exhibits the presence of an about 25 nm thick shell oxide layer around the hybrid NW. The thickness of the oxide shell is 5 nm thinner than the initial oxide thickness due to sputtering. The mapping of the Si distribution clearly shows the existence of segments which are free of Si. These segments are composed of In and As (Fig. 4.62e). Figure 4.63 shows a bright field high resolution TEM image of a Si-InAs interface within a hybrid Si-InAs-Si NW after LPE. Here the ion implantation was performed into planar NWs transferred onto a virtual substrate before ion implantation. The TEM diffraction pattern (not shown) confirms the formation of high-quality [111] oriented crystalline InAs segments. It seems that during solidification the Si part of the NW act as a crystallization seed for the InAs segments even if the lattice mismatch between InAs and Si is 11.6%. The presented hybrid NWs are potentially interesting for the new generation of nanoelectronics. The carrier mobility in InAs is about 40 times higher than that in Si which can cause a significant improvement of device performance. However, InAs is a small bandgap semiconductor (0.34 eV) which limits its application in optoelectronics. Therefore the same fabrication process was applied to integrate GaAs with Si. Glaser et al. have demonstrated RT EL from hybrid Si-Ga-GaAs-Si NWs made by ion implantation and FLA [287].

The concentration of Ga and As in the Si NWs is in the range of 15%. After ion implantation followed by FLA most of the Ga and As atoms are incorporated into GaAs single crystalline segments. The length of the GaAs segments depends mainly

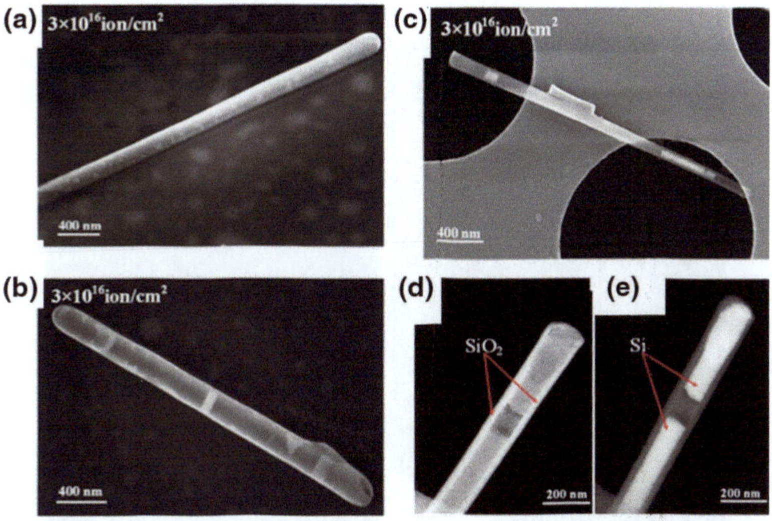

Fig. 4.62 SEM images of implanted planar deposited Si NWs (**a** and **b**), an implanted vertical Si NW (**c**) and EELS images (**d** and **e**) of the ion implanted and FLA treated Si NWs. NWs were annealed for 20 ms with an energy density of 58.1 J/cm^2 (**a**), 66.5 J/cm^2 (**b**), and 54.2 J/cm^2 (**c**). **d** and **e** present EELS plasmon mappings of the SiO_2 shell and the Si plasmon mode, respectively, taken from the as-grown Si NW presented in (**c**). The ion fluence is indicated in the images. Reprinted by permission from Springer Nature: [286]

Fig. 4.63 Bright field high resolution TEM micrographs of the interface between (111) Si and zinc blende (111) oriented InAs obtained from planar implanted NW presented in Fig. 4.62b. Reprinted by permission from Springer Nature: [286]

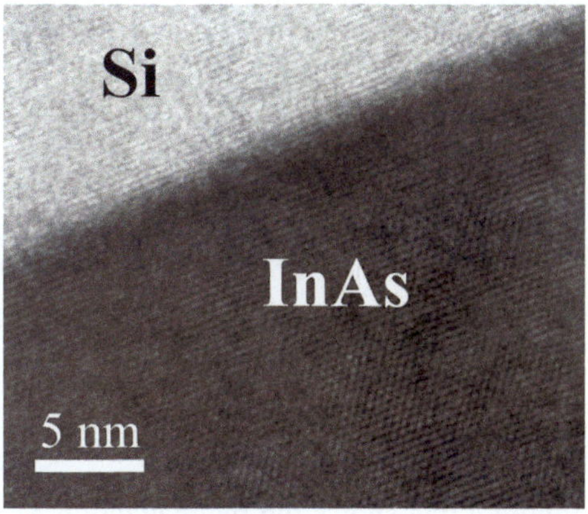

on the initial concentration of the implanted elements. But due to the different solid solubility and diffusivity of Ga and As in Si a part of Ga does not interact with As and forms metallic segments directly connected to the GaAs crystal. This metallic segment is used as a nano-contact to inject electrons into the GaAs-Si heterojunction (Fig. 4.64a). The current-voltage characteristics reveal the formation of Schottky-like diode with EL originating from the band-to-band transition in the GaAs segment. The main emission peak is located at about 700 nm and is due to impact ionization

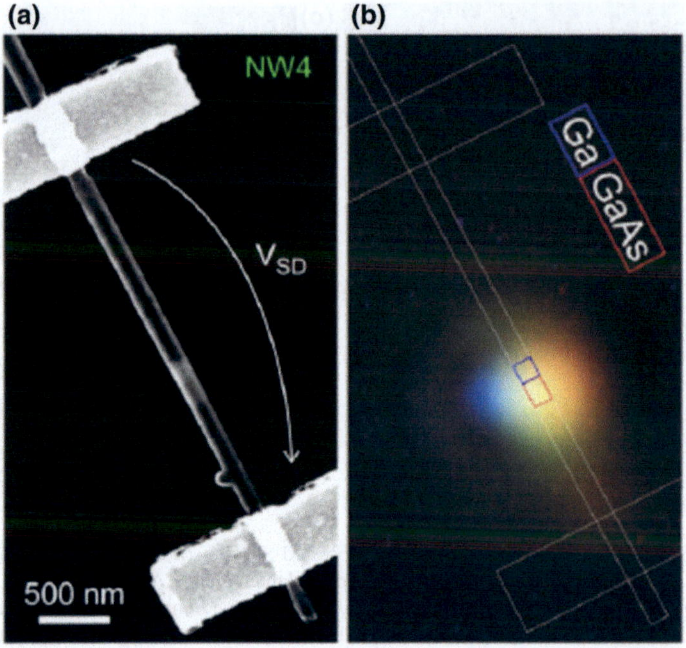

Fig. 4.64 RT EL of a Si-Ga-GaAs-Si NW heterostructure device. **a** SEM image, showing the two-terminal device (NW 4) with a Ga segment next to a 180 nm long GaAs NC. **b** Optical microscope image (1 min exposure time) of the same device under reverse bias conditions ($V_{SD} = -9.4$ V, $I_{SD} = -4$ μA) with visible EL originating from the GaAs NC. Reproduced from [287] (CC BY-NC 4.0)

in advance of the emission process. Figure 4.64b shows the optical microscope image taken from such a hybrid NW during operation with a source-drain voltage of −9.4 V and an time of 60 s. The spectral diffraction of the EL spot in the visible in Fig. 4.64b is due to artefacts of the imaging system [287].

4.4.3.3 Magnetic Silicon Nanowires

Figure 4.65a shows the magnetization as a function of temperature obtained from Si NWs implanted with Mn using a fluence of 2×10^{16} cm^{-2} which corresponds to a Mn concentration in the range of 10%. After ion implantation, the Si NWs were annealed using FLA for 20 ms. The energy density applied to the sample surface was in the range of 80–120 J cm^{-2} which corresponds to a peak temperature in the range of 1100–1350 °C. As expected, the virgin Si substrate is non-magnetic, and a similar curve was obtained from as-grown Si NWs. After ion implantation and annealing the Mn doped Si NWs become ferromagnetic. In Fig. 4.65a it is visible that three magnetic phases are formed: a low temperature phase with T_C about 50 K which

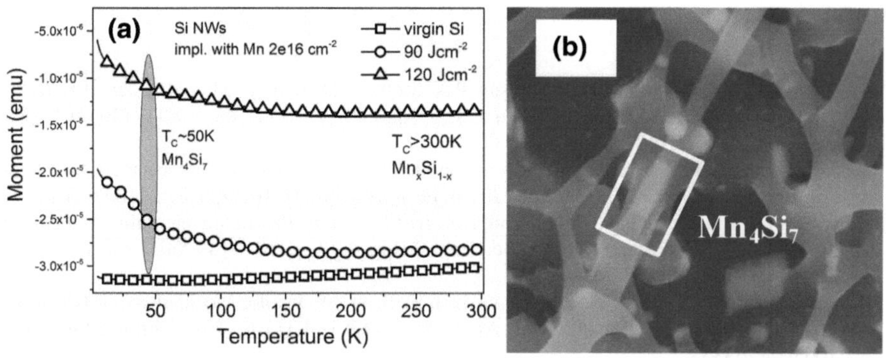

Fig. 4.65 Magnetization as a function of temperature (**a**) and SEM image of Mn implanted and annealed Si NWs (**b**). The brighter segments of Mn_4Si_7 within the Si NWs are visible due to the mass contrast

corresponds to Mn_4Si_7, another phase with T_C about 150 K which can be due to the formation of a MnSi alloy, and a high temperature phase with $T_C > 300$ K which can be caused by doping effects of Si NWs with magnetic Mn atoms.

Figure 4.65b shows the SEM image of Mn implanted and annealed Si NWs. Due to the mass contrast between Si and Mn_xSi_{1-x} alloys the magnetic phases with a high Mn content are visible as brighter parts within the Si NWs. It seems that the most stable phase is Mn_4Si_5 which segregates within Si NWs in the form of segments. A similar phenomenon has been seen in Ga and As implanted Si NWs followed by FLA [287]. The GaAs segments within the Si NWs were caused by the segregation of GaAs in the liquid phase of Si.

In conclusion, using ion implantation followed by ms-range FLA allows the integration of III–V compound semiconductors with Si technology and the formation of hybrid NWs with distinct properties which are of interest for a couple of applications like high mobility materials, light emitters or ferromagnetic materials for spintronics. The self-assembled semiconducting NW heterostructure is achieved via a novel approach, namely combining VLS growth, ion implantation, and millisecond FLA offering the possibility of large-scale production which is fully integrated with CMOS technology. A further step into this direction would be the replacement of the current NW preparation method with a top-down approach which delivers NW at defined positions.

References

1. S.C. Jain, W. Schoenmaker, R. Lindsay, P.A. Stolk, S. Decoutere, M. Willander, H.E. Maes, Transient enhanced diffusion of boron in Si. J. Appl. Phys. **91**(11), 8919–8941 (2002). https://doi.org/10.1063/1.1471941
2. E.G. Seebauer, K.W. Noh, Trends in semiconductor defect engineering at the nanoscale. Mater. Sci. Eng. R Rep. **70**(3–6), 151–168 (2010). https://doi.org/10.1016/j.mser.2010.06.007
3. L. Shao, J. Liu, Q.Y. Chen, W.-K. Chu, Boron diffusion in silicon: the anomalies and control by point defect engineering. Mater. Sci. Eng. R Rep. **42**(3–4), 65–114 (2003). https://doi.org/10.1016/j.mser.2003.08.002
4. S. Mirabella, D. De Salvador, E. Napolitani, E. Bruno, F. Priolo, Mechanisms of boron diffusion in silicon and germanium. J. Appl. Phys. **113**(3), 031101 (2013). https://doi.org/10.1063/1.4763353
5. M. Aboy, L. Santos, L. Pelaz, L.A. Marqués, P. López, Modeling od defects, dopant diffusion and clustering. J. Comput. Electron. **13**, 40–58 (2014)
6. J.F. Ziegler, J.P. Biersack, *Stopping and Range of Ions in Matter* (2003)
7. M.D. Giles, Transient phosphorus diffusion below the amorphization threshold. J. Electrochem. Soc. **138**(4), 1160 (1991)
8. F. Lanzerath et al., Boron activation and diffusion in silicon and strained silicon-on-insulator by rapid thermal and flash lamp annealings. J. Appl. Phys. **104**(4), 044908 (2008). https://doi.org/10.1063/1.2968462
9. F. Cristiano, J. Grisolia, B. Colombeau, M. Omri, B. de Mauduit, A. Claverie, L.F. Giles, N.E.B. Cowern, Formation energies and relative stability of perfect and faulted dislocation loops in silicon. J. Appl. Phys. **87**(12), 8420–8428 (2000). https://doi.org/10.1063/1.373557
10. I. Martin-Bragado, M. Jaraiz, P. Castrillo, R. Pinacho, J. Barbolla, M.M. De Souza, Mobile silicon di-interstitial: surface, self-interstitial clustering, and transient enhanced diffusion phenomena. Phys. Rev. B **68**(19) (2003). https://doi.org/10.1103/physrevb.68.195204
11. A. Claverie, B. Colombeau, B. de Mauduit, C. Bonafos, X. Hebras, G. Ben Assayag, F. Cristiano, Extended defects in shallow implants. Appl. Phys. A Mater. Sci. Process. **76**(7), 1025–1033 (2003). https://doi.org/10.1007/s00339-002-1944-0
12. A. Claverie, B. Colombeau, G. Ben Assayag, C. Bonafos, F. Cristiano, M. Omri, B. de Mauduit, Thermal evolution of extended defects in implanted Si: impact on dopant diffusion. Mat. Sci. Semicond. Process. **3**, 269–277 (2000)
13. S.H. Yeong et al., Understanding of boron junction stability in preamorphized silicon after optimized flash annealing. J. Electrochem. Soc. **155**(7), H508 (2008). https://doi.org/10.1149/1.2917901
14. R.A. Camillo-Castillo, M.E. Law, K.S. Jones, R. Lindsay, K. Maex, B.J. Pawlak, S. McCoy, Application of flash-assist rapid thermal processing subsequent to low-temperature furnace anneals. J. Vac. Sci. Technol. B Microelectr. Nanom. Struct. **24**(1), 450 (2006). https://doi.org/10.1116/1.2140003
15. L.A. Marqués, L. Pelaz, P. Castrillo, J. Barbolla, Molecular dynamics study of the configurational and energetic properties of the silicon self-interstitial. Phys. Rev. B **71**(8) (2005). https://doi.org/10.1103/physrevb.71.085204
16. C.S. Rafferty, G.H. Gilmer, M. Jaraiz, D. Eaglesham, H.J. Gossmann, Simulation of cluster evaporation and transient enhanced diffusion in silicon. Appl. Phys. Lett. **68**(17), 2395–2397 (1996). https://doi.org/10.1063/1.116145
17. S. Mirabella, D. De Salvador, E. Bruno, E. Napolitani, E.F. Pecora, S. Boninelli, F. Priolo, Mechanism of boron diffusion in amorphous silicon. Phys. Rev. Lett. **100**(15), 155901 (2008). https://doi.org/10.1103/PhysRevLett.100.155901
18. H. Kodera, Diffusion coefficients of impurities in silicon melt. Jap. J. Appl. Phys. **2**(4), 212 (1963)
19. A. Ural, P.B. Griffin, J.D. Plummer, Fractional contributions of microscopic diffusion mechanisms for common dopants and self-diffusion in silicon. J. Appl. Phys. **85**(9), 6440 (1999)

20. N.E.B. Cowern, K.T.F. Janssen, H.F.F. Jos, Transient diffusion of ion-implanted B in Si: Dose, time, and matrix dependence of atomic and electrical profiles. J. Appl. Phys. **68**(12), 6191–6198 (1990). https://doi.org/10.1063/1.346910
21. W.K. Hofker, H.W. Werner, D.P. Oosthoek, H.A.M. de Grefte, Influence of annealing on the concentration profiles of boron implantations in silicon. Appl. Phys. **2**, 265–278 (1973)
22. V.S. Lysenko, V.I. Zimenko, I.P. Tyagulskii, I.N. Osiyuk, O.V. Snitko, T.N. Sytenko, Flash-lamp annealing of Si-SiO_2 transition layer defects. Phys. Stat. Sol. (a) **87**, K175 (1985)
23. N. Mutsukura, Y. Machi, Flash annealing of ion-implanted silicon wafers with a pulsed xenon lamp and its application for solar cells. Thin Solid Films **111**, 105–119 (1984)
24. A. Usami, Y. Tokuda, M. Katayama, S. Kaneshima, T. Wada, Studies of deep-level defects in flash lamp annealing of ion-implanted silicon. J. Phys. D **19**, 1079–1089 (1986)
25. T. Sakoda et al., Characterization of millisecond-anneal-induced defects in SiON and SiON/Si interface by gate current fluctuation measurement, in *IEEE International Reliability Physics Symposium* (2010), p. 379. https://doi.org/10.1109/irps.2010.5488801
26. G. Kissinger, D. Kot, W. von Ammon, Comparison of the impact of thermal treatments on the second and on the millisecond scales on the precipitation of interstitial oxygen. ECS J. Sol. State Sci. Technol. **1**(6), P269–P275 (2012)
27. N. Matsuo, N. Yoshioka, A. Heya, Dynamics of interstitial atoms and vacancies during the crystallization of amorphous Si and Ge films by flash lamp annealing. Jap. J. Appl. Phys. **56**, 085505 (2017). https://doi.org/10.7567/JJAP.56.085505
28. S.P. McCoy, E.A. Arevalo, J.C. Gelpey, D.F. Downey, Advanced doping and millisecond annealing for ultra-shallow junctions for 65 nm and beyond, in *12th IEEE International Conference on Advanced Thermal Processing of Semiconductors* (2004), p. 99
29. M. Moras, J. Martin-Martinez, R. Rodriguez, M. Nafria, X. Aymerich, E. Simoen, Negative bias temperature instabilities induced in devices with millisecond anneal for ultra-shallow junctions. Solid-State Electr. **101**, 131–136 (2014). https://doi.org/10.1016/j.sse.2014.06.036
30. S.M. Jung, C.J. Park, H. Jeong, M.W. Shin, Effect of number of laser pulses on p+/n silicon ultra-shallow junction formation during non-melt ultra-violet laser thermal annealing. Mater. Sci. Semicond. Process. **60**, 34–39 (2017). https://doi.org/10.1016/j.mssp.2016.12.014
31. M.H. Tsai, C.-T. Wu, W.-H. Lee, Activation of boron and recrystallization in Ge preamorphization implant structure of ultra shallow junctions by microwave annealing. Jpn. J. Appl. Phys. **53**(4), 041302 (2014). https://doi.org/10.7567/jjap.53.041302
32. P. Xu et al., Ultra-shallow junctions formed using microwave annealing. Appl. Phys. Lett. **102**(12), 122114 (2013). https://doi.org/10.1063/1.4799030
33. K. Fuse, H. Tanimura, T. Aoyama, S. Kato, I. Kobayashi, Conformal SDE doping for FinFETs using an arsenic-doped Sol-Gel Coating (SGC) and flash lamp annealing (FLA), in *2017 17th International Workshop on Junction Technology* (2017)
34. J. Foggiato, W.S. Yoo, Millisecond flash annealing: applications for USJ formation and optimization of device electrical characteristics. Mater. Sci. Eng. B **124**, 219–222 (2005). https://doi.org/10.1016/j.mseb.2005.08.077
35. T. Ito et al., Low-resistance ultrashallow extension formed by optimized flash lamp annealing. IEEE Trans. Semicond. Manuf. **16**(3), 417–422 (2003). https://doi.org/10.1109/tsm.2003.815621
36. A. Jain, L. Robertson, K. Gable, Establishing advanced implant annealing for the 65 nm technology node, in *10th IEEE International Conference on Advanced Thermal Processing of Seminconductors* (2002), p. 31
37. R.B. McKnight, P.J. Timans, S.P. Tay, Z. Nenyei, RTP applications and technology options for the sub-45 nm nodes, in *12th IEEE International Conference on Advanced Thermal Processing of Seminconductors* (2004), p. 3
38. K.T. Nishinohara, T. Ito, T. Itani, K. Suguro, Impact of flash lamp annealing on 20-nm-gate-length metal oxide silicon field effect transistors. Jpn. J. Appl. Phys. **42**(Part 2, No. 10A), L1126–L1129 (2003). https://doi.org/10.1143/jjap.42.l1126
39. K.T. Nishinohara, T. Ito, K. Suguro, Improvement of performance deviation and productivity of MOSFETs with gate length below 30 nm by flash lamp annealing. IEEE Trans. Semicond. Manuf. **17**(3), 286–291 (2004). https://doi.org/10.1109/tsm.2004.831931

40. M. Abe, K. Fuse, S. Kato, T. Aoyama, I. Kobayashi, Dopant activation control using a soak pulse in Flash Lamp Annealing, in *21st International Conference on Ion Implantation Technology* (2016)
41. B. Mizuno et al., Plasma doping, in *7th International Conference on Solid-State and Integrated Circuits Technology*, Proceedings, vols. 1–3, (2004), p. 423
42. Y. Sasaki et al., New method of plasma doping with in-situ Helium pre-amorphization. Nucl. Instr. Methods B **237**(1–2), 41–45 (2005). https://doi.org/10.1016/j.nimb.2005.04.109
43. M.S. Ameen, L.M. Rubin, M.A. Harris, C. Huynh, Properties of ultralow energy boron implants using octadecaborane. J. Vac. Sci. Technol. B **26**(1), 373 (2008). https://doi.org/10.1116/1.2790925
44. W.S. Yoo, K. Kang, Electrical activation of ultra-shallow B and BF_2 implanted silicon by flash anneal. Nucl. Instr. Methods B **237**(1–2), 12–17 (2005). https://doi.org/10.1016/j.nimb.2005.04.107
45. B. Mizuno et al., Plasma doping and subsequent rapid thermal processing for ultra shallow junction formation, in *13th IEEE International Conferecne on Advanced Thermal Processing of Seminconductors* (2005), p. 45
46. J. Foggiato, W.S. Yoo, Implementation of flash technology for ultra shallow junction formation: challenges in process integration. J. Vac. Sci. Technol. B **24**(1), 515 (2006). https://doi.org/10.1116/1.2151903
47. S. Gennaro, D. Giubertoni, M. Bersani, J. Foggiato, W.S. Yoo, R. Gwilliam, Nonconventional flash annealing on shallow indium implants in silicon. J. Vac. Sci. Technol. B **24**(1), 473 (2006). https://doi.org/10.1116/1.2132321
48. C.M. Polley et al., Exploring the limits of N-Type ultra-shallow junction formation. ACS Nano **7**(6), 5499–5505 (2013). https://doi.org/10.1021/nn4016407
49. J.O. Borland, 32 nm node USJ implant & annealing options, in *15th IEEE International Conference on Advanced Thermal Processing of Seminconductors* (2007), p. 181
50. T. Gebel et al., Flash lamp annealing with millisecond pulses for ultra-shallow boron profiles in silicon. Nucl. Instr. Methods B **186**, 287–291 (2002). https://doi.org/10.1016/s0168-583x(01)00900-4
51. W. Lerch et al., Advanced activation trends for boron and arsenic by combinations of single, multiple flash anneals and spike rapid thermal annealing. Mat. Sci. Eng. B **154–155**, 3–13 (2008). https://doi.org/10.1016/j.mseb.2008.08.017
52. W. Skorupa et al., Advanced thermal processing of semiconductor materials in the msec-range, in *13th IEEE International Conference Advanced Thermal Processing of Semiconductors* (2005), p. 53. https://doi.org/10.1109/rtp.2005.1613684
53. H. Tsujii et al., Design guideline for halo condition on CMOSFETs utilizing FLA, in *5th International Workshop on Junction Technology* (2005), p. 107
54. Diffusion in silicon, germanium and their alloys, in *Landolt-Börstein*. New Series, vol. III-33A (Springer, Berlin)
55. E.J.H. Collart et al., Effects of implant temperature and millisecond annealing on dopant activation and diffusion, in *AIP Conference Proceedings on Ion Implantation Technology 2012*, vol. 1496, ed. by L. Pelaz, I. Santos, R. Duffy, F. Torregrosa, K. Bourdelle (2012), pp. 95–98
56. T. Aoyama, S. Kato, K. Yamaguchi, T. Onizawa, Y. Nara, Y. Ohji, Study of boron activation by flash lamp annealing from a view of depth dependence in silicon substrate, in *Extended Abstracts 2008 International Workshop on Junction Technology* (2008)
57. T. Ito et al., Flash lamp anneal technology for effectively activating lon implanted Si, in *Extended Abstracts of the 2001 International Conference on Solid State Devices and Materials* (2001), p. 182
58. T. Ito et al., 10–15 nm ultrashallow junction formation by flash-lamp annealing. Jpn. J. Appl. Phys. **41**(Part 1, No. 4B), 2394–2398 (2002). https://doi.org/10.1143/jjap.41.2394
59. W. Lerch et al., Advanced activation of ultra-shallow junctions using flash-assisted RTP. Mat. Sci. Eng. B **124**, 24–31 (2005). https://doi.org/10.1016/j.mseb.2005.08.047

60. D. Panknin, E. Wieser, R. Klabes, H. Syhre, Dose dependence of the flash lamp annealing of arsenic-implanted silicon. Phys. Status Solidi A **77**(2), 553–559 (1983). https://doi.org/10.1002/pssa.2210770219
61. C.T.M. Kwok, R.D. Braatz, S. Paul, W. Lerch, E.G. Seebauer, Mechanistic benefits of millisecond annealing for diffusion and activation of boron in silicon. J. Appl. Phys. **105**(6), 063514 (2009). https://doi.org/10.1063/1.3079524
62. D.-C. Ahn, B.-H. Lee, M.-H. Kang, J. Hur, T. Bang, Y.-K. Choi, Impact of crystalline damage on a vertically integrated junctionless nanowire transistor. Appl. Phys. Lett. **109**(18), 183108 (2016). https://doi.org/10.1063/1.4965851
63. T. Feudel, et al., Process integration issues with spike, flash and laser anneal implementation for 90 and 65 nm technologies, in *14th International Conference on Advanced Thermal Processing of Semiconductors* (2006)
64. C.H. Poon, A. See, Y. Tan, M. Zhou, G. Gui, Improved boron activation with reduced preheating temperature during flash annealing of preamorphized silicon. J. Electrochem. Soc. **155**(2), H59 (2008). https://doi.org/10.1149/1.2806168
65. H. Ohta et al., High performance sub-35 nm bulk CMOS with hybrid gate structures of NMOS dopant confinement layer (DCL)/PMOS; Ni-FUSI by using flash lamp anneal (FLA) in Ni-silicidation, in *Symposium on VLSI Technology* (2008), p. 115
66. C.Y. Cheng et al., Origins of flash lamp annealing induced p–n junction leakages in a 45 nm p-MOSFET with strained SiGe source/drain. J. Phys. D Appl. Phys. **42**(9), 095107 (2009). https://doi.org/10.1088/0022-3727/42/9/095107
67. R. Illgen et al., A comparative study of non-melt laser spike annealing and flash lamp annealing in terms of transistor performance and pattern effects on SOI-CMOSFETs for the 32 nm node and below, in *10th International Conference on Ultimate Integration of Silicon* (2009)
68. F. Ootsuka et al., Ultra-low thermal budget CMOS process for 65 nm-node low-operation-power applications, in *IEEE International Electron Devices Meeting Technical Digest* (2003), pp. 27–7
69. F. Ootsuka et al., Ultralow-thermal-budget CMOS process using flash-lamp annealing for 45 nm metal/high-k FETs. IEEE Trans. Electron Devices **55**(4), 1042–1049 (2008). https://doi.org/10.1109/ted.2008.917546
70. T. Sanuki et al., High-performance 45 nm node CMOS transistors featuring flash lamp annealing (FLA), in *IEEE International Electron Devices Meeting* (2007), p. 281
71. T. Sanuki et al., High density and fully compatible embedded DRAM cell with 45 nm MOS technology (CMOS6), in *Symposium on VLSI Technology, Digest of Technical Papers* (2005), p. 142
72. H.W. Cheng, C.H. Hwang, K.A. Chao, Y.M. Li, Effect of flash lamp annealing and laser spike annealing on random dopant fluctuation of 15-nm metal-oxide-semiconductor devices. J. Nanosci. Nanotechnol. **12**(3), 2462–2466 (2012). https://doi.org/10.1166/jnn.2012.5807
73. J.C. Ho et al., Wafer-scale, Sub-5 nm junction formation by monolayer doping and conventional spike annealing. Nano Lett. **9**(2), 725–730 (2009). https://doi.org/10.1021/nl8032526
74. L. Ye, M.P. de Jong, T. Kudernac, W.G. van der Wiel, J. Huskens, Doping of semiconductors by molecular monolayers: monolayer formation, dopant diffusion and applications. Mater. Sci. Semicond. Process. **62**, 128–134 (2017)
75. S.R. McKibbin, C.M. Polley, G. Scappucci, J.G. Keizer, M.Y. Simmons, Low resistivity, super-saturation phosphorus-in-silicon monolayer doping. Appl. Phys. Lett. **104**(12), 123502 (2014). https://doi.org/10.1063/1.4869111
76. H.B. Normann et al., Phosphorus in-diffusion from a surface source by millisecond flash lamp annealing for shallow emitter solar cells. Appl. Phys. Lett. **102**(13), 132108 (2013). https://doi.org/10.1063/1.4800781
77. H.N. Riise et al., Formation of shallow boron emitters in crystalline silicon using flash lamp annealing: Role of excess silicon interstitials. Appl. Phys. Lett. **107**(2), 022105 (2015). https://doi.org/10.1063/1.4926661
78. B. Kalkofen, A.A. Amusan, M.S.K. Bukhari, B. Garke, M. Lisker, H. Gargouri, E.P. Burte, Use of B_2O_3 films grown by plasma-assisted atomic layer deposition for shallow boron doping in silicon. J. Vac. Sci. Technol. A **33**, 031512 (2015). https://doi.org/10.1116/1.4917552

79. W.P. Voorthuijzen, M.D. Yilmaz, A. Gomez-Casado, P. Jonkheijm, W.G. van der Wiel, J. Huskens, Direct patterning of covalent organic monolayers on silicon using nanoimprint lithography. Langmuir **26**(17), 14210–14215 (2010). https://doi.org/10.1021/la101445n
80. J. O'Connell, S. Biswas, R. Duffy, J.D. Holmes, Chemical approaches for doping nanodevice architectures. Nanotechnology **27**(34), 342002 (2016). https://doi.org/10.1088/0957-4484/27/34/342002
81. Y. Shimizu et al., Behavior of phosphorous and contaminants from molecular doping combined with a conventional spike annealing method. Nanoscale **6**(2), 706–710 (2014). https://doi.org/10.1039/c3nr03605g
82. M. Seifner, M. Sistani, F. Porrati, G. Di Prima, P. Pertl, M. Huth, A. Lugstein, S. Barth, Direct synthesis of hyperdoped germanium nanowires. ACS Nano **12**, 1236 (2018)
83. S. Zhou, X. Pi, Z. Ni, Q. Luan, Y. Jiang, C. Jin, T. Nozaki, D. Yang, Boron- and phosphorus-hyperdoped silicon nanocrystals. Part. Part. Syst. Charact. **32**, 213–221 (2015)
84. S. Zhou, F. Liu, S. Prucnal, K. Gao, M. Khalid, C. Baehtz, M. Posselt, W. Skorupa, M. Helm, Hyperdoping silicon with selenium: solid vs. liquid phase epitaxy. Sci. Rep. **5**, 8329 (2015)
85. J.P. Mailoa et al., Room-temperature sub-band gap optoelectronic response of hyperdoped silicon. Nat. Comm. **5**, 3011 (2014)
86. E. Ertekin, M.T. Winkler, D. Recht, A.J. Said, M.J. Aziz, T. Buonassisi, J.C. Grossman, Insulator-to-metal transition in selenium-hyperdoped silicon: observation and origin. Phys. Rev. Lett. **108**, 026401 (2012)
87. K.J. Morse, R.J.S. Abraham, H. Riemann, N.V. Abrosimov, P. Becker, H.-J. Pohl, M.L.W. Thewalt, S. Simmons, A photonic platform for donor spin qubits in silicon. Sci. Adv. **3**, e1700930 (2017)
88. I. Umezu et al., Emergence of very broad infrared absorption band by hyperdoping of silicon with chalcogens. J. Appl. Phys. **113**, 213501 (2013)
89. F. Liu et al., Realizing the insulator-to-metal transition in Se-hyperdoped Si via non-equilibrium material processing. J. Phys. D Appl. Phys. **50**, 415102 (2017)
90. Y. Berencen, S. Prucnal, F. Liu, I. Skorupa, R. Huebner, L. Rebohle, S. Zhou, H. Schneider, M. Helm, W. Skorupa, Room-temperature short-wavelength infrared Si photodetector. Sci. Rep. **7**, 43688 (2017)
91. M. Wang et al., Extended infrared photoresponse in Te-hyperdoped Si at room temperature. Phys. Rev. Appl. **10**, 024054 (2018)
92. F. Liu, S. Prucnal, R. Hübner, Y. Yuan, W. Skorupa, M. Helm, S. Zhou, Suppressing the cellular breakdown in silicon supersaturated with titanium. J. Phys. D Appl. Phys. **49**, 245104 (2016)
93. F. Liu et al., On the insulator-to-metal transition in titanium-implanted silicon. Sci. Rep. **8**, 4164 (2018)
94. https://www.semiconductors.org/main/2015_international_technology_roadmap_for_semiconductors_itrs/. Accessed 26 Nov 2018
95. R.L. Cohen, J.S. Williams, L.C. Feldman, K.W. West, Thermally assisted flash annealing of silicon and germanium. Appl. Phys. Lett. **33**, 751 (1978)
96. J.S. Hovis, R.J. Hamers, C.M. Greenlief, Preparation of clean and atomically flat germanium(001) surfaces. Surf. Sci. **440**, L815–L819 (1999)
97. K. Osada, T. Fukunaga, K. Shibahara, Ge shallow junction formation by As implantation and flash lamp annealing, in *Proceedings of Technical Program: 2009 International Symposium on VLSI Technology, Systems and Applications* (2009)
98. H. Tanimura et al., 10 nm-deep n+/p and p plus /n Ge junctions with high activation formed by ion implantation and flash lamp annealing (FLA), in *2016 16th International Workshop on Junction Technology* (2016)
99. Tanimura et al., Nanometer-deep junctions with high doping concentration for Ge SDEs using solid source doping and flash lamp annealing, in *2017 17th International Workshop on Junction Technology* (2017)
100. D. Skarlatos et al., Phosphorous diffusion in $N2^{+}$ -implanted germanium during flash lamp annealing: influence of nitrogen on ge substrate damage and capping layer engineering. ECS J. Solid State Sci. Technol. **6**(7), P418–P428 (2017). https://doi.org/10.1149/2.0201707jss

101. M. Posselt et al., P implantation into preamorphized germanium and subsequent annealing: Solid phase epitaxial regrowth, P diffusion, and activation. J. Vac. Sci. Technol. B **26**, 430 (2008)
102. P. Taheri, H.M. Fahad, M. Tosun, M. Hettick, D. Kiriya, K. Chen, A. Javey, Nanoscale junction formation by gas-phase monolayer doping. ACS Appl. Mater. Interfaces. **9**, 20648–20655 (2017)
103. G. Scappucci, W.M. Klesse, L.A. Yeoh, D.J. Carter, O. Warschkow, N.A. Marks, D.L. Jaeger, G. Capellini, M.Y. Simmons, A.R. Hamilton, Bottom-up assembly of metallic germanium. Sci. Rep. **5**, 12948 (2015)
104. J. Kim, S.W. Bedell, D.K. Sadana, Multiple implantation and multiple annealing of phosphorus doped germanium to achieve n-type activation near the theoretical limit. Appl. Phys. Lett. **101**, 112107 (2012)
105. S. Prucnal, F. Liu, M. Voelskow, L. Vines, L. Rebohle, D. Lang, Y. Berencén, S. Andric, R. Böttger, M. Helm, S. Zhou, W. Skorupa, Ultra-doped n-type germanium thin films for sensing in the mid-infrared. Sci. Rep. **6**, 27643 (2016)
106. S.-H. Huang, F.-L. Lu, W.L. Huang, C.H. Huang, C.W. Liu, The $\sim 3 \times 10^{20}$ cm^{-3} electron concentration and low specific contact resistivity of phosphorus-doped Ge on Si by in-situ chemical vapor deposition doping and laser annealing. IEEE Electron. Dev. Lett. **36**, 1114 (2015)
107. C. Wündisch, M. Posselt, B. Schmidt, V. Heera, T. Schumann, A. Mücklich, R. Grötzschel, W. Skorupa, T. Clarysse, E. Simoen, H. Hortenbach, Millisecond flash lamp annealing of shallow implanted layers in Ge. Appl. Phys. Lett. **95**, 252107 (2009)
108. G. Mattoni, W.M. Klesse, G. Capellini, M.Y. Simmons, G. Scappucci, Phosphorus molecules on Ge(001). ACS Nano **7**, 11310–11316 (2013)
109. S. Prucnal et al., Nanoscale n^{++}-p junction formation in GeOI probed by tip-enhanced Raman spectroscopy and conductive atomic force microscopy. J. Appl. Phys. **125**, 245703 (2019). https://doi.org/10.1063/1.5080289
110. K. Ohdaira, S. Ishii, N. Tomura, H. Matsumura, Microstructure of polycrystalline silicon films formed through explosive crystallization induced by flash lamp annealing. Jpn. J. Appl. Phys. **50**, 04DP01 (2011)
111. T. Gebel, L. Rebohle, R. Fendler, W. Hentsch, W. Skorupa, M. Voelskow, W. Anwand, R.A. Yankov, Millisecond annealing with flash lamps: tool and process challenges, in *14th International Conference on Advanced Thermal Processing of Semiconductors* (2006), pp. 47–55
112. A. Satta, A. D'Amore, E. Simoen, W. Anwand, W. Skorupa, T. Clarysse, B. Van Daele, T. Janssens, Formation of germanium shallow junction by flash annealing. Nucl. Instr. Meth. Phys. Res. B **257**, 157–160 (2007)
113. H. Bracht, S. Schneider, R. Kube, Diffusion and doping issues in germanium. Microelectron. Eng. **88**, 452–457 (2011)
114. E. Simoen, A. Satta, A. D'Amore, T. Janssens, T. Clarysse, K. Martens, B. De Jaeger, A. Benedetti, I. Hoflijk, B. Brijs, M. Meuris, W. Vandervorst, Ion-implantation issues in the formation of shallow junctions in germanium. Mater. Sci. Semicond. Process. **9**, 634–639 (2006)
115. K. Usuda, Y. Kamata, Y. Kamimuta, T. Mori, M. Koike, T. Tezuka, High-performance tri-gate poly-Ge junction-less p- and n-MOSFETs fabricated by flash lamp annealing process. IEEE 2014 IEDM
116. S. Prucnal et al., Enhancement of carrier mobility in thin Ge layer by Sn co-doping. Semicond. Sci. Technol. **31**, 105012 (2016)
117. R. Milazzo, G. Impellizzeri, D. Piccinotti, D. De Salvador, A. Portavoce, A. La Magna, G. Fortunato, D. Mangelinck, V. Privitera, A. Carnera, E. Napolitani, Low temperature deactivation of Ge heavily n-type doped by ion implantation and laser thermal annealing. Appl. Phys. Lett. **110**, 011905 (2017)
118. M. Posselt, A. Gabriel, Atomistic simulation of amorphous germanium and its solid phase epitaxial recrystallization. Phys. Rev. B **80**, 045202 (2009)

119. F. Edelman, Y. Komem, M. Bendayan, R. Beserman, Initial crystallization stage of amorphous germanium films. J. Appl. Phys. **72**, 5153 (1992)
120. B.C. Johnson, P. Gortmaker, J.C. McCallum, Intrinsic and dopant-enhanced solid-phase epitaxy in amorphous germanium. Phys. Rev. B **77**, 214109 (2008)
121. L. Nikolova et al., Nanocrystallization of amorphous germanium films observed with nanosecond temporal resolution. Appl. Phys. Lett. **97**, 203102 (2010). https://doi.org/10.1063/1.3518069
122. J. Liu, L.C. Kimerling, J. Michel, Monolithic Ge-on-Si lasers for large-scale electronic–photonic integration. Semicond. Sci. Technol. **27**, 094006 (2012)
123. M. El Kurdi, G. Fishman, S. Sauvage, P. Boucauda, Band structure and optical gain of tensile-strained germanium based on a 30 band k·p formalism. J. Appl. Phys. **107**, 013710 (2010)
124. S. Prucnal et al., In situ ohmic contact formation for n-type Ge via non-equilibrium processing. Semicond. Sci. Technol. **32**, 115006 (2017)
125. E.A. Ekimov, V.A. Sidorov, E.D. Bauer, N.N. Melnik, N.J. Curro, J.D. Thompson, S.M. Stishov, Superconductivity in diamond. Nature **428**, 542 (2004)
126. E. Bustarret et al., Superconductivity in doped cubic silicon. Nature **444**, 465 (2006)
127. T. Herrmannsdörfer et al., Superconductivity in thin-film germaniumin the temperature regime around 1 K. Supercond. Sci. Technol. **23**, 034007 (2010)
128. V. Heera et al., Heavily Ga-doped germanium layers produced by ion implantation and flash lamp annealing: Structure and electrical activation. J. Appl. Phys. **107**, 053508 (2010)
129. R. Skrotzki, T. Herrmannsdörfer, V. Heera, J. Fiedler, A. Mücklich, M. Helm, J. Wosnitza, The impact of heavy Ga doping on superconductivity in germanium. Low Temp. Phys. **37**, 877 (2011)
130. V. Heera et al., Depth-resolved transport measurements and atom-probe tomography of heterogeneous superconducting Ge: Ga films. Supercond. Sci. Technol. **27**, 055025 (2014)
131. D. Panknin, W. Skorupa, H. Wirth, M. Voelskow, A. Mücklich, W. Anwand, G. Brauer, O. Gonzalez-Varona, A. Perez-Rodriguez, J.M. Morante, Ion beam doping of 6H-SiC for high concentration p-Type layers. Solid State Phenom. **69–70**, 391–396 (1999)
132. D. Panknin, H. Wirth, A. Mücklich, W. Skorupa, Electrical and microstructural properties of highly boron-implantation doped 6H–SiC. J. Appl. Phys. **89**(6), 3162–3167 (2001)
133. C.-M. Zetterling, *Process Technology for Silicon Carbide Device*. EMIS Processing Series 2 (INSPEC The Institution of Electrical Engineers, London 2002)
134. S.G. Davis, D.F. Anthrop, A.W. Searcy, Vapor pressure of silicon and the dissociation pressure of silicon carbide. J. Chem. Phys. **34**, 659 (1961)
135. V. Heera, D. Panknin, W. Skorupa, P-type doping of SiC by high dose Al implantation—problems and progress. Appl. Surf. Sci. **184**, 307–316 (2001)
136. D. Panknin, H. Wirth, W. Anwand, G. Brauer, W. Skorupa, High concentration doping of 6H-SiC by Ion Implantation: flash versus furnace annealing. Mater. Sci. Forum **338–342**, 877–800 (2000)
137. D. Panknin, T. Gebel, W. Skorupa, Flash lamp annealing of implantation doped p- and n-type 6H-SiC. Mater. Sci. Forum **353–356**, 587–590 (2001)
138. H. Wirth, D. Panknin, W. Skorupa, E. Niemann, Efficient p-type doping of 6H-SiC: Flash-lamp annealing after aluminium implantation. Appl. Phys. Lett. **74**(7), 979–981 (1999)
139. W. Anwand, G. Brauer, J. Kuriplach, W. Skorupa, Slow positron beam investigations of defects caused by B^+ implantation into epitaxial 6H-SiC. Mater. Sci. Forum **445–446**, 36–38 (2004)
140. J. Stoemenos, D. Panknin, M. Eickhoff, V. Heera, W. Skorupa, Improved 3C-SiC films epitaxially grown on Si by flash lamp processing. J. Electrochem. Soc. **151**(2), G136–G143 (2004)
141. D. Panknin, J. Stoemenos, M. Eickhoff, V. Heera, N. Voroutzis, G. Krötz, W. Skorupa, Improvement of the 3C-SiC/Si interface by flash lamp annealing. Mater. Sci. Forum **353–356**, 151–154 (2001)
142. W. Skorupa et al., Flash lamp supported deposition of 3C-SiC (FLASiC)—a promising technique to produce high quality cubic SiC layers. Mater. Sci. Forum **457–460**, 175–180 (2004)

143. M.P. Smith, R.A. McMahon, M. Voelskow, W. Skorupa, J. Stoemenos, G. Ferro, Process control and melt depth homogenization for SiC-on-Si structures during flash lamp annealing by carbon implantation. J. Appl. Phys. **100**(9), 094909 (2006). https://doi.org/10.1063/1.2359684
144. S. Felch, G. Higashi, Use of carbon co-implantation with millisecond anneal to produce ultra-shallow junctions, US Patent US 20080023732A1 (2008)
145. D. Panknin, P. Godignion, N. Mestres, E. Polychroniadis, J. Stoemenos, G. Ferro J. Pezoldt, W. Skorupa, Formation of 3C-SiC films embedded in SiO_2 by sacrificial oxidation. Mater. Sci. Forum **457–460**, 1515–1518 (2004)
146. S. Prucnal, K. Gao, W. Anwand, M. Helm, W. Skorupa, S. Zhou, Temperature stable 1.3 μm emission from GaAs. Opt. Express **20**, 26075–26081 (2012)
147. S. Prucnal et al., Band-gap narrowing in Mn-doped GaAs probed by room-temperature photoluminescence. Phys. Rev. B **92**, 224407 (2015)
148. S. Zhou, Dilute ferromagnetic semiconductors prepared by the combination of ion implantation with pulse laser melting. J. Phys. D Appl. Phys. **48**, 263001 (2015)
149. I. Vurgaftman, J.R. Meyer, Band parameters for nitrogen-containing semiconductors. J. Appl. Phys. **94**, 3675 (2003)
150. J. Toivonen, Growth and properties of GaAsN structures. Dissertation, Helsinki University of Technology, 2003
151. K. Gao, S. Prucnal, W. Skorupa, M. Helm, S. Zhou, Formation and photoluminescence of $GaAs_{1-x}N_x$ dilute nitride achieved by N-implantation and flash lamp annealing. Appl. Phys. Lett. **105**, 012107 (2014)
152. K. Gao, S. Prucnal, W. Skorupa, M. Helm, S. Zhou, Origin and enhancement of the 1.3 um luminescence from GaAs treated by ion-implantation and flash lamp annealing. J. Appl. Phys. **114**, 093511 (2013)
153. C.V. Reddy, S. Fung, C.D. Beling, Nature of the bulk defects in GaAs through high-temperature quenching studies. Phys. Rev. B **54**, 11290–11297 (1996)
154. J. Furdyna, Diluted magnetic semiconductors. J. Appl. Phys. **64**, R29–R56 (1988)
155. T. Dietl, H. Ohno, F. Matsukura, J. Cibert, D. Ferrand, Zener model description of ferromagnetism in Zinc-blende magnetic semiconductors. Science **287**, 1019–1022 (2000)
156. H. Ohno, A window on the future of spintronics. Nat. Mater. **9**, 952–954 (2010)
157. S. Zhou et al., Precise tuning of the Curie temperature of (Ga, Mn)As-based magnetic semiconductors by hole compensation: Support for valence-band ferromagnetism. Phys. Rev. B **94**, 075205 (2016)
158. C. Zener, Interaction between the d shells in the transition metals. Phys. Rev. **81**, 440–444 (1951)
159. T. Dietl, Ferromagnetic semiconductors. Semicond. Sci. Technol. **17**, 377–392 (2002)
160. M. Khalid et al., Synthesis and characterization of MnAs and MnP nanoclusters embedded in III–V semiconductors. Mater. Res. Express **1**, 026105 (2014)
161. H.-D. Geiler, E. Glaser, G. Götz, M. Wagner, Explosive crystallization in silicon. J. Appl. Phys. **59**(9), 3091–3099 (1986)
162. C. Grigoropoulos, M. Rogers, S.H. Ko, A.A. Golovin, B.J. Matkowsky, Explosive crystallization in the presence of melting. Phys. Rev. B **73**, 184125 (2006)
163. K. Murakami, O. Eryu, K. Takita, K. Masuda, Explosive crystallization starting from an amorphous-silicon surface region during long-pulse laser irradiation. Phys. Rev. Lett. **59**, 2203–2206 (1987)
164. D. Reichel, W. Skorupa, W. Lerch, J.C. Gelepy, Temperature measurement in rapid thermal processing with focus on the application to flash lamp annealing. Crit. Rev. Solid State Mater. Sci. **36**(2), 102 (2011)
165. K. Ohdaira, T. Fujiwara, Y. Endo, S. Nishizaki, H. Matsumura, Explosive crystallization of amorphous silicon films by flash lamp annealing. J. Appl. Phys. **106**(4), 044907 (2009). https://doi.org/10.1063/1.3195089
166. K. Ohdaira, T. Nishikawa, K. Shiba, H. Takemoto, H. Matsumura, Polycrystalline Si films with unique microstructures formed from amorphous Si films by non-thermal equilibrium flash lamp annealing. Physica Status Solidi **7**(3–4), 605–607 (2010)

167. M.A. Herman, W. Richter, H. Sitter, *Epitaxy—Physical Principles and Technical Implementation*. Springer Series in Materials Science (Springer, Berlin, Heidelberg, 2004)
168. M. Smith, R. McMahon, M. Voelskow, D. Panknin, W. Skorupa, Modelling of flash-lamp-induced crystallization of amorphous silicon thin films on glass. J. Cryst. Growth **285**, 249–260 (2005)
169. B. Pécz, L. Dobos, D. Panknin, W. Skorupa, C. Lioutas, N. Vouroutzis, Crystallization of amorphous-Si films by flash lamp annealing. Appl. Surf. Sci.**242**, 185–191 (2005)
170. K. Ohdaira, The control of the film stress of Cat-CVD a-Si films and its impact on explosive crystallization induced by flash lamp annealing. Thin Solid Films **575**, 21–24 (2015)
171. K. Yamakawa, K. Mai, O. Arisumi, T. Arikado, M. Yoshioka, T. Owada, K. Okumura, Novel $Pb(Ti,Zr)O_3$ (PZT) crystallization technique using flash lamp for ferroelectric RAM (FeRAM) embedded lsis and one transistor type FeRAM devices. Jpn. Soc. Appl. Phys. **41** (1/4B), 2630–2634 (2002)
172. R.A. McMahon, M.P. Smith, K.A. Steffen, M. Voelskow, W. Anwand, W. Skorupa, Flash-lamp annealing of semiconductor materials—applications and process models. Vacuum **81**, 1301–1305 (2007)
173. D.-H. Kim, B.-K. Kim, H.J. Kim, S. Park, Crystallization of amorphous silicon thin-film on glass substrate preheated at 650 C using Xe arc flash of 400 μs. Thin Solid Films **520**, 6581–6588 (2012)
174. G. Ferro, E.K. Polychroniadis, D. Panknin, W. Skorupa, J. Stoemenos, Y. Monteil, Effect of the crystallization conditions on the epitaxial relationship of Si deposited on 3C-SiC(100). Mater. Sci. Forum **527–529**, 1563–1566 (2006)
175. K. Ohdaira, N. Tomura, S. Ishii, H. Matsumura, Lateral crystallization velocity in explosive crystallization of amorphous silicon films induced by flash lamp annealing. Electrochem. Solid-State Lett. **14**(9), H372–H374 (2011)
176. K. Ohdaira, A method to evaluate explosive crystallization velocity of amorphous silicon films during flash lamp annealing. Can. J. Phys. **92**(7/8), 718–722 (2014)
177. K. Ohdaira, H. Matsumura, Flash-lamp-induced explosive crystallization of amorphous germanium films leaving behind periodic microstructures. Thin Solid Films **524**, 161–165 (2012)
178. K. Ohdaira, K. Sawada, N. Usami, S. Varlamov, H. Matsumura, Large-grain polycrystalline silicon films formed through flash-lamp-induced explosive crystallization. Jpn. J. Appl. Phys. **51**, 10NB15 (2012)
179. K. Ohdaira, H. Matsumura, Liquid-phase explosive crystallization of electron-beam-evaporated a-Si films induced by flash lamp annealing. J. Cryst. Growth **362**, 149–152 (2013)
180. K. Ohdaira, S. Ishii, N. Tomura, H. Matsumura, Polycrystalline silicon films with nanometer-sized dense fine grains formed by flash-lamp-induced crystallization. J. Nanosci. Nanotechnol. **12**(1), 591–595 (2012)
181. K. Ohdaira, The control of the film stress of Cat-CVD a-Si films and its impact on explosive crystallization induced by flash lamp annealing. Thin Solid Films **575**, 21–24 (2015)
182. K. Ohdaira, T. Nishikawa, K. Shiba, H. Takemoto, H. Matsumura, Drastic suppression of the optical reflection of flash-lamp-crystallized poly-Si films with spontaneously formed periodic microstructures. Thin Solid Films **518**, 6061–6065 (2010)
183. W. Skorupa, T. Schumann, L. Rebohle, Millisecond thermal processing using flash lamps for the advancement of thin layers and functional coatings. Surf. Coat. Technol. **314**, 169–176 (2016)
184. K. Ohdaira, N. Tomura, S. Ishii, H. Matsumura, Flash-lamp-crystallized polycrystalline silicon films with high hydrogen concentration formed from Cat-CVD a-Si films. Thin Solid Films **519**, 4459–4461 (2011)
185. K. Ohdaira, Defect termination of flash-lamp-crystallized large-grain polycrystalline silicon films by high-pressure water vapor annealing. Jpn. J. Appl. Phys. **52**, 04CR11 (2013)
186. R. Ishihara, P.C. van der Wilt, B.D. van Dijk, A. Burtsev, J.W. Metselaar, C.I.M. Beenakker, Advanced excimer-laser crystallization process for single-crystalline thin film transistors. Thin Solid Films **427**(1–2), 77–85 (2003). https://doi.org/10.1016/s0040-6090(02)01250-6

187. T. Henke, J.W. Bartha, L. Rebohle, U. Merkel, R. Hübner, M. Albert, W. Skorupa, Formation of regularly arranged large grain silicon islands by using embedded micro mirrors in the flash crystallization of amorphous silicon. J. Appl. Phys. **115**(3), 034301 (2014). https://doi.org/10.1063/1.4861398
188. S.-M. Jung et al., Three dimensionally stacked NAND flash memory technology using stacking single crystal Si layers on ILD and TANOS structure for beyond 30 nm node. Int. Electron. Dev. Meet. 503 (2006), https://doi.org/10.1109/iedm.2006.346902
189. S.-Y. Oh, C.-G. Ahn, J.-H. Yang, W.-J. Cho, M.-G. Jang, A two-layer stacked polycrystalline silicon thin film transistor complementary metal oxide semiconductor inverters using laser crystallized channel with high-k and metal gate on Si. Jpn. J. Appl. Phys. **47**(4), 3091–3094 (2008). https://doi.org/10.1143/jjap.47.3091
190. Y. Sonoda, K. Ohdaira, Effect of antireflection coating on the crystallization of amorphous silicon films by flash lamp annealing. Jpn. J. Appl. Phys. **56**(4S), 04CS10 (2017). https://doi.org/10.7567/jjap.56.04cs10
191. R.B. Iverson, R. Reif, Recrystallization of amorphized polycrystalline silicon films on SiO_2: temperature dependence of the crystallization parameters. J. Appl. Phys. **62**(5), 1675–1681 (1987). https://doi.org/10.1063/1.339591
192. J.-W. Choi et al., Rapid activation of phosphorous-implanted polycrystalline Si thin films on glass substrates using flash-lamp annealing. ECS J. Solid State Sci. Techn. **3**(11), P391–P395 (2014)
193. J.S. Im, H.J. Kim, On the super lateral growth phenomenon observed in excimer laser-induced crystallization of thin Si films. Appl. Phys. Lett. **64**, 2303 (1994). https://doi.org/10.1063/1.111651
194. K.B. Kadyrakunov, E.V. Nidaev, A.E. Plotnikov, L.S. Smirnov, I.G. Melnik, M.V. Makeev, Flash lamp annealing of ion-implanted polycrystalline silicon. Phys. Stat. Sol. (a) **75**, 483 (1983)
195. J.S. Im, H.J. Kim, M.O. Thompson, Phase transformation mechanisms involved in excimer laser crystallization of amorphous silicon films. Appl. Phys. Lett. **63**(14), 1969–1971 (1993). https://doi.org/10.1063/1.110617
196. F. Terai, S. Matunaka, A. Tauchi, C. Ichimura, T. Nagatomo, T. Homma, Xenon flash lamp annealing of poly-Si thin films. J. Electrochem. Soc. **153**(7), H147 (2006). https://doi.org/10.1149/1.2200291
197. D. Toet, P.M. Smith, T.W. Sigmon, T. Takehara, C.C. Tsai, W.R. Harshbarger, M.O. Thompson, Laser crystallization and structural characterization of hydrogenated amorphous silicon thin films. J. Appl. Phys. **85**(11), 7914–7918 (1999). https://doi.org/10.1063/1.370607
198. S. Saxena, D.C. Kim, J.H. Park, J. Jang, Polycrystalline silicon thin-film transistor using Xe flash-lamp annealing. IEEE Electron Dev. Lett. **31**, 1242–1244 (2010)
199. S. Saxena, J. Jang, Protrusions of super grains formed by ultrashort Xe flash-lamp annealing of amorphous silicon and its effect on the performances of thin-film transistors. IEEE Trans. Electron Dev. **58**, 2638–2643 (2011)
200. T. Mudgal, K. Bhadrachalam, P. Bischoff, D. Cormier, R.G. Manley, K.D. Hirschman, Communication—CMOS thin-film transistors via Xe flash-lamp crystallization of patterned amorphous Si. ECS J. Solid State Sci. Technol. **6**(12), Q179–Q181 (2017). https://doi.org/10.1149/2.0021802jss
201. P.C. van der Wilt, B.D. van Dijk, G.J. Bertens, R. Ishihara, C.I.M. Beenakker, Formation of location-controlled crystalline islands using substrate-embedded seeds in excimer-laser crystallization of silicon films. Appl. Phys. Lett. **79**(12), 1819–1821 (2001). https://doi.org/10.1063/1.1402641
202. K. Ohdaira, K. Shiba, H. Takemoto, T. Fujiwara, Y. Endo, S. Nishizaki, Y.R. Jang, H. Matsumura, Precursor Cat-CVD a-Si films for the formation of high-quality poly-Si films on glass substrates by flash lamp annealing. Thin Solid Films **517**(12), 3472–3475 (2009). https://doi.org/10.1016/j.tsf.2009.01.075
203. T. Watanabe, K. Ohdaira, The formation of poly-Si films on flat glass substrates by flash lamp annealing. Thin Solid Films **595**, 235–238 (2015). https://doi.org/10.1016/j.tsf.2015.08.003

204. K. Ohdaira, T. Fujiwara, Y. Endo, K. Shiba, H. Takemoto, H. Matsumura, Selection of material for the back electrodes of thin-film solar cells using polycrystalline silicon films formed by flash lamp annealing. Jpn. J. Appl. Phys **49**(4), 04DP04 (2010). https://doi.org/10.1143/jjap.49.04dp04
205. D. Sato, K. Ohdaira, Effect of starting point formation on the crystallization of amorphous silicon films by flash lamp annealing, Jpn. J. Appl Phys. **57**(4) (2018). https://doi.org/10.7567/jjap.57.04fs05
206. K. Ohdaira, T. Nishikawa, S. Ishii, N. Tomura, H. Matsumura, Carrier transport properties of flash-lamp-crystallized poly-Si films, in *IEEE 35th Photovoltaics Specialists Conference* 2010, p. 3726
207. K. Ohdaira, H. Takemoto, K. Shiba, H. Matsumura, Drastic improvement of minority carrier lifetimes observed in hydrogen-passivated flash-lamp-crystallized polycrystalline silicon films. Appl. Phys. Express **2**, 061201 (2009). https://doi.org/10.1143/apex.2.061201
208. K. Gao, S. Prucnal, A. Mücklich, W. Skorupa, S. Zhou, Fabrication of $Si_{1-x}Ge_x$ alloy on silicon by Ge-Ion-implantation and short-time-annealing. Acta Phys. Polon. A **123**(5), 858–859 (2012)
209. Y. Shiraki, N. Usami, *Silicon-Germanium (SiGe) Nanostructures—Production, Properties and Applications in Electronics*. Woodhead Publishing in Materials (Woodhead Publishing Limited, Philadelphia, New Delhi, 2011)
210. M. Voelskow, W. Skorupa, J. Pezoldt, T. Kups, Controlled localised melting in silicon by high dose germanium implantation and flash lamp annealing. Nucl. Instrum Method Phys Res B **267**, 1269–1272 (2009)
211. M. Voelskow, R. Yankov, W. Skorupa, J. Pezoldt, T. Kups, Buried melting in germanium implanted silicon by millisecond flash lamp annealing. Appl. Phys. Lett. **93**, 151903 (2008)
212. M. Voelskow, I. Stoimenos, L. Rebohle, W. Skorupa, The formation of near surface SiGe layers with combined high-dose ion implantation and flash-lamp annealing. Physics Status Solidi C **8**(3), 960–963 (2011)
213. I.Z. Mitrovic, O. Buiu, S. Hall, D.M. Bagnall, P. Ashburn, Review of SiGe HBTs on SOI. Solid State Electron. **49**(9), 1556–1567 (2005)
214. D. Bolze, B. Heinemann, J. Gelpey, S. McCoy, W. Lerch, Millisecond annealing of high-performance SiGe HBTs, in *17th IEEE International Conference on Advanced Thermal Processing of Semiconductors*. RTP (2009)
215. F. La Via, A. Severino, R. Anzalone, C. Bongiorno, G. Litrico, M. Mauceri, M. Schöler, P. Schuh, P. Wellmann, From thin film to bulk 3C-SiC growth: understanding the mechanism of defect reduction. Mater. Sci. Semicond. Process. **78**, 57–68 (2018)
216. G. Ferro, D. Panknin, J. Stoemenos, C. Baloud, J. Camassel, E. Polychroniadis, Y. Monteil, W. Skorupa, Investigation of thick 3C-SiC films re-grown on thin 35 nm "Flash Lamp Annealed" 3C-SiC layers. Mater. Sci. Forum **457–460**, 313–316 (2004)
217. G. Ferro, 3C-SiC heteroepitaxial growth on silicon: the quest for holy grail. Crit. Rev. Solid State Mater. Sci. **40**(1), 56–76 (2015)
218. D. Panknin, J. Stoemenos, M. Eickhoff, V. Heera, M. Voelskow, W. Skorupa, The beneficial role of flash lamp annealing on the epitaxial growth of the 3C–SiC on Si. Appl. Surf. Sci. **184**, 377–382 (2001)
219. W. Skorupa, W. Anwand, D. Panknin, M. Voelskow, R.A. Yankov, T. Gebel, Advanced thermal processing of semiconductor materials in the millisecond range. Vacuum **78**, 673–677 (2005)
220. M. Voelskow, W. Skorupa, M.P. Smith, R.A. McMahon, Homogenization of the melting depth in SiC on Si structures during flash lamp irradiation. Appl. Phys. Lett. **87**, 241901 (2005)
221. J. Pezoldt, F.M. Morales, Th Stauden, Ch. Förster, E. Polychroniadis, J. Stoemenos, D. Panknin, W. Skorupa, Growth acceleration in FLASiC assisted short time liquid phase epitaxy by melt modification. Mater. Sci. Forum **527–529**, 295–298 (2006)
222. S. Furukawa, T. Miyasato, Quantum size effects on the optical band gap of microcrystalline Si:H. Phys. Rev. B **38**(8), 5726–5729 (1988). https://doi.org/10.1103/PhysRevB.38.5726
223. L.T. Canham, Silicon quantum wire array fabrication by electrochemical and chemical dissolution of wafers. Appl. Phys. Lett. **57**(10), 1046–1048 (1990). https://doi.org/10.1063/1.103561

224. V. Lehmann, U. Gösele, Porous silicon formation: a quantum wire effect. Appl. Phys. Lett. **58**(8), 856 (1991)
225. G. Franzò, V. Vinciguerra, F. Priolo, The excitation mechanism of rare-earth ions in silicon nanocrystals. Appl. Phys. A **69**, 3–12 (1999). https://doi.org/10.1007/s003399900087
226. P.G. Kik, A. Polman, Exciton–erbium interactions in Si nanocrystal-doped SiO_2. J. Appl. Phys. **88**(4), 1992 (2000)
227. S. Tiwari, F. Rana, K. Chan, L. Shi, H. Hanafi, Single charge and confinement effects in nano-crystal memories. Appl. Phys. Lett. **69**(9), 1232–1234 (1996). https://doi.org/10.1063/1.117421
228. J. von Borany, T. Gebel, K.-H. Stegemann, H.-J. Thees, M. Wittmaack, Memory properties of Si^+ implanted gate oxides: from MOS capacitors to nvSRAM. Solid-State Electr. **46**, 1729–1737 (2002)
229. M. Zacharias, J. Heitmann, R. Scholz, U. Kahler, M. Schmidt, J. Bläsing, Size-controlled highly luminescent silicon nanocrystals: A SiO/SiO_2 superlattice approach. Appl. Phys. Lett. **80**(4), 661–663 (2002). https://doi.org/10.1063/1.1433906
230. J. Heitmann, F. Müller, M. Zacharias, U. Gösele, Silicon nanocrystals: size matters. Adv. Mater. **17**(7), 795 (2005)
231. F. Priolo, T. Gregorkiewicz, M. Galli, T.F. Krauss, Silicon nanostructures for photonics and photovoltaics. Nature nanotechn. **9**, 19 (2014). https://doi.org/10.1038/NNANO.2014.271
232. K. Dohnalova, T. Gregorkiewicz, K. Kusova, Silicon quantum dots: surface matters. J. Phys. Condens. Matter **26**(17), 173201 (2014). https://doi.org/10.1088/0953-8984/26/17/173201
233. L. Rebohle, J. von Borany, H. Fröb, W. Skorupa, Blue photo- and electroluminescence of silicon dioxide layers ion-implanted with group IV elements. Appl. Phys. B **71**, 131 (2000)
234. G.A. Kachurin, I.E. Tyschenko, K.S. Zhuravlev, N.A. Pazdnikov, V.A. Volodin, A.K. Gutakovsky, A.F. Leier, W. Skorupa, R.A. Yankov, Visible and near-infrared luminescence from silicon nanostructures formed by ion implantation and pulse annealing. Nuc. Instr. Meth. B **122**, 571–574 (1997)
235. G.A. Kachurin, S.G. Cherkova, D.V. Marin, R.A. Yankov, M. Deutschmann, Formation of light-emitting Si nanostructures in SiO_2 by pulsed anneals. Nanotechnology **19**(35), 355305 (2008). https://doi.org/10.1088/0957-4484/19/35/355305
236. D. Hiller, S. Gutsch, A.M. Hartel, P. Löper, T. Gebel, M. Zacharias, A low thermal impact annealing process for SiO2-embedded Si nanocrystals with optimized interface quality. J. Appl. Phys. **115**(13), 134311 (2014). https://doi.org/10.1063/1.4870819
237. N. Yoshioka, A. Heya, N. Matsuo, K. Kohama, K. Ito, Effect of SiO_x capping film on crystallization of Ge film by flash lamp annealing, in *2017 24th International Workshop on Active-Matrix Flatpanel Displays and Devices* (2017)
238. A. Kanjilal, L. Rebohle, M. Voelskow, M. Helm, W. Skorupa, Controlling blue-violet electroluminescence of Ge-rich Er-doped SiO2 layers by millisecond annealing using flash lamps. J. Appl. Phys. **107**(2), 023114 (2010). https://doi.org/10.1063/1.3296252
239. A. Kanjilal, S. Prucnal, L. Rebohle, M. Voelskow, M. Helm, W. Skorupa, Comparison of the room temperature 1.53 μm Er photoluminescence from flash lamp and furnace annealed Er-doped Ge-rich SiO_2 layers. J. Appl. Phys. **107**(11), 113523 (2010). https://doi.org/10.1063/1.3437652
240. L. Rebohle, J. von Borany, R.A. Yankov, W. Skorupa, I.E. Tyschenko, H. Fröb, K. Leo, Strong blue and violet photoluminescence and electroluminescence from germanium-implanted and silicon-implanted silicon-dioxide layers. Appl. Phys. Lett. **71**, 2809 (1997)
241. A. Kanjilal, L. Rebohle, M. Voelskow, W. Skorupa, M. Helm, Enhanced blue-violet emission by inverse energy transfer to the Ge-related oxygen deficiency centers via Er^{3+} ions in metal-oxide semiconductor structures. Appl. Phys. Lett. **94**(5), 051903 (2009). https://doi.org/10.1063/1.3077169
242. L. Rebohle, A. Kanjilal, W. Skorupa, M. Helm, The inverse energy transfer between Ge nanocrystals and erbium in SiO_2 and its dependence on microstructure. Opt. Mater. **33**(7), 1075–1078 (2011). https://doi.org/10.1016/j.optmat.2010.08.019

243. S. Seidel, L. Rebohle, S. Purcnal, D. Lehninger, R. Hübner, V. Klemm, W. Skorupa, J. Heitmann, Microstructure and charge trapping in ZrO_2- and Si_3N_4-based superlattice layer systems with Ge nanoparticles. Thin Solid Films **645**, 124 (2018). https://doi.org/10.1016/j.tsf.2017.10.029
244. J.A. del Alamo, Nanometre-scale electronics with III–V compound semiconductors. Nature **479**(7373), 317–323 (2011). https://doi.org/10.1038/nature10677
245. http://www.ioffe.ru/SVA/NSM/. Accessed 11 October 2017
246. S.M. Sze, K. Ng Kwok, *Physics of Semiconductor Devices*, 3rd edn. (Wiley-Interscience, Hoboken, N.J., 2007)
247. O. Moutanabbir, U. Gösele, Heterogeneous integration of compound semiconductors. Annu. Rev. Mater. Res. **40**(1), 469–500 (2010). https://doi.org/10.1146/annurev-matsci-070909-104448
248. P. Demeester, A. Ackaert, G. Coudenys, I. Moerman, L. Buydens, I. Pollentier, P. Van Daele, Relaxed lattice-mismatched growth of III–V semiconductors. Prog. Cryst. Growth Charact. **22**, 53–141 (1991)
249. P. Vennéguès, B. Beaumont, V. Bousquet, M. Vaille, P. Gibart, Reduction mechanisms for defect densities in GaN using one- or two-step epitaxial lateral overgrowth methods. J. Appl. Phys. **87**(9), 4175–4181 (2000). https://doi.org/10.1063/1.373048
250. T.A. Langdo, C.W. Leitz, M.T. Currie, E.A. Fitzgerald, A. Lochtefeld, D.A. Antoniadis, High quality Ge on Si by epitaxial necking. Appl. Phys. Lett. **76**(25), 3700–3702 (2000). https://doi.org/10.1063/1.126754
251. J.S. Park, J. Bai, M. Curtin, B. Adekore, M. Carroll, A. Lochtefeld, Defect reduction of selective Ge epitaxy in trenches on Si(001) substrates using aspect ratio trapping. Appl. Phys. Lett. **90**(5), 052113 (2007). https://doi.org/10.1063/1.2435603
252. Y. Xia, P. Yang, Y. Sun, Y. Wu, B. Mayers, B. Gates, Y. Yin, F. Kim, H. Yan, One-dimensional nanostructures: synthesis, characterization and applications. Adv. Mater. **15**(5), 535 (2003)
253. T. Li, M. Mastro, A. Dadgar (eds.), *III–V compound semiconductors: integration with silicon-based microelectronics* (CRC Press, Boca Raton, Fla., 2011)
254. C.W. White et al., Encapsulated semiconductor nanocrystals formed in insulators by ion beam synthesis. Nucl. Instr. Methods B **141**, 228–240 (1998)
255. A. Tchebotareva, J.L. Brebner, S. Roorda, P. Desjardins, C.W. White, Structural properties of InAs nanocrystals formed by sequential implantation of In and As ions in the Si (100) matrix. J. Appl. Phys. **92**(8), 4664–4671 (2002). https://doi.org/10.1063/1.1507822
256. F. Komarov, L. Vlasukova, W. Wesch, A. Kamarou, O. Milchanin, S. Grechnyi, A. Mudryi, A. Ivaniukovich, Formation of InAs nanocrystals in Si by high-fluence ion implantation. Nucl. Instr. Methods B **266**(16), 3557–3564 (2008). https://doi.org/10.1016/j.nimb.2008.06.010
257. F. Komarov, L. Vlasukova, O. Milchanin, W. Wesch, E. Wendler, J. Zuk, I. Parkhomenko, Ion-beam synthesis and characterization of narrow-gap A^3B^5 nanocrystals in Si: effect of implantation and annealing regimes. Mat. Sci. Eng. B **178**, 1169–1177 (2013)
258. M.A. Sortica, B. Canut, M. Hatori, J.F. Dias, N. Chauvin, O. Marty, Optical and structural properties of InAs nanoclusters in crystalline Si obtained through sequential ion implantation and RTA. Physica Status Solidi (a) **212**(12), 2686–2691 (2015). https://doi.org/10.1002/pssa.201532448
259. F. Komarov, L. Vlasukova, O. Milchanin, A. Komarov, W. Wesch, A.K. Togambayeva, Effect of implantation and annealing regimes on ion-beam synthesis of InAs nanocrystals. Lith. J. Phys. **49**(1), 105–110 (2009). https://doi.org/10.3952/lithjphys.49112
260. A.F. Komarov, F.F. Komarov, O.V. Milchanin, L.A. Vlasukova, I.N. Parkhomenko, V.V. Mikhaiov, M.A. Mokhovikov, S.A. Miskevich, Formation of InAs nanoclusters in silicon by high-dose ion implantation: experimental data and simulation results. Tech. Phys. **60**(9), 1335 (2015)
261. R. Khelifi, M. Frégnaux, Y. Le Gall, D. Muller, G. Schmerber, D. Mathiot, Ion beam synthesis of embedded III–As nanocrystals in silicon substrate. Phys Status Solidi (c) **12**(1–2), 55–59 (2015). https://doi.org/10.1002/pssc.201400107

262. S. Prucnal, M. Turek, A. Drozdziel, K. Pyszniak, S.Q. Zhou, A. Kanjilal, W. Skorupa, J. Zuk, Formation of InAs quantum dots in silicon by sequential ion implantation and flash lamp annealing. Appl. Phys. B **101**(1–2), 315–319 (2010). https://doi.org/10.1007/s00340-010-4140-5
263. S. Prucnal et al., Optical and microstructural properties of self-assembled InAs quantum structures in silicon. Cent. Eur. J. Phys. **9**(2), 338 (2011). https://doi.org/10.2478/s11534-010-0107-8
264. S. Prucnal et al., n-InAs nanopyramids fully integrated into silicon. Nano Lett. **11**(7), 2814–2818 (2011). https://doi.org/10.1021/nl201178d
265. G. Astromskas, K. Storm, O. Karlström, P. Caroff, M. Borgström, L.-E. Wernersson, Doping Incorporation in InAs nanowires characterized by capacitance measurements. J. Appl. Phys. **108**(5), 054306 (2010). https://doi.org/10.1063/1.3475356
266. M. Brötzmann, U. Vetter, H. Hofsäss, BN/ZnO heterojunction diodes with apparently giant ideality factors. J. Appl. Phys. **106**(6), 063704 (2009). https://doi.org/10.1063/1.3212987
267. S. Prucnal et al., InP nanocrystals on silicon for optoelectronic applications. Nanotechnology **23**(48), 485204 (2012). https://doi.org/10.1088/0957-4484/23/48/485204
268. S. Prucnal, W. Skorupa, Millisecond-range liquid-phase processing of silicon-based hetero-nanostructures, in *Subsecond Annealing of Advanced Materials*, ed. by W. Skorupa, H. Schmidt. Springer Series in Materials Science 192 (Springer, Heidelberg, 2011), pp. 189–210
269. S. Prucnal, M.O. Liedke, S. Zhou, M. Voelskow, A. Mücklich, M. Turek, J. Zuk, W. Skorupa, Conductivity type and crystal orientation of GaAs nanocrystals fabricated in silicon by ion implantation and flash lamp annealing. Nucl. Instr. Methods B **312**, 104–109 (2013). https://doi.org/10.1016/j.nimb.2013.07.014
270. R.C. Newman, The upper limits of useful n- and p-type doping in GaAs and AlAs. Mat. Sci. Eng. B **66**, 39–45 (1999)
271. E.P. Donovan, F. Spaepen, D. Turnbull, J.M. Poate, D.C. Jacobson, Heat of crystallization and melting point of amorphous silicon. Appl. Phys. Lett. **42**(8), 698–700 (1983). https://doi.org/10.1063/1.94077
272. M.O. Thompson, G.J. Galvin, J.W. Mayer, P.S. Peercy, J.M. Poate, D.C. Jacobson, A.G. Cullis, N.G. Chew, Melting temperature and explosive crystallization of amorphous silicon during pulsed laser irradiation. Phys. Rev. Lett. **52**(26), 2360–2363 (1984). https://doi.org/10.1103/PhysRevLett.52.2360
273. R. Wutzler, L. Rebohle, S. Prucnal, R. Hübner, S. Facsko, R. Böttger, M. Helm, W. Skorupa, III–V nanocrystal formation in ion-implanted Ge and Si via liquid phase epitaxy during short-time flash lamp annealing. Mater. Sci. Semicond. Process. **42**, 166–169 (2016). https://doi.org/10.1016/j.mssp.2015.07.075
274. F.C. Frank, J.H. van der Merwe, One-dimensional dislocations. II. Misfitting monolayers and oriented overgrowth. Proc. R. Soc. Lond. A **198**(1053), 216–225 (1949)
275. R. Wutzler, Integration of III–V compound semiconductors in silicon via ion beam implantation and flash lamp annealing. Dissertation, Technical University of Dresden, 2016
276. S. Prucnal et al., III–V/Si on silicon-on-insulator platform for hybrid nanoelectronics. J. Appl. Phys. **115**(7), 074306 (2014). https://doi.org/10.1063/1.4865875
277. R. Wutzler, L. Rebohle, S. Prucnal, J. Grenzer, R. Hübner, R. Böttger, W. Skorupa, M. Helm, Formation of $In_xGa_{1-x}As$ nanocrystals in thin Si layers by ion implantation and flash lamp annealing. New J. Phys. **19**(6), 063019 (2017). https://doi.org/10.1088/1367-2630/aa66a5
278. R. Wutzler, L. Rebohle, S. Prucnal, F.L. Bregolin, R. Hübner, M. Voelskow, M. Helm, W. Skorupa, Liquid phase epitaxy of binary III–V nanocrystals in thin Si layers triggered by ion implantation and flash lamp annealing. J. Appl. Phys. **117**(17), 175307 (2015). https://doi.org/10.1063/1.4919775
279. L. Rebohle, R. Wutzler, S. Prucnal, R. Hübner, Y.M. Georgiev, A. Erbe, R. Böttger, M. Glaser, A. Lugstein, M. Helm, W. Skorupa, Local formation of InAs nanocrystals in Si by masked ion implantation and flash lamp annealing. Phys. Status Solidi C **14**, 1700188 (2017). https://doi.org/10.1002/pssc.201700188

280. H. Schmid, M.T. Björk, J. Knoch, S. Karg, H. Riel, W. Riess, Doping limits of grown in situ doped silicon nanowires using phosphine. Nano Lett. **9**, 173 (2009)
281. M.S. Seifner, M. Sistani, F. Porrati, G. Di Prima, P. Pertl, M. Huth, A. Lugstein, S. Barth, Direct synthesis of hyperdoped germanium nanowires. ACS Nano **12**, 1236–1241 (2018)
282. K. Saito, E. Tokumitsu, T. Akatsuka, M. Miyauchi, T. Yamada, M. Konagai, K. Takahashi, Characterization of p-type GaAs heavily doped with carbon grown by metalorganic molecular-beam epitaxy. J. Appl. Phys. **64**, 3975 (1988)
283. A. Jochannes, S. Noack, W. Wesch, M. Glaser, A. Lugstein, C. Ronning, Anomalous plastic deformation and sputtering of ion irradiated silicon nanowires. Nano Lett. **15**, 3800–3807 (2015)
284. C. Ronning, C. Borschel, S. Geburt, R. Niepelt, Ion beam doping of semiconductor nanowires. Mater. Sci. Eng. R Rep. **70**, 30–43 (2010)
285. Y. Berencen et al., CMOS-compatible controlled hyperdoping of silicon nanowires. Adv. Mater. Interfac **5**, 1800101 (2018)
286. S. Prucnal et al., III–V semiconductor nanocrystal formation in silicon nanowires via liquid-phase epitaxy. Nano Res. **7**, 1769–1776 (2014)
287. M. Glaser et al., Synthesis, morphological, and electro-optical characterizations of metal/semiconductor nanowire heterostructures. Nano Lett. **16**, 3507–3513 (2016)

Chapter 5
Beyond Semiconductors

Abstract This chapter gives a brief overview about flash lamp annealing of non-semiconductor materials, although the line to semiconductors is flexible. The first section is devoted to dielectric thin films with the focus on high-k materials and rare earth doping of silicon dioxide. The next section discusses the use of flash lamp annealing for monocrystalline, polycrystalline and thin film solar cells. The following section deals with transparent conducting oxides including zinc oxide, titanium oxide, and indium tin oxide. Finally, the section about flexible substrates gives a short introduction to the state of the art and discusses the interaction of ink with the environment and the substrate if inkjet printing is combined with intense pulsed light annealing, a synonym for flash lamp annealing.

5.1 Dielectric Thin Films

5.1.1 High-k Materials

One of the numerous measures taken to continue the miniaturization in microelectronics was the introduction of high-k materials in order to replace the downscaling of the gate oxide thickness by an upscaling of the dielectric constant. However, the introduction of new materials has far-reaching consequences for the fabrication techniques. With respect to flash lamp annealing (FLA) the question is if such a type of annealing is harmless to or even beneficial for high-k materials. However, the individual material properties sensitively depend on the deposition and annealing parameters as well as on the choice of the gate metal stack. Annealing can have, among others, the following impacts: (i) it can crystallize the amorphous layer, and the crystalline phase may depend on the specific annealing conditions. In fact, there is an ongoing discussion whether high-k materials should be nanocrystalline or amorphous. Crystalline oxides have grain boundaries which potentially increase leakage currents, but their k value is probably higher [1]. (ii) In most cases there is an ultrathin SiO_x interfacial layer which is beneficial with respect to interface defect densities, but increases the equivalent oxide thickness. The formation and development of this interlayer depends on the annealing parameters, e.g. on the oxygen content of the

L. Rebohle et al., *Flash Lamp Annealing*, Springer Series in Materials Science 288,
https://doi.org/10.1007/978-3-030-23299-3_5

annealing ambient. (iii) Annealing should not allow diffusion from the metal gate, and (iv) it should be able to anneal out defects.

Among the different approaches HfO_2 is one of the most promising candidates due to its large bandgap (5.5–6.0 eV), its relatively high dielectric constant (22–25), the high electric breakdown field (3.9–6.7 MV/cm), a high thermal stability and a large heat of formation [2]. Kamiyama et al. investigated the crystallization behaviour of HfO_2 and found that a 5 nm thick layer crystallizes after rapid thermal annealing (RTA) at 650 °C 5 s, whereas the threshold for FLA with a pulse length of 0.8 ms and a preheating temperature of 400 °C was at an energy density of about 20 J cm^{-2} [3]. Cross-sectional transmission electron microscopy performed by these authors revealed that FLA leads to a slight increase of the thickness of the interfacial SiO_x layer, but also to a higher k value. As a consequence, FLA was able to improve the equivalent oxide thickness by 5% without any increase of the leakage current. A more recent study focuses on the importance of annealing ambient during FLA [4]. At first, the authors demonstrated that a reduction of the residual oxygen concentration leads to a slightly thinner interfacial thickness with a better thickness homogeneity. At second, the admixture of NH_3 leads to nitridation of HfO_2 which reduces the defect density and a potential diffusion through the HfO_2. Finally, FLA was not able to crystallize a 1 nm thick HfO_2 layer, whereas the same FLA conditions lead to crystallization if the thickness increases to 3 nm. Moreover, crystallization by FLA results in a mixed cubic and monoclinic crystal structure with a dominance of the dense cubic phase with a higher k value.

However, HfO_2 has a low crystallization temperature, a fairly high oxygen diffusivity, and tends to interfacial reactions in the gate stack system [2]. This is why Hf-based alloys like $HfSiO_x$, HfSiON or $HfAlO_x$ are used to increase the crystallization temperature or to provide a diffusion barrier for oxygen. The impact of FLA on $HfAlO_x$ in the gate-last metal-insulator-semiconductor field-effect transistor device fabrication process was investigated in [5]. As sketched in Fig. 5.1a, the high-k stack is made of a 1 nm thick SiO_2 layer and an $HfAlO_x$ layer on Si. If unprotected during FLA, the interfacial SiO_2 layer may react with the $HfAlO_x$ layer which results in a higher defect density and a higher positive bias temperature instability. In this case the flash light is mainly absorbed within the Si substrate, and the critical SiO_2-$HfAlO_x$ interface is exposed to the flash light. If a Si capping layer is used (Fig. 5.1b), the main part of the flash light is absorbed in the capping layer, and the $HfAlO_x$ layer is heated up via thermal conduction from the capping layer. As a result, an interfacial reaction with the SiO_2 layer is suppressed and the positive bias temperature instability is reduced. Furthermore, the beneficial role of FLA for the processing of full-metal-gate HfSiON transistors was discussed in [6].

FLA was also applied to other high-k materials. In [7] the FLA-induced crystallization of thin $LaLuO_3$ or $LaScO_3$ layers grown by molecular beam deposition was investigated. Whereas in case of 3 nm thick layers no sign of crystallization was detected, FLA at 1100 °C with 3 ms pulses and a preheating of 570 °C for 40 s leads to the first appearance of small crystallites in a 6 nm thick layer. At 1200 °C, larger nanocrystals in an amorphous matrix can be found in these layers. In case of 10 nm thick $LaLuO_3$ layers it was found that FLA pulse lengths of 1 or 3 ms are better

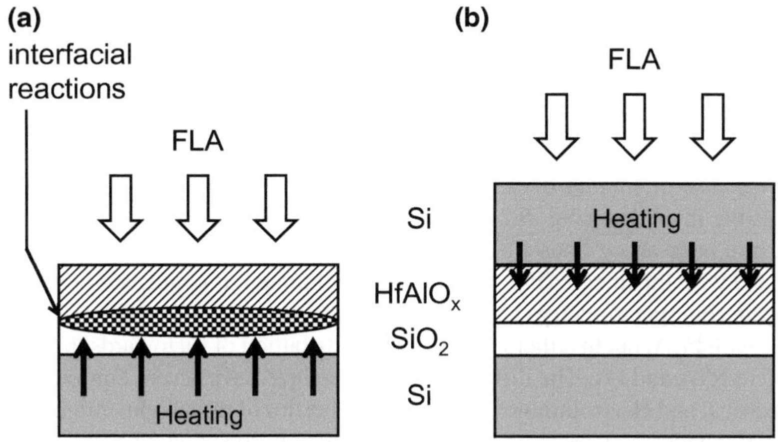

Fig. 5.1 Schematic sketch of a $HfAlO_x$-SiO_2-Si stack which is directly exposed to FLA (**a**) or protected by a Si capping layer (**b**). If protected by a capping layer, interdiffusion and positive bias temperature instability can be reduced. After [5]

than 20 ms, and that the optimum anneal temperature is around 1000 °C at a preheating of 700 °C [8]. This result was obtained by analysing the leakage current and determining the equivalent oxide thickness from CV measurements. Finally, there are reports about the successful use of FLA applied to thin fresnoite ($Ba_2TiSi_2O_8$) [9], lead zirconate titanate ($Pb(Ti, Zr)O_3$) [10] and yttria-stabilized zirconia layers [11]. These materials have high k-values, but are better known for their suitability for piezoelectric applications or for embedded ferroelectric random-access memory devices.

5.1.2 Rare Earth Doping of SiO_2

There is another class of thin dielectrics which were subjected to FLA in the recent past: thin SiO_2 layers with a typical thickness of 100 nm doped by rare earth (RE) ions for Si-based light emission. In this case the motivation to use FLA is similar to that in case of group-IV nanoclusters: FLA should activate ions and anneal out defects, but should avoid the formation of large clusters. The luminescence is usually due to trivalent RE ions which should be either finely dispersed in the bulk or be located at the surface of RE or RE oxide nanoclusters. This especially applies to Eu which features enhanced diffusion in SiO_2 compared to other RE elements [12, 13]. The impact of different annealing timescales on the cluster formation of Eu is depicted in Fig. 5.2. The structure comprises a Si substrate (bottom), a 100 nm thick SiO_2 layer implanted with Eu (peak concentration 1.5 at. %), and a 100 nm SiON capping layer (top). As seen in Fig. 5.2a, amorphous clusters with sizes around 3–4 nm already appear in the as-implanted state and are assumed to be formed by implantation-induced diffusion processes. It is not exactly known if these clusters are composed of Eu, EuO, Eu_2O_3, or a mixture of them, but based on the formation enthalpy of Eu oxides it is assumed that Eu tends to cluster in an oxidized form [14]. The application of FLA at 1000 °C for 20 ms does not cause significant changes (Fig. 5.2b). In contrast, RTA at 1000 °C for 6 s causes the formation of larger clusters at the expense of smaller ones due to Ostwald ripening and a significant diffusion towards the interfaces (Fig. 5.2c). Finally, after furnace annealing (FA) at 1000 °C for 30 min large amorphous clusters up to 20 nm in size and SiO_2 interfaces heavily decorated with Eu emerge (Fig. 5.2d). This development is confirmed by Rutherford backscattering spectrometry [15] showing that the implantation profile of Eu does not change with FLA, but that there is a strong redistribution of Eu towards the interfaces in case of RTA and FA. The different microstructures have severe consequences for the electrical and electroluminescence (EL) properties of these light emitters. The EL output of devices processed by FLA is higher than corresponding devices annealed by FA [16], and the aging during operation is dominated by positive charge trapping instead of a negative one [17]. This behaviour was explained by the defect shell model [12, 18] in which a defect shell around RE clusters is postulated and in which strong electron trapping is associated with large RE clusters.

FLA is also beneficial for other RE ions. In case of Gd-implanted SiO_2 layers the highest EL intensities are obtained by FLA for 20 ms and 1100 °C, followed by FA for 30 min, 900 °C, and RTA for 6 s, 1100 °C [19]. Once again, the lower intensity in case of FA is believed to be due to the formation of larger Gd_2O_3 clusters. If co-doped with F, the EL intensity can be further enhanced due to the beneficial role of F for defect passivation [20]. This enhancement equally applies to devices annealed by FA, but on a lower EL intensity level. Another interesting phenomenon is the co-doping of Gd with K in combination with FLA [21]. In this case K strongly alters the charge trapping mechanism during operation, which led to a lifetime improvement by one order of magnitude at the expense of EL intensity which decreases by a factor of 2–5 depending on the K concentration. Due to the high diffusivity of K, higher thermal budgets as in case of FA hold the risk of complete out-diffusion [22]. Finally, FLA was used to recover Tb-implanted metal-oxide-semiconductor devices which were stressed under constant current injection [23]. In detail, such light emitters were stressed until the EL intensity was below 80% of their initial values, followed by a mild FLA process of 400 °C. In contrast to FA 400 °C, the original EL intensity recovered. As one degradation mechanism is the blockage of RE luminescence centres by electron capture, photons from the ultraviolet (UV) or blue-violet part of the flash lamp spectrum can release such electrons resulting in the recovery of the EL ability. Unfortunately this method cannot be applied several times as the recovered light emitter undergoes accelerated aging due to other degradation processes. In general, more details about RE-implanted MOS devices for Si photonics can be found in [13].

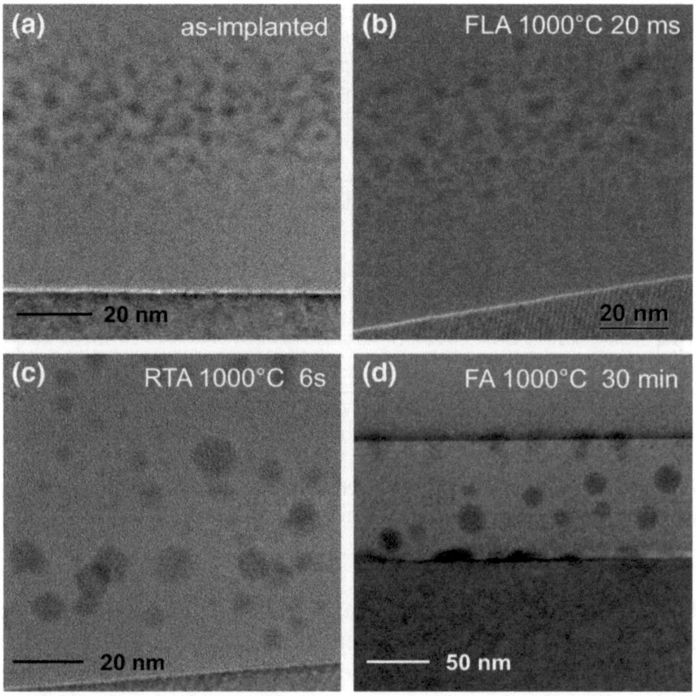

Fig. 5.2 Bright field cross section TEM images showing the gate oxide layer of Eu-implanted metal-oxide-semiconductor (MOS) structures together with its interfaces to Si (bottom) and SiON (up). Please note the different magnification in (**d**). After [13]

5.2 Photovoltaic Applications

The global photovoltaic (PV) solar industry has been grown by 68% over the last decade. At present, the main effort in the solar cell industry is directed to the "costs per watt" reduction of the solar panel fabrication. In 2013 it was postulated, that by 2020 the average price per kWh obtained from PV systems should be reduced to 0.07 €/kWh. But already in 2018 the levelized costs of electricity obtained by large PV utilities in Germany was much below 0.04 €/kWh (Fig. 5.3). For large PV systems installed on the roofs the lowest price for new installations was below the threshold price of the coal lignite (Fig. 5.3).

The main reduction of the solar cell costs was possible due to the reduction of the overall wafer thickness, an increase of the solar cell efficiency and/or the simplification of the production process. Solar cell panels can be divided into three main technologies: monocrystalline Si, polycrystalline Si, and thin-film solar cells. The most efficient and the most stable are monocrystalline silicon panels with an efficiency above 22% and a lifespan of more than 25 years. But also the fabrication costs (material and the manufacturing steps) are the highest. Nevertheless, the monocrys-

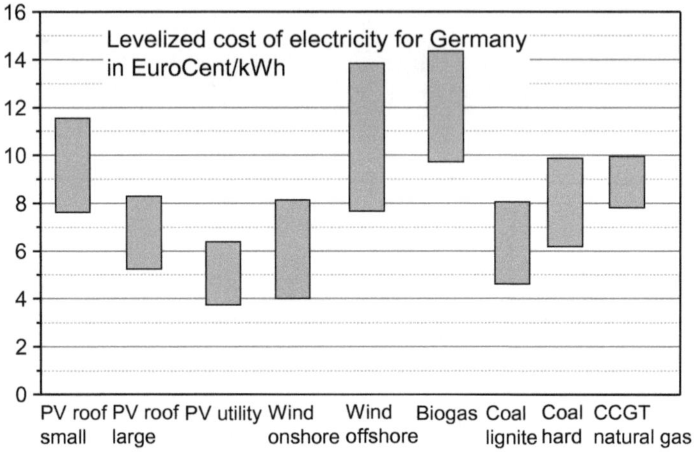

Fig. 5.3 Comparison of the levelized cost of electricity for some newly built renewable and fossil-fuel based power stations in EuroCent per kWh in Germany (March 2018) [24]

talline solar cells can compete with cheaper materials due to the high efficiency. Polycrystalline solar panels have an efficiency in the range of 13–18% and a lifespan similar to that of monocrystalline Si. The manufacturing of polycrystalline panels is simpler making the technology less expensive. Due to the lower efficiency and the lower price polycrystalline panels are mainly installed in big open areas where space plays a secondary role. Thin film solar cells are made by depositing photovoltaic materials onto a suitable substrate (glass, metal foils, polymer etc.). The thickness of the photoactive layer is in the range of few tens of nanometres to few micrometres. The energy efficiency in the laboratory for the best thin film solar cells varies between few percent for amorphous Si and more than 20% for copper-indium-gallium selenide (CIGS) and perovskites solar cells. The thin films solar cells are much cheaper than mono- or polycrystalline silicon cells but the application of thin film PV is limited due to the low efficiency per area for the *large scale* production. Thin film solar cells require huge free areas to produce a commercially significant wattage.

5.2.1 Mono- and Polycrystalline Solar Cells

As the PV industry is very conservative progress at the industry level comes in very small steps based on the gentle evolution of the fabrication processes. The conventional doping of Si solar cells relays on the diffusion of dopants during a high temperature treatment. Si solar cells can be divided into two groups either due to their type of conductivity (p-type and n-type solar cells) or due to their crystallinity (mono- or polycrystalline solar cells). The first solar cells were made of p-type Si because it is much more resistive to cosmic radiation. Unfortunately, p-type Si is highly contaminated with oxygen which significantly reduces the maximum energy conversion efficiency. At the panel level p-type solar cells show an average solar cell efficiency in the range of 17–21% with 22.2% world record solar cells made by Sunpower. Recently, the Institute for Solar Energy Research Hamelin (ISFH) and the Leibniz Universität Hannover demonstrated a prototype of a crystalline p-type silicon solar cell with an energy conversion efficiency of $(26.10 \pm 0.31)\%$ under one sun. The solar cell is 4 cm^2 large with rear-side metallization. It uses a passivating electron-selective n^+-type polysilicon on oxide junction at the minus contact of the cell and a hole-selective p^+-type polysilicon on oxide junction at the plus contact [25]. In the case of n-type Si solar cells the 2018 world record energy conversion efficiency is 26.63% achieved in a Si wafer with a size of 180 cm^2 [26]. The efficiency limit in Si solar cells is defined by the Shockley–Queisser limit and is in the range of 29%, which means that the record values achieved at the lab scale are already very close to the theoretical limit of a single junction solar cell.

Nowadays, the most efficient Si solar cells are made of n-type Si and rear-side metallization where the p-n junction is made by selective ion implantation. The use of ion implantation instead of conventional diffusion reduces the fabrication steps and improves the junction quality. Figure 5.4 shows the comparison between the conventional fabrication process of Si solar cells with diffusion doping (left) and a novel concept (right), called SUNFLASH in the following, where the p-n junction is made by plasma immersion ion implantation (PIII) and millisecond range FLA. In the case of conventional processes the fabrication of solar cells requires 9 basic steps. The replacement of the diffusion process by ion implantation and conventional RTA for p-n junction formation reduces the fabrication steps by two: the etching of phosphorous silicate glass for p-type solar cells and edge isolation. The use of FLA instead of conventional RTA can reduce the process even further if FLA is applied after the metallization. The combination of PIII and FLA enables a precise control of the doping level, the charge carrier distribution and the emitter thickness—parameters which are crucial for the solar cell performance. The implementation of the advanced technology used in current microelectronics into PV manufacturing followed by the reduction of the solar cell production steps can significantly reduce the cost.

The one-flash-one-solar-cell (OFOCell) concept was developed at Helmholtz-Zentrum Dresden Rossendorf [27] and enables a significant reduction of the overall thermal budget required for crystalline silicon solar cell manufacturing. It relays on the one-time high temperature treatment in the millisecond time scale. During the FLA process implanted dopants will be electrically activated forming a low resistance front ohmic contact, and the back surface field layer forms due to Al diffusion into Si. The OFOCell concept allows the use of inexpensive metals for the contact formation of c-Si solar cells, e.g. nickel instead of silver. Furthermore, since the anti-reflecting coating layer is deposited after metallization, it can be freely chosen. Moreover, implementation of the ultra-short high-temperature recrystallization by FLA or laser annealing (LA) together with silicon thin film ink-jet printing on flexible substrates will substantially increase the production capacity of thin film solar cells.

FLA was also successfully used to form high quality p-n junctions in solar grade Si. This material is dirty silicon highly contaminated with nitrogen and transition metal

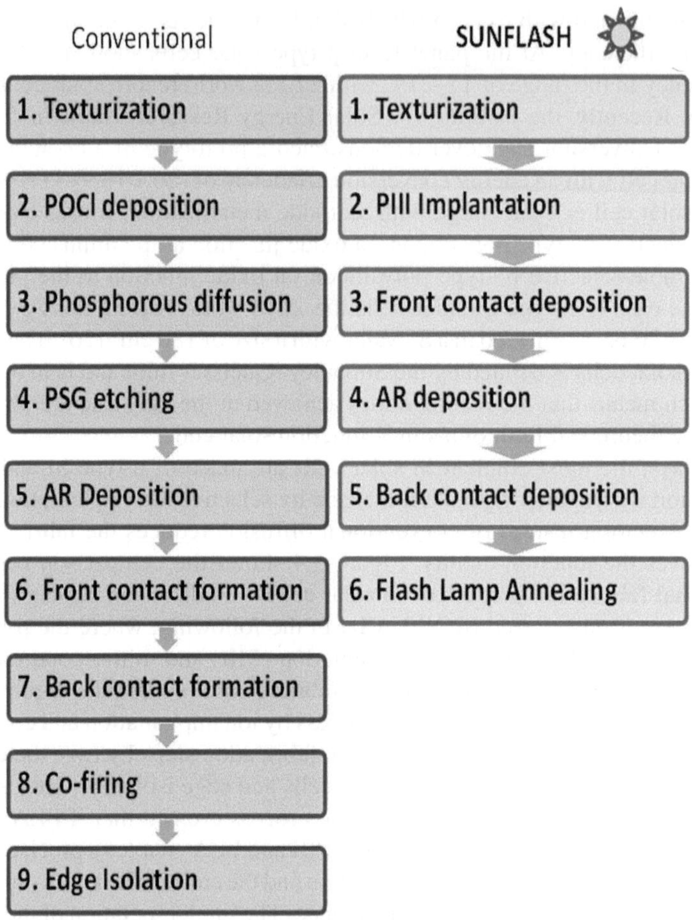

Fig. 5.4 Schematic illustration of the conventional fabrication process of Si solar cells and the SUNFLASH concept

impurities which are mainly responsible for the degradation of solar cell efficiency. The conventional purification process of solar grade silicon takes place during P diffusion. Most metals diffuse into the surface region where they are trapped within the phosphorous silicate glass layer which will be later removed by etching. If ion implantation is used as a doping technique the high temperature purification is not possible. Therefore, metal impurities must be kept away as far as possible from the p-n junction region. Prucnal et al. have shown that using ultra-short FLA the dopants implanted into solar grade silicon can be electrically activated without simultaneously affecting the distribution of transition metals [28]. The minority carrier diffusion length in FLA samples is more than one order of magnitude larger in comparison with samples annealed by RTA and FA. FLA is long enough to activate dopants and to recrystallize the implanted layer but too short to activate impurity diffusion in Si. Ion

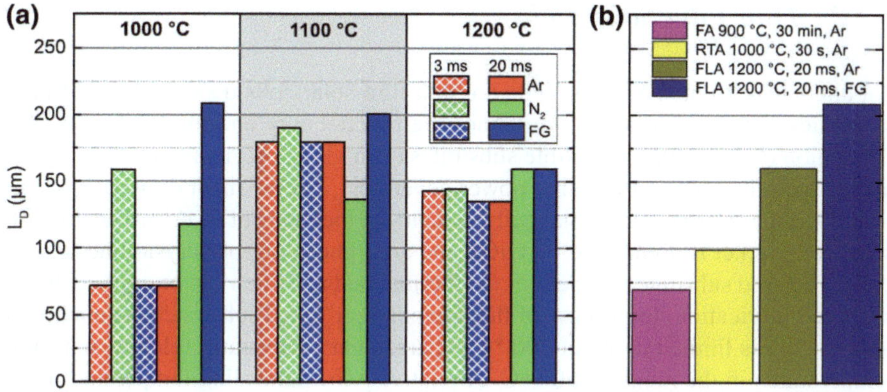

Fig. 5.5 Average minority carrier diffusion lengths as a function of temperature after FLA in different ambients and at different temperatures (**a**), and its dependence on different annealing techniques (**b**). Reprinted from [29], with the permission of AIP Publishing

implantation followed by FLA leads to the formation of high quality p-n junctions conserving the superior properties of bulk Si. Therefore, the fabrication of solar cells via a strong non-equilibrium processing is mainly recommended for monocrystalline Si. In the case of polycrystalline silicon solar cells, the energy conversion efficiency is not only reduced by the presence of impurities, but also by grain boundaries. Bregolin et al. have shown that even during millisecond FLA the grain boundaries in solar grade silicon can be efficiently passivated by hydrogen when samples are annealed in forming gas atmosphere (95% of N_2 and 5% of H_2) [29]. Figure 5.5 shows the average minority carrier diffusion length after FLA in different ambients and at different temperatures (a) as well as after applying different annealing techniques (b).

In general, samples annealed in forming gas show the best performance (except for 3 ms at 1000 °C in N_2). The real average minority carrier diffusion length could not be measured because it is longer than the wafer thickness (here standard solar cell wafers with a thickness of 200 μm were used). The main function of forming gas is to prevent out-diffusion of H_2 which was co-implanted with P during the PIII process. In addition, H_2 from the flash chamber can diffuse into the bulk of the solar grade Si sample. The main defect centres at grain boundaries are broken bonds and vacancies. Both are recombination centres for electrons and can be efficiently deactivated by hydrogen. Therefore, samples annealed in forming gas exhibit a much higher performance compared to samples annealed in N_2 or Ar as N_2 can passivate only the Si surface and Ar does not interact with the annealed sample at all.

In summary, the non-equilibrium processing has a huge potential for high performance PV. It was shown that the junction depths and defects can be properly engineered. The precise control of the junction depths is important for the UV and blue response of the Si solar cells as shallower junctions increase the light harvesting in this spectral range.

5.2.2 Thin Film Solar Cells

An alternative to bulk Si solar cells are thin film solar cells deposited on different substrates. The type of substrate defines the functionality and the final costs. The deposition of thin films on flexible substrates (thin glass, polymers, metal foils etc.) allows the fabrication of portable power generators, but the main drawback of thin film PV is the relatively low solar cell efficiency in the range of 6–12%. The thin film solar cell market is dominated by CIGS and amorphous or poly-crystalline Si solar cells. The used substrate also defines the thermal budget which can be applied during solar cell fabrication. In the case of flexible polymer-based substrates the maximum temperature is limited to about 300 °C, whereas glass substrates tolerate annealing temperatures in the range of 500 °C. A relatively low annealing temperature has a strong negative influence on the PV performance (crystal structure, activation efficiency of dopants etc.). But the maximum temperature can be significantly increased if the annealing time is very short. In such a case the total thermal budget deposited into the system is the same or even smaller than using long term conventional annealing. Moreover, the use of intense pulsed light annealing (IPL), synonymous to FLA, instead of conventional annealing gives the possibility of a selective heating of the PV system by adjusting the light spectrum of the pulse in such a way that the main absorption is within the PV active layer. In such a case the peak temperature within the absorber layer can be much higher than the substrate temperature, provided that the thermal conduction of the substrate is sufficiently low like glass. This allows the recrystallization of the absorber layer without affecting the substrate. Figure 5.6 shows an example of a Roll-to-Roll (R2R) process where the IPL (here FLA) can be used for both the recrystallization of the thin Si film and the dopant activation. In principle, any kind of absorber can be deposited on flexible and temperature-sensitive substrates, e.g. by the very efficient ink-jet printing. The doping can be realized either ex-situ by PIII or already in-situ in the ink, and FLA is applied at the end of the process. Using only one single flash in the millisecond range time scale nearly all energy is deposited within the absorber and the substrate is protected from thermal degradation.

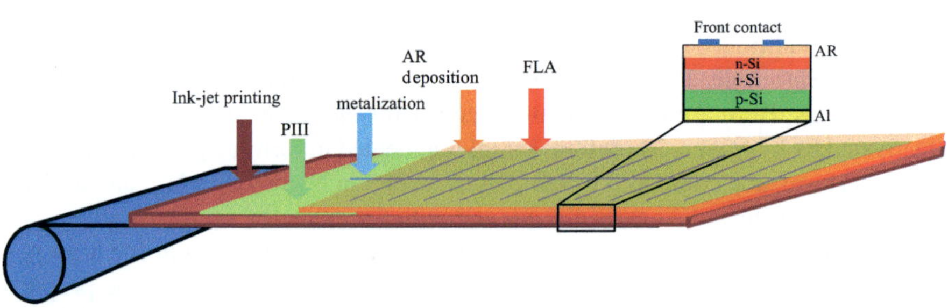

Fig. 5.6 R2R process for the thin film solar cell fabrication

Prucnal et al. have shown that the FLA can be used to significantly improve the crystalline and optical properties of the CIGS absorber [30]. After FLA for 20 ms the near band gap photoluminescencephotoluminescence (PL) intensity increases by one order of magnitude. Moreover, the use of 20 ms FLA for the sintering of CIGS avoids the secondary phase formation within the absorber which is the main reason for the degradation of the solar cell efficiency [31]. A similar effect to CIGS was obtained by utilizing IPL for recrystallization and sintering of copper zinc tin sulphide, where In, Ga and Se has been replaced by Zn, Sn and S [32]. Unfortunately, the copper zinc tin sulphide layer was not stable during IPL. Even an energy density as low as 3.9 J cm^{-2} caused crack formation. Furthermore, multi-pulse annealing led to blistering of the copper zinc tin sulphide layer. The detailed study reveals that this blistering was accompanied by the formation of SnS nanoparticles (NPs) which grow with increasing number of pulses or increasing pulse energy. In the case of Si-based thin film solar cells FLA is mainly used for the recrystallization of the Si layer. The mechanism of recrystallization of Si during FLA was already described in detail in Sect. 4.3. A more exotic use of FLA/IPL is the recrystallization of a polymeric bulk-heterojunction photoactive layer [33]. The white IPL causes the self-organization of the polymeric donor into an ordered structure which increases the power conversion efficiency by three times. The short circuit current is more than two times higher and the fill factor increases from 0.35 to 0.58. The significant improvement of solar cell parameters after IPL is mainly due to the crystallization of the absorber.

FLA or IPL were successfully used to fabricate perovskite solar cells [34–37]. The perovskite solar cells are probably the fastest developing section of PV in the last years. During the last ten years the conversion energy or solar cell efficiency increases from 3.8 to 22.1% in the laboratory [38]. Unfortunately, perovskite solar cells are unstable and degrade very fast. The lifetime is usually given in days which limits the potential applications of perovskite PVs. A significant lifetime improvement was achieved by the encapsulation of solar cells in Teflon. The power conversion efficiency of the Teflon encapsulated perovskite solar cells exhibited considerably stable performances retaining 95% of the initial power conversion efficiency after 30 d. IPL was also used to improve the power conversion efficiency of perovskite solar cells. Lavery et al. have sintered $CH_3NH_3PbI_3$ based solar cells using IPL for 2 ms with an energy of 2 kJ per pulse [34]. The maximum power conversion efficiency is 12.3% and the developed process can be easily adopted for the R2R technology. Another interesting approach is the use of $CuInSe_2$ nanocrystals which offer the possibility to overcome the Shockley–Queisser limit by multiexciton generation. Recently, such layers have been fabricated by spray deposition, and FLA was used to decompose the organic shell of the nanocrystals and to sinter them [39].

In summary, the IPL/FLA is a powerful tool for the non-equilibrium thermal processing of solar cell absorbers. The possibility of tuning the flash light supports the selective heating of the absorber layer without affecting the substrate. Therefore, the substrate used for thin film solar cells can be flexible with a relatively low melting or degradation temperature. The integration of IPL/FLA with R2R processes ensures a high throughput at moderate costs. In the previous paragraphs just a few examples of solar cell absorbers have been given whose internal properties were significantly improved by millisecond range IPL annealing. In principle, any kind of material can be effectively annealed by IPL/FLA if it can absorb most of the UV/VIS light of the flash lamp.

5.3 Transparent Conducting Oxides

Transparent conductive oxides (TCO) become one of the most studied materials, both from the point of view of application and that of basic research. The main applications of TCOs are related with their internal optical and electrical properties as well as their suitability for low-cost, highly efficient production. TCOs are basic materials for thin film PV, flat-panel displays, printed electronics and, in some cases, for light emitters. It was demonstrated that thin film transistors (TFT) made of TCOs can compete with those made of amorphous Si (a-Si) not only due to their device performance but also due to their production costs [40]. The TCOs can be directly printed on temperature sensitive substrates like paper or polymers when the maximum temperature is kept below 200 °C. From the application point of view the most important TCOs include zinc oxide (ZnO), aluminium-, fluorine-, gallium- or indium-doped ZnO (AZO, FZO, GZO or IZO, respectively), titanium dioxide (TiO_2), indium tin oxide (ITO), tin oxide (SnO_2) and other transition metal oxides. Most TCOs can be fabricated at relatively low thermal budget using different deposition techniques like molecular beam epitaxy, chemical vapour deposition, atomic layer deposition (ALD), pulsed laser deposition, magnetron sputtering, the sol-gel process using liquid precursors or ink-jet printing. Due to the huge variability of fabrication methods the final optical, electronic and structural properties differs significantly from layer to layer and from laboratory to laboratory. In this section we will focus on TCOs deposited at low thermal budget whose internal properties are modified by post-grown non-equilibrium thermal processing like FLA. The utilization of ms-range FLA allows to recrystallize TCOs on temperature sensitive substrates and/or achieve doping/alloying much above the equilibrium solid solubility limit. Moreover, the phase formation of different crystallographic orientation of synthetized oxides can be controlled by using FLA.

5.3.1 Zinc Oxide

ZnO together with TiO_2 is present in our daily life not only in electronic devices, but is also added to food, paint, antibacterial coatings and photocatalytical materials. The salient characteristics such as non-toxicity, low cost, easy synthesis, outstanding optoelectronic properties and high stability make ZnO and its derivatives attractive for applications in optoelectronics, sensing, PV and in health-related applications. A highly doped ZnO is an attractive candidate to replace the much more expensive ITO in the microelectronic and PV industry. Due to the high exciton binding energy of ~60 meV leading to efficient excitonic transitions at room temperature (RT), ZnO can be used for efficient light emitters and light converters. To this day, the main drawback of ZnO is the lack of p-type doping. Therefore, difficulties with p-type ZnO are a major obstacle for the fabrication of light emitting and laser diodes with a low turn-on-voltage and p-n junction photo-detectors. On the other hand, ZnO nanostructures were shown to be extremely promising for the application as gas sensors and antibacterial coatings due to the large surface to volume ratio [41]. Moreover, together with TiO_2, ZnO is one of the most attractive photocatalytic materials [42]. The electron-hole pairs produced under intense light illumination are responsible for the photo-degradation of organic pollutants due to oxidation or reduction reactions. It was shown that in comparison with non-polar ZnO the photocatalytic effect is stronger on the polar than on the non-polar surface of ZnO [43]. Zhang et al. have presented that a proper surface defect engineering leads to significantly higher photocatalytic performance [44]. 3D-nanostructures with properly functionalized surfaces are the most promising candidates for such an application. Prucnal et al. have shown that the surface of ZnO can be tailored by ultra-short FLA [45].

A 100 nm thick ZnO layer was deposited on a Si substrate by reactive pulsed laser deposition. During millisecond range FLA of these ZnO films in argon atmosphere Zn interstitials and oxygen vacancies are produced whereas annealing in oxygen suppresses the formation of O vacancies and increases the concentration of oxygen interstitials. This allows the control of the type and the density of surface states which is crucial for efficient light emitting devices and the photocatalytic effect in ZnO. In general, the formation energy for a certain defect type in ZnO depends on the relative position of the Fermi level [46]. During conventional annealing like RTA or FA, and independent on the atmosphere, the Fermi energy level position continuously moves towards either the conduction band minimum or the valence band maximum due to the concentration change of donor- or acceptor-like defects, respectively. This continuously changes the defect formation energy during thermal treatment. Therefore, the control of defects during conventional annealing is challenging. In contrast, ultra-short non-equilibrium thermal annealing may present a novel method to reduce the influence of the shift of the Fermi energy level on the defect formation during annealing. Moreover, it was shown that by applying different annealing parameters (annealing time and energy density) and annealing atmospheres it is possible to control the optoelectronic properties of ZnO films (Fig. 5.7a). The post-deposition NH_3 plasma treatment of the ZnO surface completely quenches the defect related green luminescence. The near-band gap emission from ZnO and the green luminescence can be efficiently modified using oxygen and hydrogen passivation. Defects in ZnO also determine the electrical properties of the material. Zn interstitials and O vacancies are shallow and deep donors, respectively, while Zn vacancies and O interstitials are deep acceptors. Moreover, the point defects in ZnO may form complex defects with H which is usually responsible for n-type conductivity. By utilizing FLA for ZnO processing in N_2 or O_2 atmosphere it is possible to control the electrical conductivity of ZnO thin films (Fig. 5.7b). Non-intentionally doped ZnO after annealing in N_2 has a high n-type conductivity with an average electron concentration in the range of $2–5 \times 10^{19}$ cm^{-3}. Annealing in O_2 atmosphere produces highly resistive ZnO where

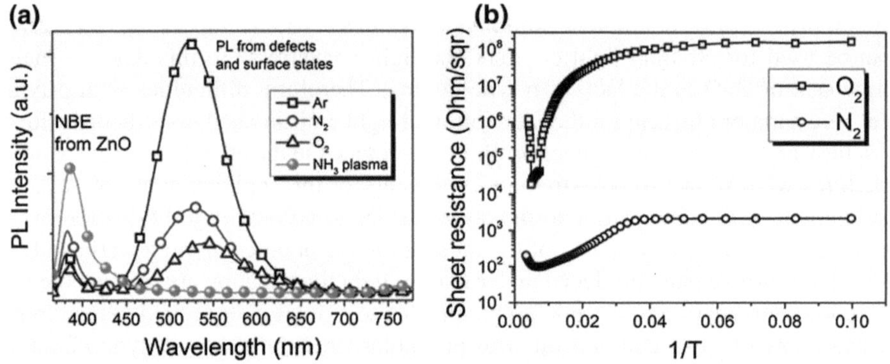

Fig. 5.7 RT PL spectra obtained from polycrystalline ZnO annealed by ms-range FLA under different atmosphere (**a**). The sheet resistance as a function of reciprocal temperature obtained from ZnO film annealed in O_2 and N_2 (**b**)

the electrons are fully compensated by holes. The average carrier concentration is in the range of 10^{13} cm^{-3}.

The electrical properties of ZnO can be easily controlled by intentional doping. In general, n-type ZnO is achieved by doping with group III elements. In order to avoid the segregation of dopants and the formation of secondary phases like Al_2O_3 within the ZnO matrix the time of the post-deposition annealing must be reduced to a minimum. Gebel et al. have shown that FLA using an 1.3 ms pulse does not only decrease the resistivity but significantly improves the transmittance of Al doped ZnO (AZO), too [47]. Lindberg et al. studied the influence of ms-range FLA on the activation efficiency of Al in ZnO and the defect interaction with electrically active Al atoms [48]. In this study, Al doped ZnO was made by magnetron sputtering on a Si substrate with Al concentrations of up to 3%. It was shown that FLA increases the activation of Al and promotes the formation of electrically non-active Zn vacancy clusters. These clusters are electrically isolated from the matrix and do not affect the film resistivity. Using 3 ms FLA pulses the lowest achieved resistivity was in the range of 2×10^{-4} Ωcm which is close to state-of-the-art value. Simultaneously, using temperature dependent deep level transient spectroscopy, the deep defect state density at the Si/AZO interface was found to be significantly reduced during FLA.

Due to the wide band gap and the high conductivity a ZnO layer can be used as a conductive anti-reflection coating for thin film solar cells. Reck et al. investigated the influence of a ZnO coating on the efficiency of copper–indium–gallium–sulphide films (CIGSu) [49]. Solar cells with CIGSu have a smaller external efficiency than those made of CIGS due to the much lower open circuit voltage. In CIGSu, the open circuit voltage can be significantly increased by Zn co-doping [50]. In principle, ZnO deposited on CIGSu followed by post-deposition annealing can act both as a transparent conductive electrode and as a source of Zn which can diffuse into the CIGSu layer during annealing. The post-deposition annealing must be short and powerful enough to control the diffusion of Zn and the redistribution of Ga within

CIGSu. Having this in mind FLA with sub-millisecond pulses appears to be the most suitable annealing method for this purpose. After FLA for 250 μs the diffusion of Zn is fully supressed and the absorption efficiency of CIGSu significantly improves [49].

As mentioned above, ZnO is also attractive for light emission devices. It was shown that the internal optical properties of ZnO can be easily modified by doping with RE. Different REs being in the +3 oxidation state exhibit sharp emission lines in the broad spectral range from the UV to the near infrared (NIR). The introduction of individual REs into a ZnO matrix allows the realization of monochromatic light sources. In contrast, the proper mixing of different RE's together (e.g. Ce, Eu and Tb) can act as a white phosphor. Unfortunately, the solid solubility limit of REs in ZnO is very low, and REs are fast diffusers with a strong tendency to form clusters and to segregate at the surface of ZnO. Therefore the fabrication of REs doped ZnO is challenging. Ratajczak et al. used ion implantation to dope ZnO with different REs [51, 52]. After conventional annealing at a temperature around 900 °C most of the implanted elements are agglomerated at the sample surface where they form different RE oxides. Only the use of FLA preserves the depth distribution of implanted RE's and leads to the incorporation of most of the RE atoms into a lattice site of ZnO. Unfortunately, by replacing Zn in the matrix the RE atoms also change their oxidation state from +3 to +2, although most of the REs are optically not active in the +2 oxidation state. In the case of Yb doped ZnO, after FLA a weak NIR PL was observed, which means that part of the Yb ions incorporated into ZnO are in the +3 oxidation state [52].

5.3.2 *Titanium Oxide*

TiO_2 is one of the most important transition metal oxides for water splitting. Due to its unique optical properties TiO_2 is also used for anti-reflecting coatings for solar concentrators. Like ZnO, TiO_2 is a wide band gap functional oxide whose internal optical and electrical properties can be tuned by controlling the phase composition starting with the pure anatase phase ($E_g = 3.2$ eV) via mixed anatase/rutile phases to the pure rutile phase ($E_g = 3.0$ eV). The formation of different crystallographic phases of TiO_2 is determined by doping and annealing. The anatase phase in non-intentionally doped TiO_2 is stable up to 500 °C. Annealing of TiO_2 in the temperature range from 500 to 700 °C causes the formation of mixed phases. After annealing above 700 °C the rutile phase becomes stable. The unique optical properties of TiO_2 like photoactivity are directly connected with the anatase phase. Unfortunately, the wide band gap of the anatase phase of 3.2 eV limits the harvesting of sun light for the catalytic effect to the UV spectral range. Therefore, the main attention is focused on the stabilization of the anatase phase simultaneously with the band gap narrowing by alloying TiO_2 with non-metal anions like N or metallic cations like Cr [53]. Due to the very low equilibrium solid solubility limit of N in TiO_2 the achievable band gap narrowing is also limited. The enhancement of visible light absorption in TiO_2:N with increasing N concentration is due to the inter-band absorption via N-related localised levels within the band gap [54]. On the other hand, Cr can be incorporated into TiO_2 in any concentration. It was shown that the photoactivity of TiO_2 doped with Cr is significantly enhanced [55]. But the main challenge is to stabilize the anatase phase in TiO_2:Cr and to suppress the diffusion of Cr during post-deposition annealing. Gago et al. investigated the influence of FLA on the Cr distribution and phase formation in TiO_2 thin films deposited by magnetron sputtering [53]. Considering the practical applications of TiO_2 the production methods have to be cost-effective including the use of low thermal budgets. It was shown that FLA, performed for 20 ms with an energy density up to 65 J cm^{-2} on samples deposited at RT, stabilizes the anatase phase and suppresses the out-diffusion of Cr. For a low Cr content (<6%) the FLA treatment significantly improves the crystal structure of anatase TiO_2:Cr. For an high Cr content (>6%) the rutile phase is stabilised, but Cr exists in TiO_2 mainly in the +6 oxidation state (Cr^{6+}) which is very toxic. The FLA treatment does not promote the anatase phase in Cr highly-doped TiO_2 but reduces the toxicity of TiO_2:Cr by converting Cr^{6+} observed in the as-deposited sample to Cr^{4+} being an evidence for the rutile phase formation.

Another approach to promote and stabilize the anatase phase in TiO_2 was developed by Neubert [56]. He has shown that the phase formation is strongly affected by the growth conditions during magnetron sputtering. By varying the deposition conditions, i.e. partial pressure and temperature, either anatase or rutile phase can be promoted already during the growth. He demonstrated that layers deposited at a low oxygen partial pressure and with a high concentration of oxygen deficiency centres catalyses the anatase phase formation. Such a system is stable up to 600 °C if using ms-range FLA in vacuum. In contrary, the growth of TiO_2 in an oxygen-rich atmosphere, where Ti-vacancies dominate, promotes the rutile phase.

5.3.3 *Indium Tin Oxide*

Nowadays the most common TCO used in the mass-production for flat-panel displays is ITO. ITO has an optical band gap in the range of 3.8 ± 0.2 eV depending on the crystal structure and indium content. Simultaneously, the electrical resistivity is much below 10^{-5} Ω cm which is not achievable by other TCOs like AZO or TiO_2. To this day only graphene can compete with ITO and can be considered to replace ITO in the mass production [57]. The electrical and optical properties of ITO are defined by the crystal structure, the composition and the conductivity (carrier concentration). However, a compromise must be always made between optical and electrical properties of the deposited film. Thus, an increase of the film thickness and the charge carrier concentration decreases the transmittance of the ITO film but increases its conductivity. Moreover, the crystallographic phase of ITO strongly affects the transparency. Poly- or single crystalline ITO is much more transparent than their amorphous counterpart. Using conventional annealing the recrystallization temperature of ITO is in the range of 230 °C. Similar to other TCOs, magnetron

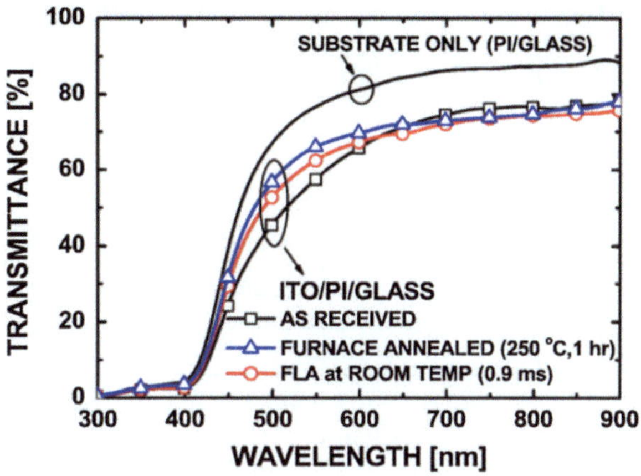

Fig. 5.8 Transmittance of the ITO films on glass before and after FLA. Reprinted from [63], Copyright (2017), with permission from Elsevier

sputtering is the most common technique to deposit thin ITO films on both flexible and non-flexible substrates. The deposition can be performed at elevated temperature providing crystalline ITO or at RT (amorphous film) followed by post-deposition annealing. In order to reduce production costs and to avoid substrate heating, FLA is the most commonly used technique to recrystallize amorphous ITO. Recently, ITO layers on glass have also been fabricated by electron beam physical vapour deposition and FLA [58].

It was shown that ITO can also be deposited on flexible glass with a thickness in the range of 100 μm (Fig. 5.8) or on other large area transparent substrates with superior optical and electrical properties utilizing sputtering techniques followed by ms-range FLA [59, 60, 63]. Both the optical and electrical properties of ITO annealed by FLA are comparable to samples annealed by conventional FA. Panckow et al. investigated the dependence of the optical and electrical properties of ITO films on the O_2/Ar ratio and the post-growth annealing conditions. Figure 5.9 shows that independent of the O_2/Ar ratio both transmittance and conductivity of the deposited films significantly improve after annealing. Samples grown at low O_2 pressure contain a high density of O vacancies which decreases transmittance and carrier concentration. Nevertheless, both FLA and FA remove most defects. However, the obtained results also show that FLA applied to samples grown at an optimum O_2/Ar ration is more beneficial than the conventional FA [60, 61]. Moreover, the FLA can be easily integrated into R2R processes where the repetition rate of flash pulses can be in the range of 4 Hz with an energy density of 25 J/cm^2 per pulse [59].

Gebel et al. simulated the impact of different FLA conditions on ITO films on a polyethylene terephthalate (PET) foil and a thin glass substrate [62]. Due to the lower heat conductivity of PET compared to glass, larger temperature gradients can be achieved in case of the PET foil. In detail, the maximum temperature in the PET layer in a depth of 10 μm from the ITO-PET interface drops down from 870 °C at 20 ms to 640 and 460 °C for pulse lengths of 3 ms and 600 μs, respectively. In fact the temperature of the PET-foil strongly depends on the spectrum of flash lamps and

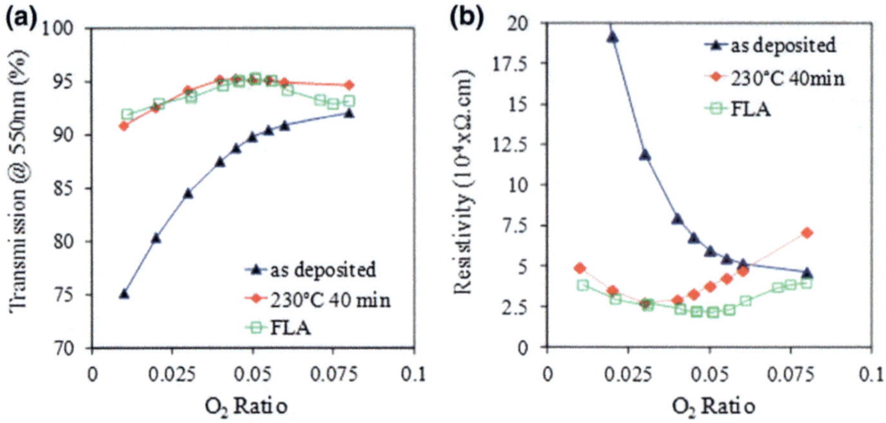

Fig. 5.9 Transmittance (**a**) and resistivity (**b**) of an ITO layer deposited on flexible glass before and after annealing using either ms-range FLA or conventional FA for 40 min at 230 °C. Samples were deposited by using different O_2/Ar ratio. Reprinted from [61], Copyright (2017), with permission from Elsevier

the absorption within the ITO film. To be able to absorb more light within the top ITO layer the spectral weight of the flash lamps must be in the UV range. Another approach can be realized by the use of highly transparent polyimide substrate where the flash light is absorbed mainly by the deposited film [63].

5.3.4 Other TCOs and Beyond

The final application determines the required properties of the designed metal oxide. The band gap and the electrical properties of TCOs can be tuned by using different metal oxides. The use of wide band gap TCOs is very important for PV as a higher transparency in the UV increases the open circuit voltage. Therefore, pure In_2O_3 is even more favourable than conventional ITO [64]. Moreover, the hydrogenation of In_2O_3 significantly increases the carrier mobility which enhances the solar cell efficiency. The carrier mobility is also affected by the microstructural properties of the oxide film. Thermal processing in the PV requires low thermal budget processes, i.e. fast processes with low energy consumption and low-cost non-vacuum processing. Hence ms-range FLA became the most promising annealing technique. By using FLA the deposited films are recrystallized, hydrogen has no time to diffuse out during annealing and the electron mobility increases from 48 to 112 $cm^2\ V^{-1}\ s^{-1}$ after FLA [64]. In combination with ZnO, In_2O_3 can be used for the TFT s. The superior properties of such TFTs are obtained after FLA for 500 μs of In_2O_3/ZnO layers on glass [65]. Figure 5.10 shows the temperature distribution in an oxide layered system on glass after FLA for 250 μs as an example [49].

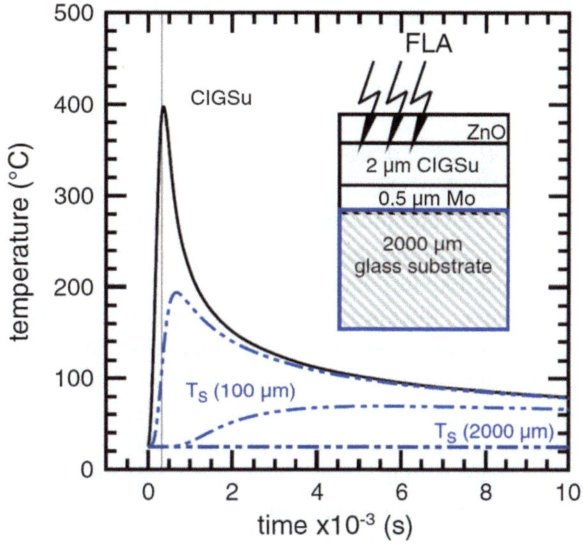

Fig. 5.10 Time-dependent temperature distribution at the surface of the multilayer system on a glass substrate. Reprinted from [49], Copyright (2014), with permission from Elsevier

Indium gallium zinc oxide is another promising material for TFTs due to its relative high carrier mobility. However, degradation under negative bias temperature illumination stress is a common problem [66]. In few recent studies, FLA was applied to TFTs made of indium gallium zinc oxide to increase carrier mobility and transparency, whereas the latter also increases the device stability under illumination [66–69]. Henke et al. have used ALD with in-situ FLA (see also Sect. 2.7.2) to increase the electrical properties of *high-k* materials like Al_2O_3 [70]. It was shown that in-situ FLA increases the Al_2O_3 layer density and the mass gain per cycle by 10 and 25%, respectively. Moreover, the FLA was used to control the nanostructure of a polymeric bulk-heterojunction photoactive layer to improve the efficiencies of polymer solar cells [33].

5.4 Metallic Films

Metallic thin films have caught users' and scientists' interest due to their promising mechanical, structural and magnetic properties. In fact, they may differ significantly from the respective bulk materials. For example, it has been found that elasticity is larger in the case of thin metallic films which in turn makes them attractive for novel applications on flexible substrates. Also they have become a major component in integrated circuit industries and printed electronics [71–73]. The following section intends to look into the ability to engineer thin metallic film properties by millisecond annealing and investigates their role in various applications, among which are printed and integrated electronics.

5.4.1 *Annealing of Thin Metal Films*

The low thickness of metallic films has a direct consequence also on millisecond annealing. In fact, the thickness may be smaller than the penetration depth of the flash [74]. Thus those films can become semi-transparent to the flash. In fact, thin metal films have been studied as a low-cost substitute for TCOs (previous section). This is to overcome certain disadvantages of TCOs regarding their suitability for flexible machining and costs by providing an alternative with fundamentally different properties like thin metal films [75].

However, thickness reduction is limited to a final thickness of (10–20) nm as the resistivity rises with decreasing thickness. The reason for this can be found in the decreasing mean free path of the electrons which now may scatter elastically off the surface of the film more often than inside the bulk material. Figure 5.11 shows this behaviour exemplarily for copper [75, 76].

In contrast to the more expensive noble metals like gold and silver copper forms oxides during annealing under atmospheric conditions. This both reduces its conductivity which requires a further post-annealing and demands elevated annealing temperatures which are applied for a comparatively long time if using conventional FA. In order to protect the often temperature-sensitive substrates from excessive heating and thus from warpage or even subsequent destruction, South Korean researchers showed that two-step flash sintering can prevent warpage of polymer substrates during millisecond annealing of copper NP ink while retaining high conductivity [77]. Furthermore, the small film thickness with respect to the flash penetration depth into the metal leads to gradient-free heating throughout its thickness. This in turn means that there is no heat sink that supports the idea of millisecond annealing which is why processing parameters need to be adapted accordingly. On the other hand researchers have found that it induces local heating at the interface between the substrate and the metal film (here: copper) giving rise to an enlarged adhesion strength of the metal [77].

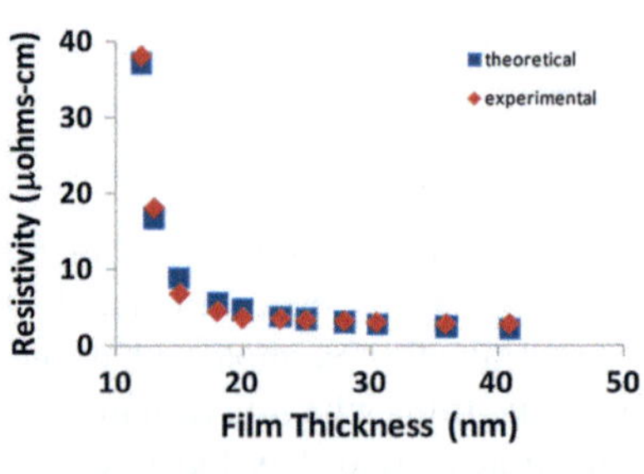

Fig. 5.11 Decreasing resistivity for increasing film thickness of copper as a consequence of a decreasing mean free path for the electrons [76]. Published under license to BioMed Central Ltd. and distributed under the terms of the Creative Commons Attribution License 2.0

5.4.2 Material Property Engineering

Millisecond annealing has been studied for material engineering, among which magnetic properties of FePt films were object of inspection. It could be shown that FLA can successfully transform the films' microstructure from its disordered A1 phase to the $L1_0$ chemically ordered phase which is closely connected to an increase in coercivity. The velocity of this transformation was found to increase for slightly Fe-rich films [78]. During preparation of such amorphous alloys from the liquid state stress is produced downgrading their elastic and thus magnetomechanical properties. It was studied whether FLA with an applied magnetic field can help dissolve these stresses and therefore reimprove these properties in contrast to isothermal annealing. Braun et al. have found that FLA can indeed reduce internal stress which causes the domain walls to move more freely and thus improves the magnetomechanical properties of the samples [79].

Another group showed that for a limited phosphorous content amorphous clusters in Fe-Pt-Si alloys can be fully dissolved to give improved magnetic properties, among which the temperature dependence of the elastic modulus. This was not found for the more stable phosphorous-rich alloy. Annealing was carried out at multiple pulses with a total time duration of less than 1.1 s [80]. The suitability for multi-pulse FLA with a single pulse length of 20 ms to recrystallize amorphous alloys has been shown. Material hardness is increased and in contrast to isothermal heating regrowth can be bound to the sample surface [81, 82].

5.4.3 Silicides

Silicides are metallic alloys with silicon being semiconductors with the high reflectivity typical for metals. They may be produced from powder or by reduction of silicon dioxide with an excess of the metal to form the desired alloy. A third method to form silicides—which is the most common one for integrated circuit design—is brought about by diffusion from a thin metal film on top of the wafer surface into the silicon substrate by thermal treatment where again FLA is needed for shallow junction formation. Figure 5.12 shows that after evaporation of the metal and deposition onto the Si substrate different silicides may form depending on the annealing temperature. Strictly speaking, however, the metal-Si interface disappears already shortly after metal deposition before an annealing temperature is applied because a thin silicide film is formed [83–85].

For their use as interconnections and metal contacts transition metal silicides need to be locally confined onto the Si substrate. In the mid-1990s Bischoff et al. presented an alternative to using oxide masks in applying a focused ion beam of cobalt onto Si with subsequent FLA. The ultra-short annealing nature of FLA in combination with the focused ion beam has allowed for feature dimensions as small as 200 nm and even

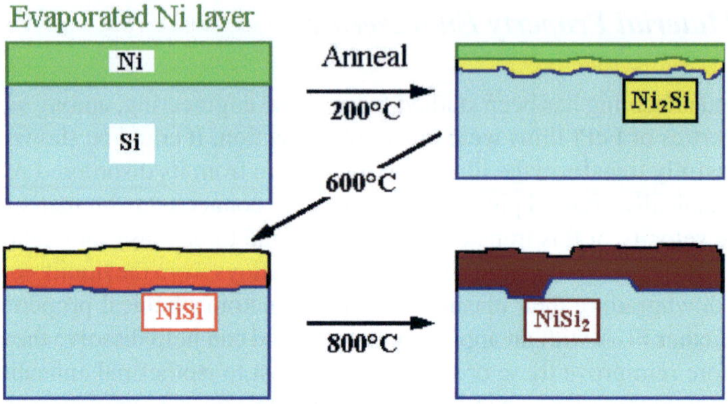

Fig. 5.12 Schematic showing nickel silicide Formation [84]

smaller structures are expected by the researchers for optimized flash parameters and reduced ion doses [86].

It has been found for platinum-nickel silicides, which are widely used for very large scale integration technology, that the silicide formation not only proceeds faster with multi-pulse FLA and a total irradiation time of 1.3–2.0 ms in contrast to isothermal annealing, but also starts earlier, namely at 100–150 °C. Furthermore, it produces only one phase (the orthorhombic one), whereas conventional FA leads to the formation of both hexagonal and orthorhombic platinum-nickel silicide phases and requires more elevated temperatures of at least 400 °C for a processing time of 30 min which in turn promotes diffusion and thus opposes shallow doping and helps agglomeration. The composition of this orthorhombic phase, however, has been found to vary depending on the metal thickness and the irradiation energy. Similar finding had been made earlier with titanium silicides [87, 88].

The heat treatment for silicide formation as explained above is traditionally done by RTA with a time interval of up to a minute. However, the thermal budget thus imposed on temperature-sensitive materials such as the silicides themselves or high-k metal gates may be viewed critically. On the other hand, it has been shown that millisecond annealing can account for dopant segregation into the silicide without changing its morphology by an excess of thermal treatment (Fig. 5.13). Investigations into FLA have further shown that an equally low sheet resistance could be achieved after multiple shots while obtaining a smooth interface between the silicide and the silicon substrate (Fig. 5.13) which is promising for low leakage currents. The final silicide structure has been found to depend crucially on the metal film thickness and the annealing temperature [85, 89–91].

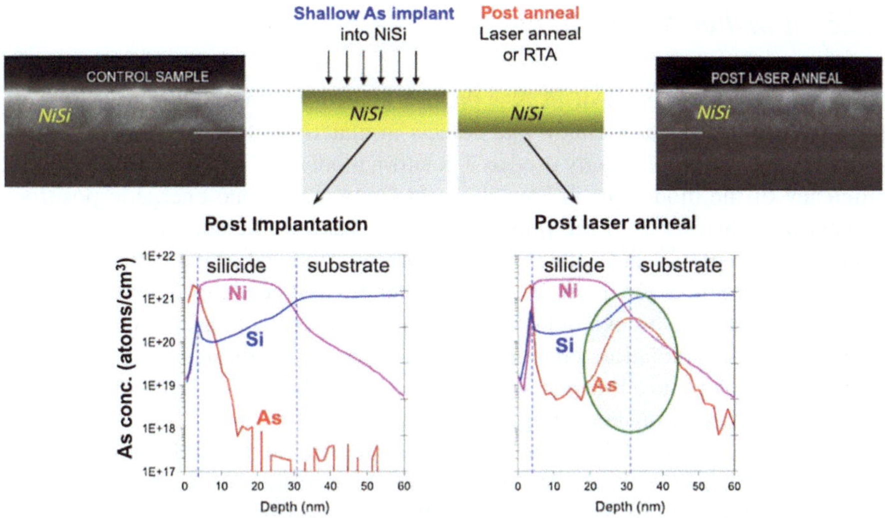

Fig. 5.13 Top left and top right: scanning electron microscope images showing a smooth interface between the Silicide and the substrate. Bottom: secondary ion mass spectroscopy measurements in As implanted nickel silicides demonstrate dopant segregation using millisecond LA. Reprinted from [91], Copyright 2016, World Scientific Publishing

5.4.4 Printed Electronics

As an alternative to the traditional ITO-based touch sensors the technology of printed electronics allows for sensors consisting of a metal mesh which can be more easily processed by inexpensive R2R processing. The mesh allows for flexible and stretchable structures, but it also requires less material thus saving costs [92]. Similar studies have been undertaken for a mesh of large-scale silver nanowires. Only one shot of ultra-short FLA at comparatively high intensity and during a time interval of 50 μs suffices to effectively sinter the nanowires and gain high conductivity and strong substrate adhesion at the same time which once more allows for R2R processing [93]. Although metals definitely play a larger role in inks for printed electronics (cf. Sect. 5.5) they are also found as a substrate alternative. For example, stainless steel has been studied as an option for flexible substrates for displays with ITO where transparency is not required. Not only can it withstand high temperatures which makes it suitable for high temperature annealing in contrast to its opponents glass and polymers, but also it is much less permeable to water or oxygen vapour as well as more stable. Similar studies involve metallic substrates as a backplane for ITO. Further information may be found in Sect. 5.3 [94–96].

5.4.5 Further Applications

A more exotic application of FLA is the use in positron annihilation spectroscopy which is used to detect vacancy-like defects in materials. Tungsten foil of a thickness of some μm is commonly used as a positron moderator for the source ^{22}Na. The efficiency of the moderator is crucial to obtain enough mono-energetic positrons, especially with respect to the potassium price. Yet its efficiency in turn depends strongly on the prior heat treatment of the moderator foil and conventional annealing at 2000 °C introduces too many technical challenges as it needs to be done under vacuum. Furthermore, although being available at low price the tungsten foil moderator is often subject to radiation damage and requires regular post-annealing. Therefore, it has been looked into inexpensive and easy-to-handle millisecond annealing for cost reduction and array annealing.

Anwand et al. studied the suitability of FLA to increase the moderator efficiency. It was found that FLA is suitable for annealing and heating up a tungsten mesh to just beneath the melting temperature which cleans the surface without introducing additional local overheating as it can be seen in Fig. 5.14. Further, the positron diffusion length could be increased as it would not have been possible by conventional means. The obtained moderator efficiency of 3×10^{-4} agrees well with the values published in literature [97].

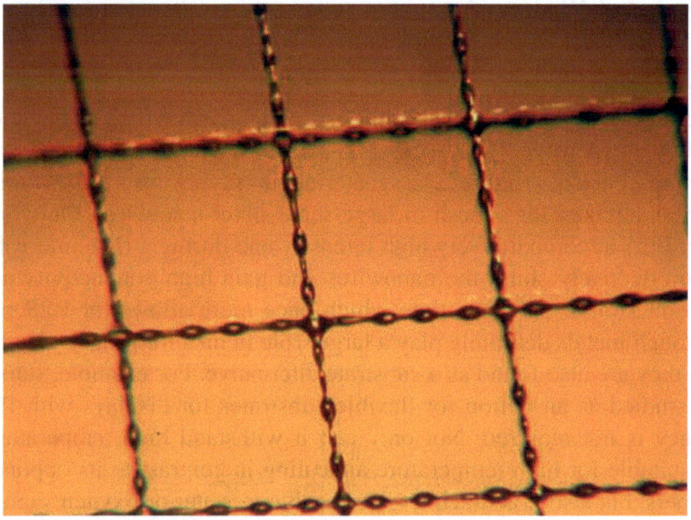

Fig. 5.14 Tungsten mesh after FLA treatment [97]. Distributed under the terms of the Creative Commons Attribution Licence 3.0

5.5 Flexible Substrates

Printed or *low-cost* electronics is an emerging field of large-scale, flexible and inexpensive technology. In terms of high power performance it may not be able to compete with conventional electronics, but it is well-suited to be spread quickly for everyday life applications like package labelling or self-updating newspaper. The fact that the flexible substrate s used in printed electronics allow for volume manufacturing on R2R printers reduces significantly production cost and thus makes printed electronics economically interesting. Furthermore, it allows for large area substrates. Instead of microelectronics' motivation to scale-down dimensions this technological path aims to scale them up. One may refer to as *macroelectronics*. Therefore, FLA—also known as photonic curing, photonic sintering, pulsed thermal processing or IPL—matches very well with printed electronics w.r.t. time, costs and volume [98–100].

Substrates, inks and architecture differ from conventional electronics which is why they get special attention in current research activities [98, 99]. Yet, traditional printing can be applied though they require adjustment to suit the new purpose. So do the substrates and inks such that the otherwise brittle materials for electronics are made printable by being brought into solution or being made dispensable. At the same time ink and substrate need to be adapted to each other in order to be combined.

Almost all components for electronic devices can be printed. These include organic electronics like organic light emitting diodes or organic solar cells, diodes and electrodes, for example source-drain-electrodes for organic field effect transistors and Schottky diodes by inkjet printing or whole integrated circuits altogether. Furthermore, RFID (radio frequency identification) antenna can be produced by screen printing for labelling and anti-theft protection [101–103].

Better performance of the so-created products is not necessarily the reason to push printed electronics, but its capability of inexpensive, planar mass production at good quality for the manufacturing of high amounts of everyday needs is outstanding. The possibility to make thin and bendable products by printing them has inspired man's phantasy for goods like a self-updating newspaper. Printed electronics is regarded as a key technology, not least due to its ability to realize quite easily manifold functions on one item. Phrases like „ubiquitous computing“ and „ambient intelligence“ have thus appeared [33].

The following chapter aims to briefly introduce into the current state of art of printed electronics, the availability and characteristics of substrates and inks and gives a guide to means of processing. Conventional means of sintering are as much looked into as the use of millisecond annealing, whereupon the focus will be on the latter, clearly.

5.5.1 State of the Art

5.5.1.1 Substrates

Printed electronics is mainly based on flexible substrates that can be handled in large volume by inexpensive R2R processing. R2R (reel-to-reel or web-fed) printing terms a process, at which a substrate which is yet to be printed is unwound from a roll, then processed (printing and sintering) before being rewound again onto another roll (Fig. 5.15). Thereby, an area of several 10,000 m^2 of substrate per hour can be processed for a variety of materials. At the same time sheet quality is not negatively affected and spatial resolutions of less than 20 μm can be achieved. There are three types of printing that are mainly used for volume manufacturing on reels: gravure, offset and flexographic printing [98]. More details will be found in Sect. 5.5.1.3.

R2R processing can only be applied by the use of flexible substrates such as polymer foil, paper or thin metallic films, whereas the latter has been closely looked at in Sect. 5.2. For advanced performance printed electronics resorts to rigid silicon or glass substrates using sheet-based inkjet or screen printing instead (Sect. 5.5.1.3) [59, 92, 104]. The right choice of substrate drives the success of the desired product. For this purpose the base material for printed electronics may be functionalized to meet the requested criteria. Among the polymeric foils PET is preferred with respect to costs as well as paper, but in comparison to the latter PET is more resistant to tearing, chemicals and humidity [105, 106].

This is closely linked to the structure of the polymers (Fig. 5.16). Thus, in the case of PET, for example, the missing OH-groups favour its water resistance and due to the high resonance energy of the aromatic ring, PET is comparatively slow to react with chemicals, for example. Further, paper causes technological challenges with regard to its roughness and absorbency. However, it is still paper with all its advantages, being bendable and foldable without to be destroyed and naturally compostable. Polyethylene naphthalate as well as polyimide show high thermal stability. Besides optical quality this is one of the major demands for e.g. display production [107, 108].

The substrates can be further improved by surface modification altering for example its tension to widen the choice of inks for this particular substrate. However, these

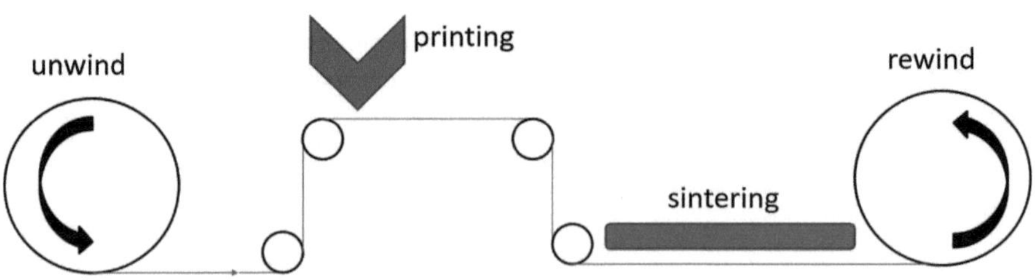

Fig. 5.15 Schematic drawing of R2R processing

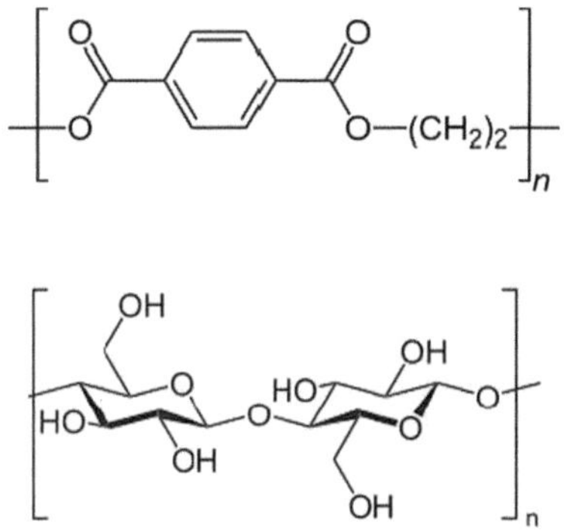

Polyethylene terephthalate (PET)

Cellulose (feedstock for paper production)

Fig. 5.16 Constitutional formulae of polyethylene terephthalate (PET) and cellulose

modifications might cause an undesired reduction in the transmission of visible light. But the key is dimensional stability. The fit of substrate, ink as well as pastes and possible coatings is crucial to the overall performance of the electronic device in the following application [109]. Flexible glass and silicon substrates—being made bendable by using thin films—are mainly preferred if high annealing/sintering temperatures are required and especially in the case of glass due to its optical properties (transparency) and chemical stability [59, 104, 109–111].

Their rigid counterparts can also be imprinted and they are used for the same characteristics as their flexible equivalents although volume manufacturing is more laborious and less financially beneficial for rigid substrates. Strictly speaking, they represent conventional substrates for printed electronics due to the handiness of printing interconnects onto existing rigid silicon substrates—like solar cells—using inexpensive, easy to install, non-contact means like inkjet or screen printing (Sect. 5.5.1.3). The demand for flexible materials grew along the way. Similarily, this holds for rigid glass substrates which are still in favour for many display manufacturers [111, 112].

5.5.1.2 Ink

Ink for printed electronics must—besides its ability to withstand tension—fulfil two major demands. On the one hand it is expected to suit its purpose as electronically functional material, but on the other hand it needs to be printable, i.e. in the liquid state. The variety of inks that follow these prerequisites can be summarized by three major groups: polymers, metal-containing ink and metal-oxides (cf. Sect. 5.1). The choice of material and thus its conductivity, functionality, suitability for customisation and finally costs are crucial aspects for the economic success of the final device [98, 110]. The biggest challenge is to convert these solid materials into liquids. A number of inks and company-owned compositions are commercially available. The manufacturer avail themselves of conventional methods such as dispersions or sol-

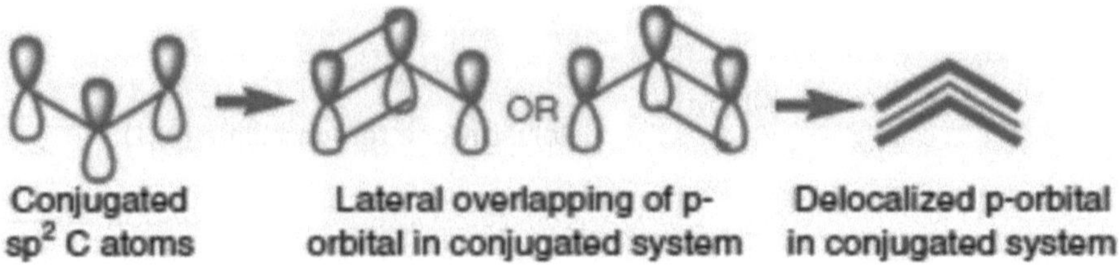

Fig. 5.17 Delocalization in conjugated system through overlapping p_z-orbitals. Reprinted from [115], Copyright (2013), with permission from John Wiley and Sons

ubles; alternatively they make use of precursors to solve NPs of otherwise almost insoluble inorganic materials like silver or silicon [98].

Through the discovery of conjugated polymers and the ability to put them into solution the first organic conducting inks were found [113]. In a simple 1-D model conduction in conjugated polymers is brought about by mutually alternating single (σ- only) and double (σ- and π-) bonds forming resonance structures of a delocalized π-system. This is possible due to the weak overlap of p_z-orbitals in π-bonds (Fig. 5.17). The phenomenon results in new optical and electronic properties among which (semi)conduction leads to their application in printed electronics. However, similarly to inorganic semiconductors, undoped they almost behave like insulators due to the lack of free electrons and the bandgap which is too large for thermal activation. Only doping—which in this case refers to redox reactions—alters the redox potential and thus leads to free electrons for charge transport [114, 115].

The possibility of numerous material combinations allows for functionality engineering especially when using polymeric inks. Inorganic ink is based on the possibility of storing of otherwise insoluble or hardly soluble inorganic (metallic or semiconducting) particles inside a soluble—mostly organic—matrix. Generally speaking, any inorganic substituent can be that way transferred into printable inks. In fact, they have certain advantages over their organic counterparts. Although their designability and suitability for functional engineering rank behind organic printed electronics they show better charge mobility and environmental/oxidation stability. Nevertheless, inorganic ink is more tedious in production and its mechanical flexibility is inferior to organic ink as well. Further, inorganic inks often require annealing at elevated temperatures [98, 116].

Due to their larger conductivity compared to conjugated polymers metal ink is often preferred for interconnects and contacts. Two major categories can be distinguished: NP ink and metal-organic decomposition inks, which covalently bond the metal atom inside an organometallic precursor. The metal may be released by heat treatment due to decomposition of the precursor [110, 115, 117, 118]. Though speaking of inorganic ink, organic constituents are yet often added to promote printability, mechanical integrity and adhesion. If used as a matrix they help reducing agglomeration by screening their inorganic core against neighbouring particles. The same job is done by soap-like particles, long alkyl chains with a polar head, in nonpolar solvents [110, 119].

Metal NPs are mostly stabilised by organic ligands in solution which have to be removed to produce a conductive pattern. Due to their size and thus due to their increased curvature with respect to larger particles NPs have merely contact to neighbouring particles or to the substrate surface. This lowers their surface energies and hence the particles's resistance to change their phase state. Consequently, they may be sintered at temperatures well below the melting point of the bulk metal [120] (melting point depression down as low as 150 °C) to remove the organic solvent and to agglomerate the NPs to highly conductive sheaths, although in practice more elevated temperatures are sometimes required for complete solvent removal. IPL is a useful tool for low-temperature sintering as its penetration depth into the material can be engineered by a suitable choice of the flash lamp spectrum saving the substrate from elevated temperatures while the organic matrix containing the metal NPs can be successfully removed [102, 110, 118].

Current research studies inexpensive alternatives to the commonly used silver, like copper, aluminium or nickel, although these less noble metals are more easily subjected to oxidation than silver. For metal-organic decomposition inks, however, the metal is already present in an oxidised state and will be reduced during sintering by the presence of a reducing atmosphere like hydrogen. Similarly, metal oxides provide easier processing due to their natural stability towards oxidation. As they are easily dissolved no stabilisers are necessary to avoid agglomeration. *However*, NP inks are more heavily loaded per ink droplet and show a lower "contact resistance" [110, 118, 121, 122].

Metal chalcogenides, such as sulfides, selenides and tellurides, are inorganic semiconductors that consist of group VI elements and usually transition metals. Due to their direct bandgap, this group is especially interesting for optoelectronic components like solar cells or displays. For printing they are synthesised as soluble precursors, dispersed quantum dots or NPs to give a print-ready ink [98]. For printed electronics, silicon plays a smaller role in comparison to its importance for conventional electronics. However, it is still utilised for solar cell, sensor and TFT production. It may be suspended in solution as NPs or it may be sprayed onto the substrate in an organic precursor which is removed by pyrolysis and FLA [123, 124]. For insulation and charge storage dielectric materials are not only indispensable to silicon-wafer-based microelectronics, it is also not straightforward to put them into solution. Only a few inks have been produced so far [98].

Also carbon and graphene [125, 126] are known to printed electronics, although their commercial importance is still limited. For carbon nanotubes this is mainly due to the difficulty in obtaining considerable amounts with suitable electronic and chemical properties. Graphene, on the other hand, has not shown outstanding performance as conductive ink, but is used for transparent conductive oxides (cf. Sect. 5.1) [98, 127]. The two-fold demand of printable, yet electronically functional ink induces certain challenges. For example, larger polymer chains usually mean increased conductivity, however, they are also usually more hardly soluble and increase solvent viscosity which in turn reduces their printability. The situation is similar with surface tension and concentration. Typical additives as known from conventional printing cannot be readily used as they may affect the ink's electrical functionality [128, 129].

5.5.1.3 Processing

The production of printed electronics (Fig. 5.18) comprises two major steps: the printing itself and the material-dependent sintering or annealing. If necessary, a tape coating is applied for encapsulation of sensitive electronics [92, 130]. However, the variety of production lines starts with the right choice of printer. According to the material, purpose and costs the printing method may vary. There are web-fed printers for mass production or sheet-fed ones for more sophisticated tasks or rigid substrates. Volume manufacturing is brought about by R2R processing. This saves costs by reducing production time due to a higher throughput and by creating less waste. Existing printing technology can be transferred, namely gravure, flexographic as well as offset printing.

In gravure printing, cavities in a gravure cylinder take on the ink from a reservoir (Fig. 5.19a) and similarly, elevations on the plate cylinder of the flexographic printer (Fig. 5.19b) transfer the ink to the substrate. During offset printing, however, plate and impression cylinder are separated by a blanket cylinder (Fig. 5.19c) to prevent damage to the second. The plate cylinder is structured by hydrophobic (print elements) and hydrophilic surfactants. It is further wetted first with water and subsequently with ink during each turn. As the latter is usually in an organic matrix it is repelled by water and only coats the hydrophobic print elements. The blanket cylinder serves as offset between the (printing) plate cylinder and the substrate to conserve the latter. An impression cylinder provides the support during printing [98, 132].

Yet, conventional sheet-based printing technologies are still in use as they are required for rigid substrates. They include inkjet printing and screen printing. Inkjet printing (Fig. 5.20a) may use pressure waves to destabilize a steady *inkjet* into droplets of identical size. In continuous mode the latter are directed onto the substrate surface by an electrical potential and are deflected according of the print pattern by electrostatic plates just like electrons in a cathode ray tube. Screen printed substrates are tinted with ink by a squeegee through a screen mesh taking the ink to its desired destination (Fig. 5.20b) [98, 132].

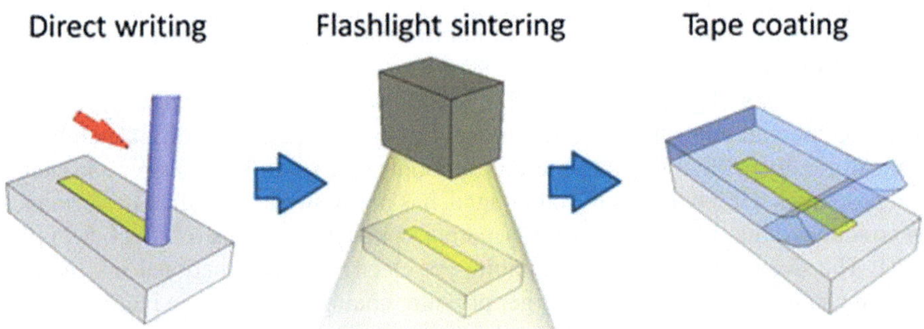

Fig. 5.18 Processing steps for printed electronics on the example of flashlight sintering. Reprinted with permission from [131]. Copyright (2014) American Chemical Society

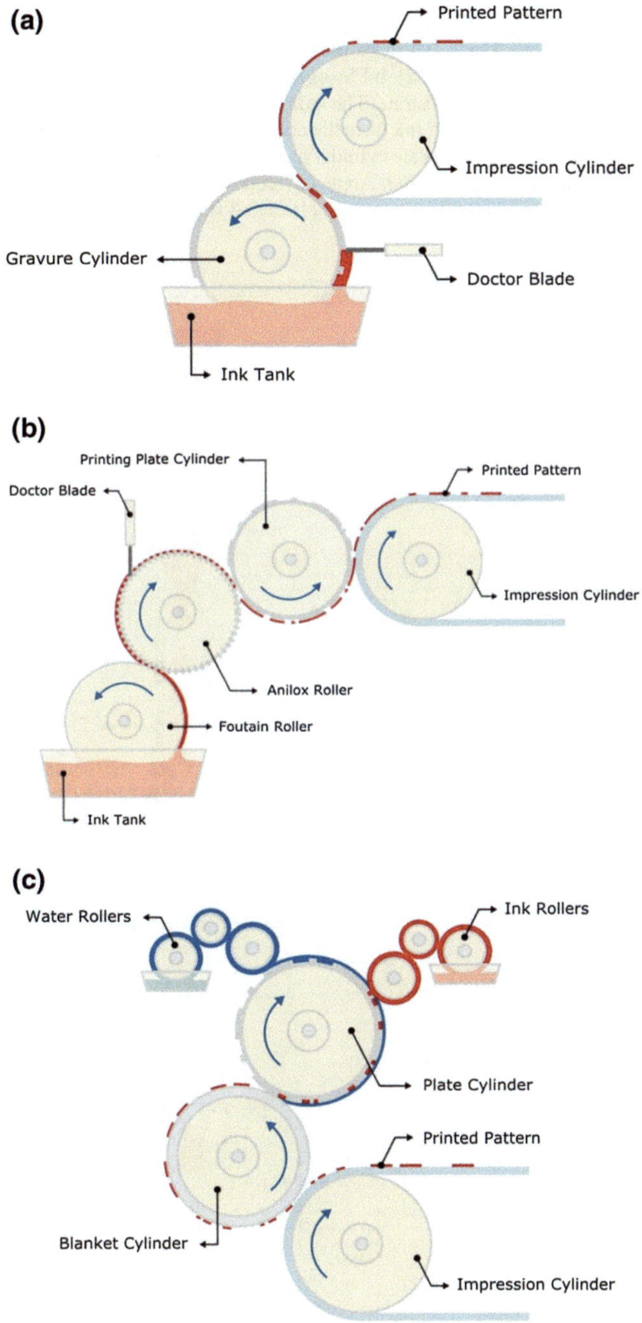
(a)
Printed Pattern
Impression Cylinder
Gravure Cylinder
Doctor Blade
Ink Tank
(b)
Printing Plate Cylinder
Doctor Blade
Printed Pattern
Impression Cylinder
Anilox Roller
Foutain Roller
Ink Tank
(c)
Water Rollers
Ink Rollers
Plate Cylinder
Printed Pattern
Blanket Cylinder
Impression Cylinder

◄**Fig. 5.19** **a** Gravure printing—An engraved cylinder takes up the ink from a tank and the residual ink is swiped off by a doctor blade. Through pressure applied by a second cylinder the engraved pattern is printed onto the substrate [132]. **b** Flexographic printing—In contrast to gravure printing elevations transfer the ink onto the substrate. The anilox roller ensures a constant uptake of ink from the tank and the doctor blade wipes off ink from the nose-pieces [132]. **c** Offset printing—Solubility in water determines the printing. The plate cylinder has hydrophilic and hydrophobic areas (printing pattern) which either attract or repel water. Areas covered with water cannot take up ink anymore. The blanket cylinder serves as an offset to ensure clean and sharp printing [132]

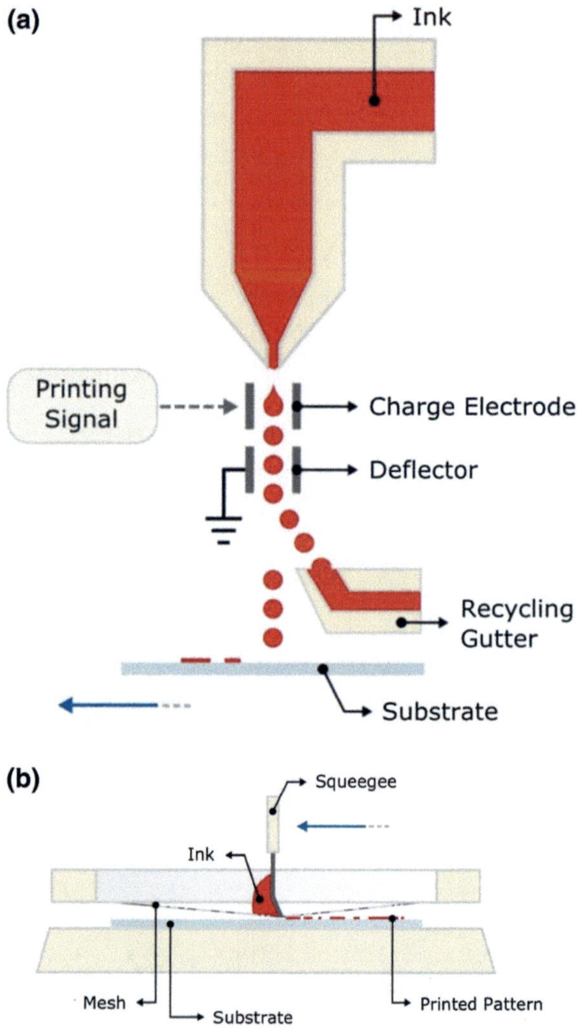

Fig. 5.20 **a** Inkjet printing in continuous mode showing the formation of droplets from an inkjet and their deflection onto the paper. Drop-on-demand is also possible [132]. **b** Screen printing—the pattern to be printed is transferred to the paper via a partially ink permeable screen. Residual ink is removed by a blade [132]

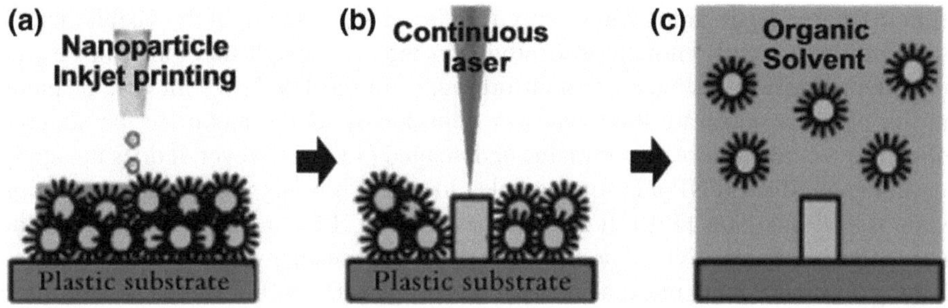

Fig. 5.21 Sketch showing selective laser sintering of NPs (**b**) after inkjet printing (**a**). Unsintered remains have to be removed by a suitable solvent (c) [118]. Copyright Kamyshny et al.; Licensee Bentham Open

After printing an annealing or sintering step is required to form conductive precipitates and to evaporate any organic solvent thus raising conductivity. Often this sintering step requires temperatures above 250 °C for a period of time of more than 30 min [110]. Although NP ink has a significantly reduced melting point due to its large surface-to-volume ratio, especially most (cost-efficient) polymers cannot handle such elevated temperatures for the above mentioned duration as a consequence of their low glass-transition temperature [107]. Therefore research into alternative sintering approaches is omnipresent. There are various suggestions to protect temperature-sensitive substrates by considering the constant demand for cost reduction in printed electronics [133].

One of them is LA by an argon ion laser that follows the printed pattern. On the one hand this reduces the thermal load of the substrate next to the laser line, but on the other hand the local thermal budget is very large and due to the Gaussian nature of the laser spot unsintered material remains along the track which needs to be removed later (Fig. 5.21). The possible feed rate of the laser is quite small (0.2 mm/s) compared to other laser applications, but increasing the speed of the laser movement has been found to reduce the achievable conductivity. Resistivities of six times that of bulk silver have been gained. Yet, it has been found to be a complementary to FLA annealed samples [118, 134–136].

A faster alternative to LA is microwave sintering. Through polarization thus creating dielectric losses materials like ceramics can be successfully sintered. This should, however, not work for metals as they behave as reflectors towards microwave radiation. However, it works for metal particles to which the microwaves couple inductively. Moreover, this coupling does not affect the polymer substrate which is transparent to microwave radiation. This mechanism can achieve processing times of a few minutes though the achievable conductivity is only a small percentage of that of the bulk material.

Even faster while at the same time effective but gentle, is the use of IPL. Moreover, it stands out for its possibility to selectively treat ink and substrate and to combine it with volume manufacturing processing [133]. IPL or FLA makes use of radiation from discharge flash lamps whose spectrum is at its maximum in the visible, which is where there is still enough absorptivity in metals, though this may not be true for polymer solvents, which often absorb better in the UV [137]. In case of metal ink on polymer substrate this comes in quite handy as the metal can be sintered efficiently while the substrate remains undamaged [138]. However, it does not apply for all metal NPs. Cu NP ink, for example, does not show significant UV absorption while Ag NP ink does [139]. It has been shown that IPL can successfully sinter the latter to give low resistivities on the order of three times the value of bulk silver and thereby outrun thermal, plasma and laser sintering [140–142]. So here UV annealing serves two purposes: evaporation of the polymeric fluid and sintering of the Ag NPs. But even for the yet cheaper though less noble Cu, which easily forms an oxide shell at ambient conditions, UV assisted FLA has shown to significantly reduce the resistivity of the sintered metal film. Moreover, its higher melting point in comparison to Ag demands for a higher sintering temperature or/and sintering time which again helps oxidation. Still the oxide can merely be broken up by thermal annealing only. Therefore, a polymer binder was used for the functionalisation of the oxide, which was effectively decomposed by the UV light [102, 143–145]. Other groups put the Cu into a complex or into a dispersion to save it from oxidation [122, 146, 147].

Hwang et al. first dried the samples at two temperatures 80 and 120 °C before flashing them [143]. Subsequently, they investigated the effect of NIR and UV assisted sintering and found a considerable decrease in resistivity for UV assisted FLA at 12,5 J/cm^2 and 30 mW/cm^2 UV irradiation from 100 to 7 $\mu\Omega$ cm due to effective decomposition of the binder which further promotes Cu oxide reduction by the decomposition products. For larger UV power, however, the conductivity declined again because the Cu film got damaged by the accompanying higher temperature during sintering. Furthermore, SEM imaging revealed smaller pores for UV assistance than for FLA only. This hindered subsequent oxidation of the metal ink when exposed to atmosphere. The sole UV share to the resistivity decrease was about 8%, NIR assistance could decrease the resistivity of the printed pattern to about 2% due to higher sintering temperatures which gave rise to a denser metal film and thus larger conductivity. Literature does not agree on certain flash parameters, but consensus prevails that the resistivity drops significantly with increasing deposited energy density up to a limit above which the printed pattern is either destroyed again due to excessive heating or may peel off because of too fast evaporation of the binder/solvent [133, 143]. Also, damage to the substrate was not found, neither for PET, polyimide or glass fibre although the gained resistivities were found to vary for different substrates depending on their individual thermal conductivity. Combination of UV and NIR assistance, however, led to a significant increase in resistivity. SEM images showed large pores in the metal film of the so-treated samples, which can explain the decline in conductivity [143].

Han et al. investigated multiple pulse sintering of Cu NPs using flash lamps in order to reduce the thermal load on the substrate while maintaining the required energy density to remove the organic matrix and sinter the printed metal film [148]. Besides pulse energy, pulse length and peak intensity they also studied the influence of the number of pulses on the sintering efficiency and found that it correlates to the sintering depth and that a maximum number exists above which the substrate may be damaged due to overheating [136]. However, heat is transferred to the substrate by the Cu NPs themselves, not by the flash light [148]. Multiple flashes may further lead to a swelling of the printed film, which was found for Ag NPs. The first pulse causes sintering of a thin Ag film and the layers below are sintered primarily by thermal conduction. The sintered surface, however, acts as a sealing to the evaporating decomposition products of the organic matrix below, which causes it to lift off [149].

For metal-organic decomposition inks UV curing, often accompanied by conventional thermal annealing, is usually applied due to the high UV absorption of the ink. This should reduce both annealing temperature and time through efficient reduction of the metal, but for complete removal of any organic residues and highest conductivity practice has shown that more elevated temperatures and prolonged sintering times are often needed [135, 138]. Straight UV curing, which is also known to printed electronics sintering, however, is—compared to photonic curing, which takes place in the millisecond regime—very slow on the order of a few seconds and thus allows for significant UV energy absorption of the polymer substrate which may not withstand the heating. FLA, however, does allow for both an increase in UV output if the current density through the lamp is raised [150, 151] as well as for thermal annealing in a short time.

Metal chalcogenides nanocrystals were processed with 160 μs flash pulse at an energy density of 2.2 J/cm^{-2}. The resulting peak external quantum efficiencies were found to be larger than 120%, which the authors contribute to multiexciton extraction. However, removal of the organic ligands led to the formation of traps which lowered the device performance [39]. FLA has also shown that it can be successfully applied to TCOs on flexible thin glass substrates. It was found that the resistivity could be reduced by almost 70% and the transmittance of the ITO films was also raised by a few % in the visible range. Comparison to conventional glass substrates showed that FLA allows for faster processing while the electrical and optical properties remain similar (Sect. 5.3) [59, 104].

Though in distinct variations even RT sintering is known. However, this requires chemical agents which may be not desired with respect to purity requirements [118, 135]. Further alternatives leave similar unsatisfactory results. The low pressure plasma used for microwave sintering, for example, is not suitable for metal ink films beyond a few μm due to a small penetration depth [118]. A clear gradient of sintered to unsintered material can be found and the resistivity remains larger than that of the bulk [137, 152]. Although a combination of the two predicts more promising results [153], photonic sintering is not only faster, it also does not require a non-oxidative atmosphere.

Indeed, electrical sintering achieved by Joule heating through an electrical current that is applied across the printed pattern, can be very fast on the order of μs, while saving the substrate from excessive heat. But for one, there is a considerably larger activation time (few ms) and movement of a possibly small electrode over comparably large substrates (eg. for volume processing) may limit the overall processing time to conventional sintering times. Further, a substantial conductivity of the ink is required *before* annealing can occur. Usually this needs to be achieved by preliminary thermal heating. Spark plasma sintering or pulsed electric current sintering—a combined method of Plasma and Joule heating—denotes a high temperature sintering although processing times are kept to a minimum. A substantial initial conductivity, however, is likewise required [118, 135, 154, 155].

So FLA shows yet again its suitability to printed electronics as a powerful tool to reduce both sintering time and temperature load of the substrate (Sect. 3.2). Its use is mainly confined to metallic and semiconducting ink where it shows its possibility to produce crystalline thin films on temperature-sensitive substrates such as polymers or paper while achieving conductivities of up to one third of the bulk material [102, 122, 133, 156].

Although UV curing is also known for organic ink IPL often does not play a significant role, which is most likely due to its mostly diminishing absorption in the visible part of the spectrum, where the main energy is emitted by the flash lamps. Indirect heating of added metal NPs is applied instead. However, it has been shown that considerable absorption exists in the visible for conjugated polymers with large polymer stiffness as well as for those containing chromophores thus leading to split-second crystallization [33, 157, 158]. More traditional means are, besides heat, the removal of the solvent by oxidation, penetration into the substrate or evaporation [159].

5.5.1.4 Electrical Response

Conductivity—or more specifically—charge carrier mobility of conventional electronics cannot be reproduced by printed electronics. Still its suitability for R2R processing helps the latter to its popularity. Likewise, an annealing concept is needed that supports these advantages of printed electronics while at the same time yields minimum sheet resistance values. FLA represents an easy to use and to handle process while providing comparable conductivity results to other sintering techniques in this area [160]. Regardless of the method of sintering the comparison of commercially available inks suggests that the printing technique itself has an impact on the resulting resistivity of the ink. For silver on polymer substrates, for example, inkjet printing was found to give lower resistivities than screen printing. This may be a function of line spacing during printing, droplet size or feed rate of the printer [160–162].

Surprisingly, the method of annealing itself—at least for the three major groups LA, FA and FLA—may not produce a significant difference on the conductivity of the printed structures. Most of them give at least three times smaller values when compared to the bulk metal. Rather, the choice of a sintering process is more crucial with respect to substrate protection and R2R suitability [136, 140]. Important process parameters for conductivity tuning during FLA include energy density and pulse length. Figure 5.22 shows that an increase in energy density, for example, leads to better physical contact between the NPs (Fig. 5.22 left) and thus resulting in a decrease of sheet resistance (Fig. 5.22 right). The reason for the increase in resistance first

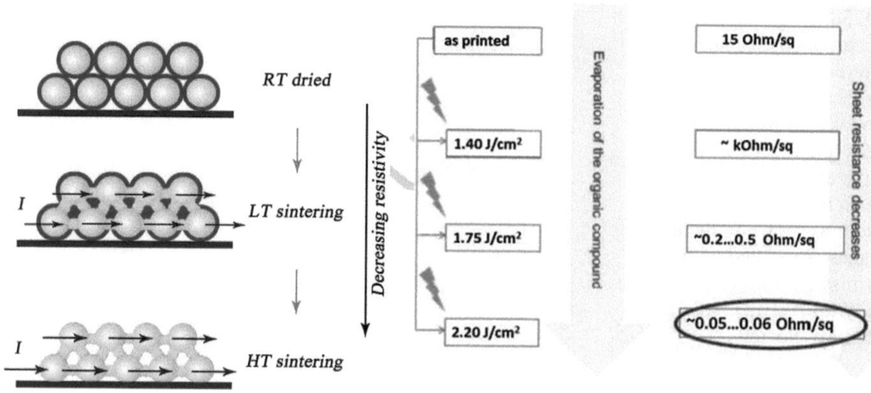

Fig. 5.22 Left: Sketch showing that increasing the temperature during annealing decreases the conductivity (RT = room temperature; LT = low temperature, HT = high temperature); Republished with permission of RSC Pub. from [163]; permission conveyed through Copyright Clearance Center, Inc. Right: Sketch showing the correlation between energy density and sheet resistance (acc. to [100])

after drying can be found when looking at the solvent. Indeed, it adds to the sample conductivity, but the biofilm formed after drying does not.

5.5.2 *Ink-Environment Interaction*

5.5.2.1 Photonic Sintering

The interaction between the incident light and the printed electronics during FLA is obvious. Regardless of the different absorption spectra of different metals, the organic matrix in which inorganic NPs may be dispersed also needs to be considered for maximum outcome after sintering. It was found that the carbon chain lengths of copper salt ligands may significantly change the ink's absorption behaviour thus changing ink colour (Fig. 5.23). Moreover, long chains have a negative effect on the resistivity after annealing [164]. A similar significant influence, however, is carried by the substrate. The optical *and* thermal properties of the substrate, namely reflectivity, absorptivity and thermal conductivity, may drastically change the electrical performance of the printed top, but can be likewise used to tune them [165].

As an example of this, photonic sintering of various silver patterns on PET led to lower sheet resistance than on polyimide foil, yet the latter did not show any cracks or other defects as they could be found on the PET foil (next section). Different optical properties of the substrate will affect the light energy that is absorbed and thus influence the resulting heat dissipation throughout the material. Further, different

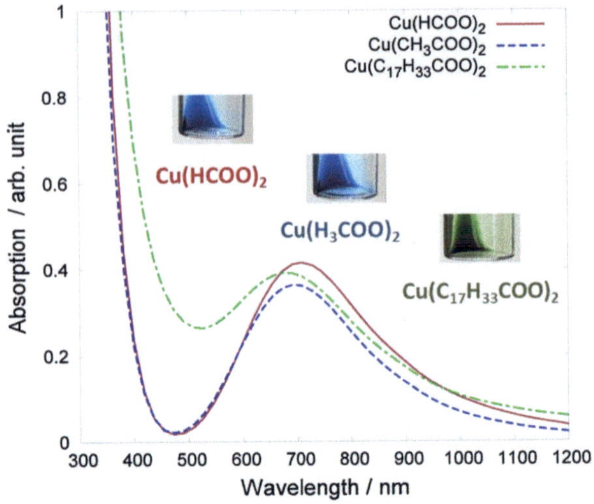

Fig. 5.23 Absorption spectra of copper salt with different ligand sizes. The number of carbon atoms increases from red to blue to green and illustrates the shift of the absorption spectrum which in turn changes the ink colour as shown in the inset images. Reprinted with permission from [164]. Copyright (2013) American Chemical Society

material properties, like a different glass-transition temperature, will influence the result of sintering, namely the resistivity of the printed patterns [133, 166].

5.5.2.2 Oxidative Stability and Thermal Stability

Though silver ink is still widely spread as ink for printed electronics, research into cheaper alternatives, meaning high conductivity at low cost like in case of copper, aluminium and nickel, is ongoing. Silver is already more than 130 times more expensive than copper, and the increasing demand will further push this tendency. However, less noble metals are more subject to oxidation causing new issues when processing at ambient temperature and decreasing the device performance or product quality.

Hybrid-type conductive inks aim to solve this dilemma. Cu/Ag hybrid nanoinks, for example, are both low cost due to the substantial amount of Cu, but highly conductive (resistivity ~4 μΩ cm) exhibiting a low oxidation affinity [133, 167]. Further research into this issue suggests introducing protective layers which can be made of various materials depending on the application. These include carbon, graphene, silica, metals or a combination of surfactants and polymers. These layers not only keep oxygen away, but also water, for a reasonable lifetime and ensure mechanical protection. The conductivity of the printed electronics is said not to be affected [103, 168]. The encapsulant usually consists of at least a similar material as that what is to be protected. Therefore, it is dispensed in a liquid and then deposited onto the material. Contact holes are etched into the encapsulation layer afterwards [92]. Finally, the right choice of FLA conditions can be used to reduce CuO nanocrystals via Cu_2O to metallic Cu forming a fully sintered, bulk-like Cu coating [144, 169].

In the case of printed solar cells oxygen may diffuse into the active layer which may have drastic consequences for the device stability. A solution presented by

Hau et al. [170] proposes an inverted geometry, namely a swap of cathode and anode. The otherwise top anode made of a TCO is now turned into a cathode by a buffer layer even though the transmissivity is thus reduced. The metal back contact which normally consists of a good conductor, like copper or aluminium [171], is now made of a noble metal like gold or silver and thus serves as an anode. Low temperature sintering which may be achieved by reducing organic additives further helps oxidation stability. Moreover, the latter is closely related to thermal stability. It may not only affect the ink's conductivity but also damage the substrate. Research into this topic of different silver patterns has shown a dependence of the critical curing temperature on organic additives, but unchanged resistance independent of the temperature and of the individual line thickness for most silver tracks. Although in some cases ink formulations showed a loss of conductivity when heated it has been assumed that this was due to cracks and agglomerations during heating [107, 110, 142, 172].

The oxygen affinity is also larger for metal NPs than for the bulk material due to its large surface-to-volume ratio although a minimal size seems to exists below which this relationship turns around. Otherwise if the metallic ink is already in an oxidised state (Sect. 5.5.1.2) additional oxidation during storage, for example, is hindered. As seen before the organic matrix or certain protective agents can have yet again a positive effect on the oxidation stability of printed structures. However, some of them may reduce the overall conductivity of the ink. Kanninen et al. studied Cu NPs in an organic ligand complex [173]. Different stabilizing ligands were studied and an increasing oxidation resistance for certain ligands depending on their chain length and metal-ligand-ratio was found [103, 173–175].

Of course, oxidation is not limited to metals. Polymers are likewise affected, especially at elevated temperatures. Higher thermal stability of polymers is though enabled through interchain interactions. The stronger those interactions the better the polymer can withstand a temperature raise and subsequent degradation by thermolysis [176]. The deformability of polymers may be—among others—explained by the so-called glass transition temperature below which the polymer appears to be brittle in its glassy state whereas above it enters its rubber state which exhibits soft and flexible behaviour [177]. Accordingly, cracks as they occur in [166] may be explained by crossing the glass transition temperature of the polymer which may be the reason why cracks have been found during annealing of PET, but not when polyimide was sintered probably due to its three times higher glass transition temperature.

5.5.3 *Ink-Substrate Interaction*

5.5.3.1 Compatibility

Materials used for printed electronics show very different Young's moduli, which is a measure for the ratio of stress to strain and thus a material property for its stiffness. In fact, values range from a few Pa for liquid metals to several 100 GPa for

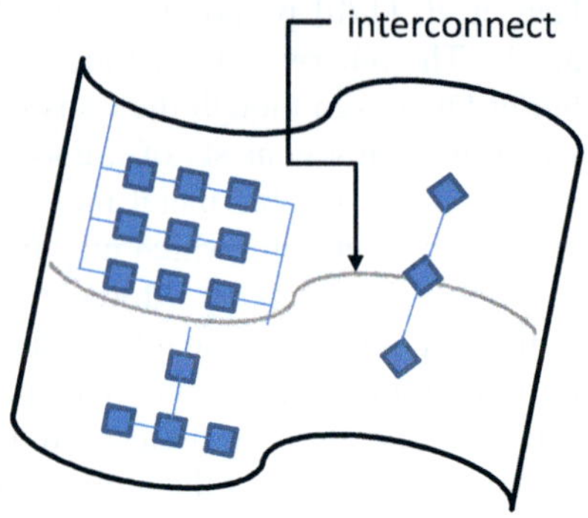

Fig. 5.24 Sketch of a printed electrical device showing how active sensors with brittle materials are placed on bendable substrates to contain its overall flexibility. After [92]

silicon, gold or graphene. Surprisingly, combining these materials, for example metal NPs in a polymer matrix, brings about new material properties through mechanical engineering [92, 178].

Similarly opposing are the demands for substrates and inks. The comparatively brittle optoelectronic devices are arranged in islands on the flexible substrate. Efforts to replace these brittle structures have already partly proven to be a success, e.g. for ITO free polymer solar cells [179]. Those parts of the substrate that are exposed to the environment and the interconnects remain stretchable through the choice of material, the thickness of the layer, special architecture (e.g. nets) and degrees of freedom to move. The islands of sensitive and non-stretchable electronics, however, are so thin such that on a local scale no bending occurs, but on global scale the whole device remains flexible (Fig. 5.24) [92, 178].

5.5.3.2 Wettability

Besides ink composition (Sect. 5.5.1.2) ink deposition is a further constraint put up towards the choice of material which in turn is a function of wettability, which describes the ability of a liquid to contact the surface of a solid due to adhesion. It is related to the contact angle which is the angle between the liquid-vapour and the liquid-solid interface. The larger the contact angle the less efficient the wetting of the liquid on the solid surface [180]. The ink's wettability has a direct influence on line resolution, thus printing quality and edge definition as well as film formation and it is affected by the chemical polarity and viscosity of the ink, the porosity of the substrate and surface energy [109].

Thus the ink's polarity shows a significant difference in terms of wetting between PET and paper [181]. This relationship has been exploited to produce printed structures that are even smaller than the individual ink droplet during inkjet printing [182]. These so-called superhydrophilic places on the substrate could be wetted by the ink with a wetting angle smaller than 5°. These results were achieved by surface modification prior to printing. The substrate was coated with hydrophilic and hydrophobic thin films making parts of the surface water-repellent and others water-adhesive [182]. Surface energy is a further factor to engineer the ink's wettability on the substrate. They found that reducing the surface energy of the substrate in turn reduces the wettability of the ink, which results in improved line resolution due to a larger contact angle. If the critical surface tension of a substrate is lower than the surface energies of the inks, good resolution can be achieved even with low viscosity inks which is even better for printing [109].

Besides line resolution wettability influences the electronic properties themselves. Researchers found a direct relationship between the density of semiconducting single-walled carbon nanotubes, their electronic properties and the contact angle of water on a SiO_2/Si substrate. The larger the wetting of the polar solvent the better the uniformity, the carrier mobility, the higher the on/off ratio, the lower the operation voltage and the smaller the hysteresis of a TFT consisting of printed Ag top-gate electrodes and Al_2O_3 as the dielectric [183]. Pulsed light sintering of the substrates has also shown to improve wetting [184]. The comparison of these findings with Fig. 5.22 suggests that IPL indeed helps advanced electrical properties.

References

1. J. Robertson, High dielectric constant gate oxides for metal oxide Si transistors. Rep. Prog. Phys. **69**(2), 327–396 (2006). https://doi.org/10.1088/0034-4885/69/2/r02
2. J.H. Choi, Y. Mao, J.P. Chang, Development of hafnium based high-k materials—a review. Mater. Sci. Eng. R Rep. **72**(6), 97–136 (2011). https://doi.org/10.1016/j.mser.2010.12.001
3. S. Kamiyama, T. Miura, Y. Nara, Ultrathin HfO_2 films treated by xenon flash lamp annealing for use as transistor gate dielectric replacements. Electrochem. Solid-State Lett. **8**(12), G367 (2005). https://doi.org/10.1149/1.2119584
4. H. Kawarazaki, A. Ueda, M. Furukawa, T. Aoyama, S. Kato, I. Kobayashi, new flash lamp annealing tool equipped with an ambient control feature suitable for high-k gate stack anneals, in *21st International Conference Ion Implantation Technology* (2016)
5. T. Matsuki et al., Area-selective post-deposition annealing process using flash lamp and Si photoenergy absorber for metal/high-kgate metal–insulator–semiconductor field-effect transistors with NiSi source/drain. Jpn. J. Appl. Phys. **45**(4B), 2939–2944 (2006). https://doi.org/10.1143/jjap.45.2939
6. F. Ootsuka et al., Ultralow-thermal-budget CMOS process using flash-lamp annealing for 45 nm metal/high-k FETs. IEEE Trans. Electron Dev. **55**(4), 1042–1049 (2008). https://doi.org/10.1109/ted.2008.917546
7. J. Lehmann, R. Hübner, J.V. Borany, W. Skorupa, T. Mikolajick, A. Schäfer, J. Schubert, S. Mantl, Millisecond flash lamp annealing for $LaLuO_3$ and $LaScO_3$ high-k dielectrics. Microelectron. Eng. **109**, 381–384 (2013). https://doi.org/10.1016/j.mee.2013.04.021
8. J. Lehmann, N. Shevchenko, A. Mücklich, J.V. Borany, W. Skorupa, J. Schubert, J.M.J. Lopez, S. Mantl, Millisecond flash-lamp annealing of $LaLuO_3$. Microelectr. Eng. **88**(7), 1346–1348 (2011), https://doi.org/10.1016/j.mee.2011.03.126
9. A. Müller, M. Lorenz, K. Brachwitz, J. Lenzner, K. Mittwoch, W. Skorupa, M. Grundmann, T. Höche, Fresnoite thin films grown by pulsed laser deposition: photoluminescence and laser crystallization. CrystEngComm **13**(21), 6377 (2011). https://doi.org/10.1039/c1ce05265a
10. K. Yamakawa, K. Imai, O. Arisumi, T. Arikado, M. Yoshioka, T. Owada, K. Okumura, Novel Pb(Ti, Zr)O_3(PZT) crystallization technique using flash lamp for ferroelectric RAM (FeRAM) embedded LSIs and one transistor type FeRAM devices. Jpn. J. Appl. Phys. **41**(4B), 2630–2634 (2002). https://doi.org/10.1143/jjap.41.2630

11. J.S. Park, D.J. Kim, W.H. Chung, Y. Lim, H.S. Kim, Y.B. Kim, Rapid, cool sintering of wet processed yttria-stabilized zirconia ceramic electrolyte thin films. Sci Rep. **7**(1), 12458 (2017). https://doi.org/10.1038/s41598-017-12438-9
12. A.N. Nazarov et al., The effect of rare-earth clustering on charge trapping and electroluminescence in rare-earth implanted metal-oxide-semiconductor light-emitting devices. J. Appl. Phys. **107**(12), 123112 (2010). https://doi.org/10.1063/1.3436591
13. L. Rebohle, W. Skorupa, *Rare-Earth Implanted MOS Devices for Silicon Photonics: Microstructural, Electrical and Optoelectronic Properties*. Springer Series in Materials Science 142 (Springer, Berlin Heidelberg, 2012)
14. L. Rebohle, J. Lehmann, S. Prucnal, A. Kanjilal, A. Nazarov, I. Tyagulskii, W. Skorupa, M. Helm, Blue and red electroluminescence of Europium-implanted metal-oxide-semiconductor structures as a probe for the dynamics of microstructure. Appl. Phys. Lett. **93**(7), 071908 (2008). https://doi.org/10.1063/1.2964176
15. L. Rebohle et al., Anomalous wear-out phenomena of europium-implanted light emitters based on a metal-oxide-semiconductor structure. J. Appl. Phys. **106**(12), 123103 (2009). https://doi.org/10.1063/1.3272781
16. A.N. Nazarov, I.P. Tyagulskyy, S.I. Tyagulskiy, L. Rebohle, W. Skorupa, J. Biskupek, U. Kaiser, Correlation between electroluminescence and charge trapping in multi-color Eu implanted Si-based light-emitting diodes. Physica E **41**(6), 902–906 (2009). https://doi.org/10.1016/j.physe.2008.08.025
17. S. Tyagulskiy, I. Tyagulskyy, A. Nazarov, V. Lysenko, L. Rebohle, J. Lehmann, W. Skorupa, Electroluminescence, charge trapping and quenching in Eu implantes SiO_2–Si structures. Microelectron. Eng. **86**, 1954–1956 (2009)
18. S. Tiagulskyi, A. Nazarov, I. Tyagulskii, V. Lysenko, L. Rebohle, J. Lehmann, W. Skorupa, Shell model for REO_x nanoclusters in amorphous SiO_2: charge trapping and electroluminescence quenching, Phys. Status Solidi C **9**(6), 1468–1470 (2012). https://doi.org/10.1002/pssc.201100780
19. S. Prucnal, J.M. Sun, A. Mücklich, W. Skorupa, Flash lamp annealing vs rapid thermal and furnace annealing for optimized metal-oxide-silicon-based light-emitting diodes. Electrochem. Solid-State Lett. **10**(2), H50 (2007). https://doi.org/10.1149/1.2404225
20. S. Prucnal, J.M. Sun, L. Rebohle, W. Skorupa, Fourfold increase of the ultraviolet (314 nm) electroluminescence from SiO2: Gd layers by fluorine coimplantation and flash lamp annealing. Appl. Phys. Lett. **91**(18), 181107 (2007). https://doi.org/10.1063/1.2803855
21. S. Prucnal, J.M. Sun, H. Reuther, W. Skorupa, C. Buchal, Strong improvement of the electroluminescence stability of SiO2: Gd layers by potassium co-implantation. Electrochem. Solid-State Lett. **10**(2), J30 (2007). https://doi.org/10.1149/1.2404293
22. S. Prucnal, J.M. Sun, H. Reuther, C. Buchal, J. Żuk, W. Skorupa, Electronegativity and point defect formation in the ion implanted SiO_2 layers. Vacuum **81**(10), 1296–1300 (2007). https://doi.org/10.1016/j.vacuum.2007.01.032
23. S. Prucnal, L. Rebohle, A.N. Nazarov, I.N. Osiyuk, I.P. Tjagulskii, W. Skorupa, Reactivation of damaged rare earth luminescence centers in ion-implanted metal–oxide–silicon light emitting devices. Appl. Phys. B **91**(1), 123–126 (2008). https://doi.org/10.1007/s00340-008-2948-z
24. Studie: Stromgestehungskosten erneuerbare Energien—März 2018, Fraunhofer ISE, 2018. Accessed 2 April 2018
25. https://isfh.de/en/26-1-record-efficiency-for-p-type-crystalline-si-solar-cells/. Accessed 2 April 2018
26. K. Yoshikawa et al., Silicon heterojunction solar cell with interdigitated back contacts for a photoconversion efficiency over 26%. Nat. Energy **2**, 17032 (2017)
27. Functional structure design of new high-performance materials via atomic design and defect engineering (ADDE). ed. by D. Rafaja, Saxonia Standortentwicklungs- und -verwaltungsgesellschaft mbH (2015)
28. S. Prucnal, B. Abendroth, K. Krockert, K. Koenig, D. Henke, A. Kolitsch, H.J. Moeller, W. Skorupa, Millisecond annealing for advanced doping of dirty-silicon solar cells. J. Appl. Phys. **111**, 123104 (2012)

29. F.L. Bregolin et al., Hydrogen engineering via plasma immersion ion implantation and flash lamp annealing in silicon-based solar cell substrates. J. Appl. Physics **115**(6), 064505 (2014). https://doi.org/10.1063/1.4865737
30. S. Prucnal et al., Influence of flash lamp annealing on the optical properties of CIGS layer. Acta Phys. Pol. **125**, 1404–1407 (2014)
31. S.R. Dhage, H. Thomas Hahn, Rapid treatment of CIGS particles by intense pulsed light. J. Phys. Chem. Sol. **71**, 1480–1483 (2010)
32. B.A. Williams, M.A. Smeaton, C.S. Holgate, N.D. Trejo, L.F. Francis, E.S. Aydil, Intense pulsed light annealing of copper zinc tin sulfide nanocrystal coatings. J. Vac. Sci. Technol. A **34**, 051204 (2016)
33. H.Y. Yang, J.-M. Hong, T.W. Kim, Y.-W. Song, W.K. Choi, J.A. Lim, Split-second nanostructure control of a polymer: fullerene photoactive layer using intensely pulsed white light for highly efficient production of polymer solar cells. ACS Appl. Mater. Interfaces **6**, 1495–1501 (2014)
34. B.W. Lavery, S. Kumari, H. Konermann, G.L. Draper, J. Spurgeon, T. Druffel, Intense pulsed light sintering of $CH_3NH_3PbI_3$ solar cells. ACS Appl. Mater. Interfaces **8**, 8419–8426 (2016)
35. T. Druffel, R. Dharmadasa, B.W. Lavery, K. Ankireddy, Intense pulsed light processing for photovoltaic manufacturing. Solar Energy Mater. Sol. Cells **174**, 359–369 (2018)
36. M.H. Zhu et al., Millisecond-pulsed photonically-annealed tin oxide electron transport layers for efficient perovskite solar cells. J. Mater. Chem. A **5**(46), 24110–24115 (2017). https://doi.org/10.1039/c7ta07969a
37. K. Ankireddy, A.H. Ghahremani, B. Martin, G. Gupta, T. Druffel, Rapid thermal annealing of $CH_3NH_3PbI_3$ perovskite thin films by intense pulsed light with aid of diiodomethane additive. J. Mater. Chem. A **6**(20), 9378–9383 (2018). https://doi.org/10.1039/c8ta01237g
38. Q. Fu, X. Tang, B. Huang, T. Hu, L. Tan, L. Chen, Y. Chen, Recent progress on the long-term stability of perovskite solar cells. Adv. Sci. **5**, 1700387 (2018)
39. C.J. Stolle et al., Multiexciton solar cells of $CuInSe_2$ nanocrystals. J. Phys. Chem. Lett. **5**(2), 304–309 (2014). https://doi.org/10.1021/jz402596v
40. S. Adachi, S. Okamura, Amorphous indium gallium zinc oxide thin-film transistors fabricated by direct transfer printing. Appl. Phys. Express **3**, 104101 (2010)
41. B.S. Kang, H.T. Wang, L.C. Tien, F. Ren, B.P. Gila, D.P. Norton, C.R. Abernathy, J. Lin, S.J. Pearton, Wide bandgap semiconductor nanorod and thin film gas sensors. Sensors **6**, 643 (2006)
42. A. Kołodziejczak-Radzimska, T. Jesionowski, Zinc oxide—from synthesis to application: a review. Materials **7**, 2833–2881 (2014)
43. A. McLaren, T. Valdes-Solis, G. Li, S.C. Tsang, Shape and size effects of ZnO nanocrystals on photocatalytic activity. J. Am. Chem. Soc. **131**, 12540–12541 (2009)
44. X. Zhang, J. Qin, Yanan Xue, P. Yu, B. Zhang, L. Wang, R. Liuc, Effect of aspect ratio and surface defects on the photocatalytic activity of ZnO nanorods. Sci. Rep. **4**, 4596 (2014)
45. S. Prucnal, K. Gao, S. Zhou, J. Wu, H. Cai, O.D. Gordan, D.R.T. Zahn, G. Larkin, M. Helm, W. Skorupa, Optoelectronic properties of ZnO film on silicon after SF6 plasma treatment and milliseconds annealing. Appl. Phys. Lett. **105**, 221903 (2014)
46. S. Prucnal, J. Wu, Y. Berencén, M.O. Liedke, A. Wagner, F. Liu, M. Wang, L. Rebohle, S. Zhou, H. Cai, W. Skorupa, Engineering of optical and electrical properties of ZnO by non-equilibrium thermal processing: The role of zinc interstitials and zinc vacancies. J. Appl. Phys. **122**, 035303 (2017)
47. T. Gebel, M. Neubert, R. Endler, J. Weber, M. Vinnichenko, A. Kolitsch, W. Skorupa, H. Liepack, Millisecond-annealing using flash lamps for improved performance of AZO layers. Mater. Res. Soc. Symp. Proc. **1287** (2011). https://doi.org/10.1557/opl.2011.1438
48. P.F. Lindberg, F.L. Bregolin, K. Wiesenhuetter, U. Wiesenhuetter, H.N. Riise, L. Vines, S. Prucnal, W. Skorupa, B.G. Svensson, E.V. Monakhov, The effect of millisecond flash lamp annealing on electrical and structural properties of ZnO:Al/Si structures. J. Appl. Phys. **119**, 185305 (2016)

49. J. Reck, S. Seeger, M. Weise, R. Mientus, J. Schulte, K. Ellmer, Flash-lamp annealing of ZnO-layers on copper–indium–gallium–sulphide layers: A spectroscopic ellipsometry study. Thin Solid Films **571**, 762–766 (2014)
50. S.S. Schmidt, D. Abou-Ras, T. Unold, T. Eisenbarth, H. Wilhelm, A. Grimm, J. Klaer, Effect of Zn incorporation into $CuInS_2$ solar cell absorbers on microstructural and electrical properties. J. Appl. Phys. **110**, 064515 (2011)
51. R. Ratajczak, C. Mieszczynski, S. Prucnal, E. Guziewicz, M. Stachowicz, D. Snigurenko, J. Gaca, M. Wojcik, R. Böttger, R. Heller, W. Skorupa, J.V. Borany, A. Turos, Structural and optical studies of Pr implanted ZnO films subjected to a long-time or ultra-fast thermal annealing. Thin Solid Films **643**, 24–30 (2017)
52. R. Ratajczak, S. Prucnal, E. Guziewicz, C. Mieszczynski, D. Snigurenko, M. Stachowicz, W. Skorupa, A. Turos, The photoluminescence response to structural changes of Yb implanted ZnO crystals subjected to non-equilibrium processing. J. Appl. Phys. **121**, 075101 (2017)
53. R. Gago, S. Prucnal, R. Perez-Casero, I. Caretti, I. Jimenez, F. Lungwitz, S. Cornelius, Structural impact of chromium incorporation in as-grown and flash-lamp-annealed sputter deposited titanium oxide films. J. Alloys Compd. **729**, 438–445 (2017)
54. M. Batzill, E.H. Morales, U. Diebold, Influence of nitrogen doping on the defect formation and surface properties of TiO_2 rutile and anatase. Phys. Rev. Lett. **96**, 026103 (2006)
55. J. Zhu, Z. Deng, F. Chen, J. Zhang, H. Chen, M. Anpo, J. Huang, L. Zhang, Hydrothermal doping method for preparation of Cr^{3+}-TiO_2 photocatalysts with concentration gradient distribution of Cr^{3+}. Appl. Catal. B **62**, 329–335 (2006)
56. M. Neubert, Die Rolle des Sauerstoffanteils in Titandioxid bei Tantal-Dotierung zur Verwendung als transparentes leitfähiges Oxid, Dissertation, TU Dresden (2014) (in German)
57. X. Wang, L. Zhi, K. Müllen, Transparent, conductive graphene electrodes for dye-sensitized solar cells. Nano Lett. **8**, 323–332 (2008)
58. J.B. Plumley, A.W. Cook, C.A. Larsen, K. Artyushkova, S.M. Han, T.L. Peng, R.A. Kemp, Crystallization of electrically conductive visibly transparent ITO thin films by wavelength-range-specific pulsed Xe arc lamp annealing. J. Mater. Sci. **53**(18), 12949–12960 (2018). https://doi.org/10.1007/s10853-018-2534-7
59. S. Weller, M. Junghähnel, Flash lamp annealing of ITO thin films on ultra-thin glass. Vak. Forsch. Prax. **27**, 29–33 (2015)
60. A.N. Panckow, C. David, J. Weber, Flash lamp annealing (FLA) of magnetron sputtered low-temperature TCO coatings. Vak. Forsch. Prax. **29**, 21–25 (2017). https://doi.org/10.1002/vipr.201700652
61. C. David, B.P. Tinkham, P. Prunici, A. Panckow, Highly conductive and transparent ITO films deposited at low temperatures by pulsed DC magnetron sputtering from ceramic and metallic rotary targets. Surf. Coat. Technol. **314**, 113–117 (2017)
62. T. Gebel, L. Rebohle, R. Fendler, W. Hentsch, W. Skorupa, M. Voelskow, W. Anwand, R.A. Yankov, Millisecond annealing with flash lamps: tool and process challenges, in *14th International Conference on Advanced Thermal Processing of Semiconductors* (2006), pp. 47–55
63. Y. Kim, S. Park, S. Kim, B.-K. Kim, Y. Choi, J.-H. Hwang, H.J. Kim, Flash lamp annealing of indium tin oxide thin-films deposited on polyimide backplanes. Thin Solid Films **628**, 88–95 (2017)
64. H. Scherg-Kurmes, S. Seeger, S. Körner, B. Rech, R. Schlatmann, B. Szyszka, Optimization of the post-deposition annealing process of high-mobility In2O3: H for photovoltaic applications. Thin Solid Films **599**, 78–83 (2016)
65. K. Tetzner, Y.-H. Lin, A. Regoutz, A. Seitkhan, D.J. Payne, T.D. Anthopoulos, Sub-second photonic processing of solution-deposited single layer and heterojunction metal oxide thin-film transistors using a high-power xenon flash lamp. J. Mater. Chem. C **5**, 11724 (2017)
66. S.J. Moon, K.M. Yu, S.H. Jeong, J.Y. Kim, B.K. Kim, H.J. Kim, E.J. Yun, B.S. Bae, Flash lamp annealing effect on stability of oxide TFT, in *Thin Film Transistors* 12, vol. 64, ed. by Y. Kuo. ECS Transactions, vol. 10 (2014), pp. 109–113
67. T.-H. Yoo, S.-J. Kwon, H.-S. Kim, J.-M. Hong, J.A. Lim, Y.-W. Song, Sub-second photo-annealing of solution-processed metal oxide thin-film transistors via irradiation of intensely pulsed white light. RSC Adv. **4**(37), 19375 (2014). https://doi.org/10.1039/c4ra01371a

68. Y. Kim, J. Kim, B. Kim, H.J. Kim, S. Kim, E. Choi, J.H. Hwang, S. Park, Application of flash lamp annealing on nitrogen-doped amorphous indium-gallium-zinc-oxide thin film transistors. ECS J. Solid State Sci. Technol. **6**(12), P778–P785 (2017). https://doi.org/10.1149/2.0041712jss
69. T.Y. Eom et al., Investigation of the evolution of nitrogen defects in flash-lamp-annealed InGaZnO films and their effects on transistor characteristics. Appl. Phys. Express **11**(6) (2018). https://doi.org/10.7567/apex.11.061104
70. T. Henke, M. Knaut, C. Hossbach, M. Geidel, M. Albert, J.W. Bartha, Growth of aluminum oxide thin films with enhanced film density by the integration of in situ flash annealing into low-temperature atomic layer deposition. Surf. Coat. Technol. **309**, 600–608 (2017)
71. D.D. Phuong, N.T. Hoa, V.V. Hung, D.Q. Khoa, H.K. Hieu, Mechanical properties of metallic thin films: theoretical approach. Eur. Phys. J. B **89**(84) (2016)
72. H. Huang, F. Spaepen, Tensile testing of free-standing Cu, Ag and Al thin films and Ag/Cu multilayers. Acta Mater. **48**(12), 3261–3269 (2000)
73. A.J. Kalkman, A.H. Verbruggen, G.C.A.M. Janssen, Young's modulus measurements and grain boundary sliding in free-standing thin metal films. Appl. Phys. Lett. **78**, 2673 (2001)
74. A. Axelevitch, B. Gorenstein, G. Golan, Investigation of optical transmission in thin metal films. Phys. Procedia **32**, 1–13 (2012)
75. M. Shtein, Thin metal films as simple transparent conductors. SPIE Newsroom (2009) https://doi.org/10.1117/2.1200912.1848
76. F. Lacy, Developing a theoretical relationship between electrical resistivity, temperature, and film thickness for conductors. Nanoscale Res. Lett. **6**, 636 (2011)
77. K. Eun, M.-W. Chon, T.-H. Yoo, Y.-W. Song, S.-H. Choa, Electromechanical properties of printed copper ink film using a white flash light annealing process for flexible electronics. Microelectron. Reliab. **55**(5), 838–845 (2015)
78. C. Brombacher, C. Schubert, M. Daniel, A. Liebig, G. Beddies, T. Schumann, W. Skorupa, J. Donges, S. Häberlein, M. Albrecht, Chemical ordering of FePt films using millisecond flash-lamp annealing. J. Appl. Phys. **111**, 023902 (2012)
79. V.V. Braun, V.V. Vavilova, V.M. Ievlev, Y.E. Kalinin, Y.K. Kovneristyi, N.A. Palii, A.V. Perov, O.V. Serbin, Effects of thermal annealing and pulsed photon processing on the magnetomechanical properties of the amorphous $Fe_{79.3}P_{18.2}V_{2.5}$ alloy. Inorg. Mater, **41**(7), 823–827 (2005)
80. M.O. Anosova, Y.V. Baldokhin, V.V. Vavilova, V.M. Ievlev, Y.E. Kalinin, V.P. Korneev, N.A. Palii, Formation of nanocomposites during annealing of amorphous Fe-P-Si alloys. Inorg. Mater. **45**(9), 993–997 (2009)
81. O.K. Belousov, V.V. Vavilova, V.M. Ievlev, Y.E. Kalinin, S.B. Kushchev, N.A. Palii, S.A Pokazan'eva, Nanocrystallization in amorphous $Al_{83}Ni_{10}La_7$ and $Al_{83.5}Ni_{9.5}La_{5.6}Si_{1.4}$ alloys during thermal annealing and flash lamp processing. Inorg. Mater. **45**(8), 873–879 (2009)
82. V.V. Vavilova, V.M. Ievlev, Y.E. Kalinin, Y.K. Kovneristyi, N.A. Palii, B.G. Sukhodolov, V.N. Timofeev, Nanoscale crystallization in the amorphous alloy $Fe_{79}P_{14.2}Si_{4.4}Mn_{2.2}V_{0.2}$ upon pulsed photon annealing. Inorg. Mater. **39**(1), 82–87 (2003)
83. V.E. Borisenko (Ed.), *Semiconducting Silicides* (Materials Science by Springer, Berlin, Heidelberg, 2000)
84. H. Föll, Defects in crystals. University of Kiel, https://www.tf.uni-kiel.de/matwis/amat/def_en/index.html. Accessed 21 Dec 2018
85. L.J. Chen, Metal silicides: an integral part of microelectronics. J Miner. Metals Mater. Soc. **57**(9), 24–31 (2005)
86. L. Bischoff, K.-H. Heinig, J. Teichert, W. Skorupa, Submicron $CoSi_2$ structures fabricated by focused ion beam implantation and local flash lamp melting. Nucl. Instrum. Methods Phys. Res. B **112**, 201–205 (1996)
87. V.M. Ievlev, S.A. Soldatenko, S.B. Kushchev, Y.V. Gorozhankin, Effect of flash lamp annealing on silid-state reactions in (111)Si-Ni-Pt heterostructures. Inorg. Mater. **44**(7), 705–712 (2008)

88. J.T. Lue, V.C. Liu, W.J. Shen, Light-flash induced metallic silicides from titanium films on silicon. Appl. Phys. Lett. **38**(5), 372–374 (1981)
89. B. Adams, D. Jennings, K. Ma, A.J. Mayur, S. Moffatt, S.G. Nagy, V. Parihar, Characterization of nickel silicides produced by millisecond anneals, in *IEEE International Conference on Advanced Thermal Processing of Semiconductors—RTP* (2007)
90. I. Nishimura, T. Karasawa, T. Owada, M. Yasuhira, T. Arikado, Formation of Ni silicide using flash lamp technology. in *Extended Abstracts of the 2002 International Conference on Solid State Devices and Materials*, Nagoya (2002), pp. 436–437
91. V. Narayanan, M.M. Frank, A.A. Demkov (ed.), *Thin films on silicon: electronic and photonic applications. materials and energy*, vol. 8 (World Scientific Publishing, 2016)
92. S. Wagner, S. Bauer, Materials for stretchable electronics. MRS Bull. **37**, 207–213 (2012)
93. J. Jiu, M. Nogi, T. Sugahara, T. Tokuno, T. Araki, N. Komoda, K. Suganuma, H. Uchida, K. Shinozaki, Strongly adhesive and flexible transparent silver nanowire conductive films fabricated with a high-intensity pulsed light technique. J. Mater. Chem. **22**(44), 23561–23567 (2012)
94. V. Cannella, M. Izu, S. Jones, S. Wagner, I.-C. Cheng, Flexible stainless-steel substrates. SID Inf. Disp. **21**(06), (2005)
95. S. Wagner, Mechanics of TFT technology on flexible substrates, in *Flexible Flat Panel Displays*, ed. by G.P. Crawford (Wiley, Chichester, 2005), pp. 263–283
96. K.R. Sarma, Flexible Displays: TFT technology: substrate options and TFT processing strategies. in *Handbook of Visual Display Technology*, ed. by J. Chen, W. Cranton, M. Fihn (Springer, Berlin Heidelberg, 2012)
97. W. Anwand, J.M. Johnson, M. Butterling, A. Wagner, W. Skorupa, G. Brauer, Flash lamp annealing of tungsten surfaces marks a new way to optimized slow positron yields. J. Phys: Conf. Ser. **443**, 012072 (2013)
98. Z. Cui (Ed.), C. Zhou, S. Qiu, Z. Chen, J. Lin, J. Zhao, C. Ma, W. Su, *Printed Electronics: Materials, Technologies and Applications* (Higher Education Press by Wiley, Singapore 2016)
99. R.H. Reuss et al., Macroelectronics: perspectives on technology and applications. Proc. IEEE **93**(7), 1239–1256 (2005)
100. L. Rebohle, S. Prucnal, W. Skorupa, A review of thermal processing in the subsecond range: semiconductors and beyond. Semicond. Sci. Technol. **31**(10) (2016), https://doi.org/10.1088/0268-1242/31/10/103001
101. H.-Y. Jin et al., Rapid sintering of TiO_2 photoelectrodes using intense pulsed white light for flexible dye-sensitized solar cells. Appl. Phys. Lett. **104**, 143902 (2014)
102. N. Marjanovic, J. Hammerschmidt, J. Perelaer, S. Farnsworth, I. Rawson, M. Kus, E. Yenel, S. Tilki, U.S. Schubert, R.R. Baumann, Inkjet printing and low temperature sintering of CuO and CdS as functional electronic layers and Schottky diodes. J. Mater. Chem. **21**, 13634 (2011)
103. S. Magdassi, M. Grouchko, A. Kamyshny, Copper nanoparticles for printed electronics: routes towards achieving oxidation stability. Materials **3**, 4626–4638 (2010)
104. M. Junghähnel, S. Weller, T. Gebel, Advanced processing of ITO and IZO thin films on flexible glass, in *SID Symposium Digest of Technical Papers P-65* (2015), 1378–1381
105. W.A. Macdonald, M.K. Looney, D.A. MacKerron, R. Eveson, R. Adam, K. Hashimoto, K. Rakos, Latest advances in substrates for flexible electronics. J. Soc. Inform. Display **15**(12), 1075–1083 (2007)
106. H. Klauk (ed.), *Organic Electronics—Materials, Manufacturing and Applications* (Wiley, Weinheim, 2006)
107. J. Perelaer, A.W.M. de Laat, C.E. Hendriks, U.S. Schubert, Inkjet-printed silver tracks: low temperature curing and thermal stability investigation. J. Mater. Chem. **18**(27), 3209–3215 (2008)
108. J. Chen, W. Cranton, M. Fihn (eds.), *Handbook of Visual Display Technology* (Springer, Berlin, Heidelberg, 2016)
109. S. Khan, L. Lorenzelli, D. Ravinder, Technologies for printing sensors and electronics over large flexible substrates: a review. IEEE Sens. J. **15**(6), 1–22 (2015)

110. J. Perelaer, P.J. Smith, D. Mager, D. Soltman, S.K. Volkman, V. Subramanian, J.G. Korvink, U.S. Schubert, Printed electronics: the challenges involved in printing devices, interconnects, and contacts based on inorganic materials. J. Mater. Chem. **20**, 8446–8453 (2010)
111. M. Caironi, Y.-Y. Noh, *Large Area and Flexible Electronics* (Wiley, Weinheim, 2015)
112. G. Crawford, *Flexible Flat Panel Displays. SID Series in Display Technology* (Wiley, Southern Gate, 2005)
113. M. Berggren, D. Nilsson, N.D. Robinson, Organic materials for printed electronics. Nat. Mater. **6**, 3–5 (2007)
114. W. Barford (ed.), *Electronic and Optical Properties of Conjugated Polymers* (Oxford University Press, New York, 2005)
115. P. Kar (ed.), *Doping in Conjugated Polymers* (Wiley, New York, 2013)
116. https://www.idtechex.com/research/reports/inorganic-and-composite-printed-electronics-2014-2024-000401.asp. Accessed 14 Sept 2018
117. D. Sangeeta, J.R. LaGraff (eds.), *Inorganic Materials Chemistry Desk Reference* (CRC Press Book, Taylor&Francis Group, Boca Raton, 2005)
118. A. Kamyshny, J. Steinke, S. Magdassi, Metal-based inkjet inks for printed electronics. The Open Applied Physics Journal **4**, 19–36 (2011)
119. S.K. Garlapati, M. Divya, B. Breitung, R. Kruk, H. Hahn, S. Dasgupta, Printed electronics based on inorganic semiconductors: from processes and materials to devices. Adv. Mater. **30**(40), 1707600 (2018)
120. S. Das, D. Cormier, S. Williams, Potential for multi-functional additive manufacturing using pulsed photonic sintering, in A.J. Shih, L.H. Wang (eds.) *43rd North American Manufacturing Research Conference, Namrc* 43, 1. Procedia Manufacturing (2015), pp. 366–377
121. H.-J. Hwang, W.-H. Chung, H.-S. Kim, In situ monitoring of flash-light sintering of copper nanoparticle ink for printed electronics. Nanotechnology **23**, 485202 (2012)
122. H.-S. Kim, S.R. Dhage, D.-E. Shim, H.T. Hahn, Intense pulsed light sintering of copper nanoink for printed electronics. Appl. Phys. A **97**, 791–798 (2009)
123. B. Büchter, F. Seidel, R. Fritzsche, J. Toader, R. Buschbeck, A. Jakob, S. Schulze, H. Freitag, H. Lang, M. Hietschold, D.R.T. Zahn, M. Mehring, Ultrasonic spray coating and flash lamp annealing of silicon nanoparticle dispersions for silicon thin film formation. J. Mater. Sci. **49**, 7979–7990 (2014)
124. B. Büchter, F. Seidel, R. Fritzsche, D. Lehmann, D. Bülz, R. Buschbeck, A. Jakob, S. Schulze, H. Freitag, H. Lang, M. Hietschold, D.R.T. Zahn, M. Mehring, Polycrystalline silicon foils by flash lamp annealing of spray-coated silicon nanoparticle dispersions. J. Mater. Sci. **50**, 6050–6059 (2015)
125. M. Akbari, H. He, J. Juuti, M.M. Tentzeris, J. Virkki, L. Ukkonen, 3D printed and photonically cured graphene UHF RFID tags on textile, wood, and cardboard substrates. Int. J. Antennas Propag. (2017). https://doi.org/10.1155/2017/7327398
126. B. Bourgeois et al., Pulsed photoinitiated fabrication of inkjet printed titanium dioxide/reduced graphene oxide nanocomposite thin films. Nanotechnology **29**(31) (2018). https://doi.org/10.1088/1361-6528/aac306
127. W. Yang, C. Wang, Graphene and the related conductive inks for flexible electronics. J. Mater. Chem. C **4**, 7193 (2016)
128. K. Meerholz, J. Heinze, Influence of chain length and defects on the electrical conductivity of conducting polymers. Synth. Met. **57**(2–3), 5040–5045 (1993)
129. E. Hrehorova, A. Pekarovicova, P.D. Fleming, Gravure printability of conducting polymer inks. NIP Dig. Fabr. Conf **3**, 107–110 (2006)
130. J.K. Mackenzie, R. Shuttleworth, A phenomenological theory of sintering. Proc. Phys. Soc. B **62**(12), 833 (1949)
131. R.-Z. Li, A. Hu, T. Zhang, K.D. Oakes, Direct writing on paper of foldable capacitive touch pads with silver nanowire inks. ACS Appl. Mater. Interfaces. **6**(23), 21721–21729 (2014)
132. P. Rosa: Minimal computation structures for visual information applications based on printed electronics. Dissertation, New University of Lisboa, 2015

133. H.S. Lim, S.J. Kim, H.W. Jang, J.A. Lim, Intense pulsed light for split-second structural development of nanomaterials. J. Mater. Chem. C. (2017). https://doi.org/10.1039/c7tc01848g
134. A. Chiolerio, G. Maccioni, P. Martino, M. Cotto, P. Pandolfi, P. Rivolo, S. Ferrero, L. Scaltrito, Inkjet printing and low power laser annealing of silver nanoparticle traces for the realization of low resistivity lines for flexible electronics. Microelectron. Eng. **88**(8), 2481–2483 (2011)
135. S. Wünscher, R. Abbel, J. Perelaer, U.S. Schubert, Progress of alternative sintering approaches of inkjet-printed metal inks and their application for manufacturing of flexible electronic devices. J. Mater. Chem. C **2**, 10232 (2014)
136. J. Niittynen, E. Sowade, H. Kang, R.R. Baumann, M. Mäntysalo, Comparison of laser and intense pulsed light sintering (IPL) for inkjet-printed copper nanoparticle layers. Sci. Rep. **5**, 8832 (2015)
137. K. Suganuma, *Introduction to Printed Electronics*. SpringerBriefs in Electrical and Computer Engineering (Springer, New York, 2014)
138. E. Saleh, F. Zhang, Y. He, J. Vaithilingam, J. Ledesma, F.R. Wildman, I. Ashcroft, R. Hague, P. Dickens, C. Tuck, 3D inkjet printing of electronics using UV conversion. Adv. Mater. Technol. **2**, 1700134 (2017)
139. V. Amendola, M. Meneghetti, Laser ablation synthesis in solution and size manipulation of noble metal nanoparticles. Phys. Chem. **11**(20), 3805–3821 (2009)
140. J. Niittynen, R. Abbel, M. Mäntysalo, J. Perelaer, U.S. Schubert, D. Lupo, Alternative sintering methods compared to conventional thermal sintering for inkjet printed silver nanoparticle ink. Thin Solid Films **556**, 453–459 (2014)
141. J.S. Kang, J. Ryu, H.S. Kim, H.T. Hahn, Sintering of inkjet-printed silver nanoparticles at room temperature using intense pulsed light. J. Electron. Mater. **40**(11), 2268–2277 (2011)
142. D. Kim, J. Moon, Highly conductive ink jet printed films of nanosilver particles for printable electronics. Electrochem. Solid-State Lett. **8**(11), J30–J33 (2005)
143. H.-J. Hwang, K.-H. Oh, H.-S. Kim, All-photonic drying and sintering process via flash white light combined with deep-UV and near-infrared irradiation for highly conductive copper nano-ink. Sci. Rep. **6**, 19696 (2016)
144. F. Paglia, D. Vak, J. v. Embden, A.S.R. Chesman, A. Martucci, J.J. Jasieniak, E.D. Gaspera, Photonic sintering of copper through the controlled reduction of printed CuO nanocrystals. ACS Appl. Mater. Interfaces **7**, 25473 (2015). https://doi.org/10.1021/acsami.5b08430
145. J. Ryu, H.-S. Kim, H.T. Hahn, Reactive sintering of copper nanoparticles using intense pulsed light for printed electronics. J. Electron. Mater. **40**(1), 42–50 (2011)
146. B.-Y. Wang, T.-H. Yoo, Y.-W. Song, D.-S. Lim, Y.-J. Oh, Cu ion ink for a flexible substrate and highly conductive patterning by intensive pulsed light sintering. ACS Appl. Mater. Interfaces. **5**, 4113–4119 (2013)
147. C.S. Choi, Y.H. Jo, M.G. Kim, H.M. Lee, Control of chemical kinetics for Sub-10 nm Cu nanoparticles to fabricate highly conductive ink below 150 °C. Nanotechnology **23**(6), 065601 (2012)
148. W.-S. Han, J.-M. Hong, H.-S. Kim, Y.-W. Song, Multi-pulsed white light sintering of printed Cu nanoinks. Nanotechnology **22**, 395705 (2011)
149. D.J. Lee, S.H. Park, S. Jang, H.S. Kim, J.H. Oh, Y.W. Song, Pulsed light sintering characteristics of inkjet-printed nanosilver films on a polymer substrate. J. Micromech. Microeng. **21**(12), 125023 (2011)
150. J.L. Emmett, A.L. Schawlow, Enhanced ultraviolet output from double-pulsed flash lamps. Appl. Phys. Lett. **2**(11), 204 (1963)
151. Y. Wu, J. Zhu, Z. Zhang, Y. Li, Radiation model of a xenon flash lamp in a laser amplifier pump cavity. High Power Laser Sci. Eng. **3**, e31 (2015). https://doi.org/10.1017/hpl.2015.34
152. I. Reinhold, C.E. Hendriks, R. Eckardt, J.M. Kranenburg, J. Perelaer, R.R. Baumann, U.S. Schubert, Argon plasma sintering of inkjet printed silver tracks on polymer substrates. J. Mater. Chem. **19**, 3384–3388 (2009)
153. J. Perelaer, R.J.M. Grouchko, A. Kamyshny, S. Magdassi, U.S. Schubert, Plasma and microwave flash sintering of a tailored silver nanoparticle ink, yielding 60% bulk conductivity on cost-effective polymer foils. Adv. Mater. **24**(29), 3993–3998 (2012)

154. G. Delaizira et al., A comparative study of Spark Plasma Sintering (SPS), Hot Isostatic Pressing (HIP) and microwaves sintering techniques on p-type Bi_2Te_3 thermoelectric properties. Mater. Res. Bull. **47**(8), 1954–1960 (2012)
155. M.L. Allen, M. Aronniemi, T. Mattila, A. Alastalo, K. Ojanperä, M. Suhonen, H. Seppä, Electrical sintering of nanoparticle structures. Nanotechnology **19**, 175201 (2008)
156. D. Tobjörk, H. Aarnio, P. Pulkkinen, R. Bollström, A. Määttänen, P. Ihalainen, T. Mäkelä, J. Peltonen, M. Toivakka, H. Tenhu, R. Österbacka, IR-sintering of ink-jet printed metal-nanoparticles on paper. Thin Solid Films **520**, 2949–2955 (2012)
157. M.S. Vezie, S. Few, I. Meager, G. Pieridou, B. Dörling, R.S. Ashraf, A.R. Goñi, H. Bronstein, I. McCulloch, S.C. Hayes, M. Campoy-Quiles, J. Nelson, Exploring the origin of high optical absorption in conjugated polymers (2016). https://doi.org/10.1038/nmat4645
158. L. Xu, L. Cheng, C. Wang, R. Peng, Z. Liu, Conjugated polymers for photothermal therapy of cancer. Polym. Chem. **5**, 1573–1580 (2014)
159. D. Gamota, P. Brazis, K. Kalyanasundaram, J. Zhang, *Printed Organic and Molecular Electronics* (Springer Science+Business Media, New York, 2004)
160. W.-H. Chung, H.-J. Hwang, S.-H. Lee, H.-Sung Kim, In situ monitoring of a flash light sintering process using silver nano-ink for producing flexible electronics. Nanotechnology **24**, 035202 (2013)
161. J.F. Salmerón, F. Molina-Lopez, D. Briand, J.J. Ruan, A. Rivadeneyra, M.A. Carvajal, L.F. Capitán-Vallvey, N.F. de Rooij, A.J. Palma, Properties and printability of inkjet and screen-printed silver patterns for RFID antennas. J. Electron. Mater. **43**(2), 604–617 (2014)
162. C. Kim, M. Nogi, K. Suganuma, Effect of ink viscosity on electrical resistivity of narrow printed silver lines, in *11th IEEE International Conference on Nanotechnology, Portland* (2011), pp. 197–200
163. W. Wu, Inorganic nanomaterials for printed electronics: a review. Nanoscale **9**(22), 7342 (2017). https://doi.org/10.1039/C7NR01604B
164. T. Araki, T. Sugahara, J. Jiu, S. Nagao, M. Nogi, H. Koga, H. Uchida, K. Shinozaki, K. Suganuma, Cu salt ink formulation for printed electronics using photonic sintering. Langmuir **29**(35), 11192–11197 (2013)
165. D. Mitra, K.Y. Mitra, M. Hartwig, R.R. Baumann, Intense pulsed light sintering of an inkjet printed silver nanoparticle ink depending on the spectral absorption and reflection of the background. J. Imagin. Sci. Technol. **60**(4), 40403-1–40403-5(5) (2016)
166. D. Weise, K.Y. Mitra, P. Ueberfuhr, R.R. Baumann, Effect of the light spectrum of various substrates for inkjet printed conductive structures sintered with intense pulsed light. AIP Conf. Proc. **1646**, 101–105 (2015)
167. W.H. Chung, Y.T. Hwang, S.H. Lee, H.S. Kim, Electrical wire explosion process of copper/silver hybrid nano-particle ink and its sintering via flash white light to achieve high electrical conductivity. Nanotechnology **27**(20), 205704 (2016). https://doi.org/10.1088/0957-4484/27/20/205704
168. N.A. Luechinger, E.K. Athanassiou, W.J. Stark, Graphene-stabilized copper nanoparticles as an air-stable substitute for silver and gold in low-cost ink-jet printable electronics. Nanotechnology **19**(44), 445201 (2008)
169. M.S. Rager, T. Aytug, G.M. Veith, P. Joshi, Low-thermal-budget photonic processing of highly conductive Cu interconnects based on CuO nanoinks: potential for flexible printed electronics. ACS Appl. Mater. Interfaces **8**(3), 2441–2448 (2016). https://doi.org/10.1021/acsami.5b12156
170. S.K. Hau, H.-L. Yip, N.S. Baek, J. Zou, K. O'Malley, A.K.-Y. Jen, Air-stable inverted flexible polymer solar cells using zinc oxide nanoparticles as an electron selective layer. Appl. Phys. Lett. **92**, 253301 (2008)
171. J.M. Marshall, D. Dimova-Malinovska (eds.), *Photovoltaic and Photoactive Materials: Properties, Technology and Applications*, NATO Science Series II (Springer Science+Business Media, Dordrecht, 2002)
172. Y. Kim, S. Park, B.-K. Kim, H.J. Kim, J.-H. Hwang, Xe-arc flash annealing of indium tin oxide thin-films prepared on glass backplanes. Int. J. Heat Mass Transf. **91**, 543–551 (2015)

173. P. Kanninen, J. Christoffer, J. Merta, K. Kontturi, Influence of ligand structure on the stability and oxidation of copper nanoparticles. J. Colloid Interface Sci. **318**(1), 88–95 (2008)
174. H. Goesmann, *Less-Noble Metal Nanoparticles* (Institute for Organic Chemistry, Karlsruhe Institute of Technology, 2014). http://www.aoc.kit.edu/2547.php
175. O.S. Ivanova, F.P. Zamborini, Size-dependent electrochemical oxidation of silver nanoparticles. J. Am. Chem. Soc. **132**, 70 (2010)
176. J.D. Yuen, J. Fan, J. Seifter, B. Lim, R. Hufschmid, A.J. Heeger, F. Wudl, High performance weak donor-acceptor polymers in thin film transistors: effect of the acceptor on electronic properties, ambipolar conductivity, mobility, and thermal stability. J. Am. Chem. Soc. **133**(51), 20799–20807 (2011)
177. C.L. Beyler, M.M. Hirschler, Thermal decomposition of polymers. in *SFPE Handbook of Fire Protection Engineering*, vol. 2, Section 1, Chapter 7 (2002), pp. 111–131
178. Z. Suo, Mechanics of stretchable electronics and soft machines. MRS Bull. **37**, 218–225 (2012)
179. D. Angmo, T.T. Larsen-Olsen, M. Jørgensen, R.R. Søndergaard, F.C. Krebs, Roll-to-roll inkjet printing and photonic sintering of electrodes for ITO free polymer solar cell modules and facile product integration. Adv. Energy Mater. **3**, 172 (2012). https://doi.org/10.1002/aenm.201200520
180. K.Y. Law, H. Zhao (eds.), *Surface Wetting: Characterization, Contact Angle and Fundamentals* (Springer International Publishing, 2016)
181. P. Ihalainen, A. Määttänen, J. Järnström, D. Tobjörk, R. Österbacka, J. Peltonen, Influence of surface properties of coated papers on printed electronics. Ind. Eng. Chem. Res. **51**, 6025–6036 (2012)
182. P.Q. Nguyen, L.P. Yeo, B.K. Lok, Y.C. Lam, Patterned surface with controllable wettability for inkjet printing of flexible printed electronics. ACS Appl. Mater. Interfaces **6**(6), 4011–4016 (2014)
183. Z. Liu, J. Zhao, W. Xu, L. Qian, S. Nie, Z. Cui, Effect of surface wettability properties on the electrical properties of printed carbon nanotube thin-film transistors on SiO_2/Si substrates. ACS Appl. Mater. Interfaces **6**(13), 9997–10004 (2014)
184. M. Benwadih, R. Coppard, K. Bonrad, A. Klyszcz, D. Vuillaume, High mobility flexible amorphous IGZO thin-film transistors with a low thermal budget ultra-violet pulsed light process. ACS Appl. Mater. Interfaces **8**(50), 34513–34519 (2016)

Index

L. Rebohle et al., *Flash Lamp Annealing*, Springer Series in Materials Science 288,
https://doi.org/10.1007/978-3-030-23299-3

MIX
Papier aus verantwortungsvollen Quellen
Paper from responsible sources
FSC® C105338

If you have any concerns about our products,
you can contact us on
ProductSafety@springernature.com

In case Publisher is established outside the EU,
the EU authorized representative is:
Springer Nature Customer Service Center GmbH
Europaplatz 3, 69115 Heidelberg, Germany

Printed by Libri Plureos GmbH
in Hamburg, Germany